Praise from the Experts

"The author is to be congratulated and appreciated for writing such a practical and useful book for both entry-level and seasoned practitioners. It provides very clear step-by-step instructions using various case examples. In addition, it also provides the explanation of essential theoretical background in each modeling technique."

Carl Lee, Ph.D.
Professor of Statistics
Data Mining Program Coordinator
Central Michigan University

"Kattamuri Sarma has written a practical guide that will be of use to modeling professionals of all backgrounds and levels of experience. In it he shows us the power of SAS Enterprise Miner to enhance the modeling process in a way that is straightforward, easy to follow, and thorough. From data scrubbing to model fitting, each step in the model building process is described in a clear manner. The book draws on the author's extensive real-world experience to illustrate the ways in which SAS Enterprise Miner can help to facilitate the modeling process."

Lee Medoff
SAS User

"The content is wonderful, clear, and thorough."

Andrea Wainwright-Zimmerman
Senior Statistical Analysis Manager

"I think this book will be very helpful to new users of SAS Enterprise Miner. It presents many examples in a way that puts the material across clearly. I especially liked the chapters on decision trees and neural networks for their detailed exposition. I also was pleased to see many cases where the author shows the SAS code behind the scenes. I think Dr. Sarma did a fine job."

Rich Perline, Ph.D.
Senior Scientist
NuTech Solutions, Inc.

Predictive Modeling *with* SAS® Enterprise Miner™

Practical Solutions for Business Applications

Kattamuri S. Sarma, Ph.D.

The correct bibliographic citation for this manual is as follows: Sarma, Kattamuri S. 2007. *Predictive Modeling with SAS® Enterprise Miner™: Practical Solutions for Business Applications*. Cary, NC: SAS Institute Inc.

Predictive Modeling with SAS® Enterprise Miner™: Practical Solutions for Business Applications

ISBN 978-1-59047-703-8

SAS Institute Inc., SAS Campus Drive, Cary, North Carolina 27513.

1st printing, September 2007

Contents

Preface

In order to make effective use of the tools provided in SAS Enterprise Miner, you need to be able to do more than just set the software in motion. While it is essential to know the mechanics of how to use each tool (or *node*, as Enterprise Miner tools are called), you should also understand the methodology behind each tool and be able to interpret the output produced by each tool and the multitude of options available with each one. Although this book will appeal to beginners because of its step-by-step, screen-by-screen introduction to the basic tasks that can be accomplished with Enterprise Miner, it also provides the depth and background needed to master many of the more complex but rewarding concepts contained within this versatile product.

The book begins by introducing the basics of creating a project, manipulating data sources, choosing the right property values for each node, and navigating through different results windows. It then demonstrates various pre-processing tools required for building predictive models before beginning its treatment of the three main predictive modeling tools—**Decision Tree**, **Neural Network**, and **Regression**. These are addressed in considerable detail, with numerous examples of practical business applications that are illustrated with tables, charts, displays, equations, and even manual calculations that let you see the essence of what Enterprise Miner is doing as it estimates or optimizes a given model. By the time you finish with this book, Enterprise Miner will no longer be a "black box": you will have an in-depth understanding of the product's inner workings. Throughout the book, the link between the output generated by Enterprise Miner and the statistical theory behind the business analysis for which Enterprise Miner is being used is explicitly shown. Even the SAS code generated by each node is examined line by line to show the correspondence between the theory and the results produced by Enterprise Miner. In many places, however, intuitive explanations are used to give you an appreciation of the way that various nodes such as **Decision Tree**, **Neural Network**, **Regression**, and **Variable Selection** operate and how different options such as Model Selection Criteria and Model Assessment are implemented. These explanations are intended not to replicate the exact steps that SAS uses internally to make these computations, but to give a good practical sense of how these tools work in general. Overall, I believe this approach will help you use the tools in Enterprise Miner with greater comprehension and confidence.

Several examples of business questions drawn from the insurance and banking industries and based on simulated, but realistic, data are used to illustrate the Enterprise Miner tools. However, the procedures discussed are relevant for any industry. Tables and graphs from the output data sets created by various nodes are also included to give you an idea of how to make custom tables from the results produced by Enterprise Miner. In the end you should have gained enough understanding of Enterprise Miner to become comfortable and innovative in adapting the applications discussed here to solve your own business problems.

Chapter Details

Chapter 1 discusses research strategy. I include general issues such as defining the target population, defining the target (or dependent) variable, collecting data, cleaning the data, and selecting an appropriate model.

Chapter 2 shows how to open Enterprise Miner, start a new project, and create data sources. It shows various components of the Enterprise Miner window and shows how to create a process flow diagram. In this chapter, I use example data sets to demonstrate in detail how to use the **Input Data**, **StatExplore**, **MultiPlot**, **Impute**, **Data Partition**, **Filter**, **Variable Selection**, **Transform Variables**, **SAS Code**, and **Drop** nodes. I also discuss the output and SAS code generated by some of these nodes. I manually compute certain statistics such as Cramer's V and compare the results with those produced by **StatExplore**.

Chapter 3 covers the **Variable Selection** and **Transform Variables** nodes in detail. In using the **Variable Selection** node you have a choice of many options, depending on the type of target and the measurement scale of the inputs. To help make things clear, I illustrate each situation with a separate data set.

Chapter 4 discusses decision trees and regression trees. First I present the general tree methodology, and, using a simple example, I manually work through the sequence of steps—growing the tree, classifying the nodes, and pruning—which is performed by the **Decision Tree** node. I then show how decision tree models are built for predicting response and risk by presenting two examples based on a hypothetical auto insurance company. The first model predicts the probability of response to a mail order campaign. The second model predicts risk as measured by claim frequency, and since claim frequency is measured as a continuous variable, the model built in this case is a regression tree. A detailed discussion of the SAS code generated by the **Decision Tree** node is included at the end of the chapter.

Chapter 5 provides an introduction to neural networks. Here I try to demystify the neural networks methodology by giving an intuitive explanation using simple algebra. I show how to configure neural networks to be consistent with economic and statistical theory and how to interpret the results correctly. The neural network architecture—input layer, hidden layers, and output layer—is illustrated algebraically with numerical examples. Although the formulas presented here may look complex, they do not require a high-level knowledge of mathematics, and patience in working through them will be rewarded with a thorough understanding of neural networks and their applications.

In this chapter, the iterative processes of estimation of the model using the training data set as well as the selection of the optimal weights for the model using the validation data set are first discussed intuitively. Next, explicit numerical examples are given to clarify each step. As in Chapter 4, two models are developed using the hypothetical insurance data—a response model with a binary target, and a risk model with (in this case) an ordinal target representing accident frequency. I examine line by line the SAS code generated by the **Neural Network** node and show the correspondence between the theory and the results produced by Enterprise Miner.

Chapter 6 demonstrates how to develop logistic regression models for targets with different measurement scales: binary, categorical with more than two categories, ordinal, and continuous (interval-scaled). Using an example data set with a binary target, I demonstrate various model selection criteria and model selection methods. I also present business applications from the banking industry involving two predictive models, one with a binary target and one with a continuous target. The model with a binary target predicts the probability of response to a mail campaign while the model with a continuous target predicts the increase in deposits that is due to an interest rate increase. This chapter also shows how to calculate the lift and capture rates of the models when the target is continuous.

In Chapter 7, I compare the results of three modeling tools—**Decision Tree**, **Neural Network**, and **Regression**—that were presented in earlier chapters. For this purpose I develop two predictive models and then take turns applying the three modeling tools to each model. The first model has a binary target and predicts the probability of customer attrition for a fictitious bank. The second model has an ordinal target, which is a discrete version of a continuous variable, and predicts risk (as measured by loss frequency) for a fictitious auto insurance company. This chapter also provides a method of computing the lift and capture rates of these models using the expected value of the target variable.

Chapter 8 shows how to calculate profitability for each of the ten deciles created when a data set of prospective customers is scored using the output of the modeling process. It then shows how to use these profitability estimates to address questions such as how to choose an optimum cut-off point for a mailing campaign. Here my objective is to introduce the notion of the marginal cost and marginal revenue associated with risk and response and to show how they can be used to make rational quantitative decisions in the marketing sphere.

How to Use the Book

- To get the most out of this book, open Enterprise Miner and follow the sequence of tasks performed in each chapter, using either the data sets stored on the CD included with this book or, even better, your own data sets.
- Work through the manual calculations as well as the mathematical derivations presented in the book to get an in-depth understanding of the logic behind different models.
- To learn predictive modeling, read the general explanation and the theory, and then follow the steps given in the book to develop models using either the data sets provided on the CD or your own data sets. Try variations of what is done in the book to strengthen your understanding of the topics covered.
- If you already know Enterprise Miner and want to get a good understanding of decision trees and neural networks, focus on the examples and detailed derivations given in Chapters 4, 5, and 6. These derivations are not as complex as they appear to be.

Prerequisites

- Elementary algebra and basic training (equivalent to one to two semesters of course work) in statistics covering inference, hypothesis testing, probability, and regression
- Familiarity with measurement scales of variables—continuous, categorical, ordinal, etc.
- Experience with Base SAS software and some understanding of simple SAS macros and macro variables

Acknowledgments

I would like to thank a number of people who helped me at various stages during the writing of this book.

At the beginning of this project, I had many discussions about practical aspects of modeling techniques, especially neural networks and decision trees, with Ravindra Sarma, who was at that time a graduate student at MIT working on a neural networks project. In addition, he read the manuscript at several stages and contributed editorial assistance, as well as many in-depth questions and insightful comments that greatly improved the text.

While I was working on SAS Enterprise Miner projects, I turned several times to SAS Technical Support. In addition to always answering my questions about how to run the software, they often provided informative explanations of the methodology behind it. I learned a great deal by sending questions to them and getting solid answers. I do not have a list of all the people I talked with in SAS Technical Support, so I would like to recognize them as a group.

I would like to thank Randal Blank, a friend and a senior associate at KPMG LLP, for his constant encouragement and editorial help. Chris Monroe, a colleague from my days at AT&T, provided valuable editorial assistance on several sections.

In addition, I would like to thank seven reviewers from SAS—Fang Chen, Brent Cohen, David Duling, Leonardo Auslender, Mary Grace Crissey, R. Wayne Thompson, and Rob Agnelli—and two outside reviewers, Goutam Chakraborty and Michael Szenberg.

I could not have written this book without the encouragement and support of Julie Platt and Patsy Poole at SAS Publishing. Julie was the supervisor of the project, and Patsy worked closely with me and was very helpful with organization and editing. Other members of the SAS Publishing team who made indispensable contributions were Caroline Brickley, who edited the book, Mary Beth Steinbach, who was the managing editor and reviewed the final version of the book, and Candy Farrell, the electronic production specialist whose wizardry was applied to the production of the book. I express my heartfelt thanks to all of them.

I especially would like to thank my wife, Lokamatha Sarma, for putting up with me while I spent many weekends working on this book.

Research Strategy

1.1 Introduction

This chapter discusses the planning and organization of a predictive modeling project. Planning involves tasks such as these:

- defining and measuring the target variable in accordance with the business question
- collecting the data
- comparing the distributions of key variables between the modeling data set and the target population to verify that the sample adequately represents the target population
- defining sampling weights if necessary
- performing data-cleaning tasks that need to be done prior to launching SAS Enterprise Miner

Alternative strategies for developing predictive models using Enterprise Miner are discussed at the end of this chapter.

1.2 Measurement Scales for Variables

Because many of the steps above will involve a discussion of the data and types of variables in our data sets, it is important that I first define the measurement scales for variables that are used in this book. In general, I have tried to follow the definitions given by Alan Agresti[1]:

- A *categorical variable* is one for which the measurement scale consists of a set of categories.
- Categorical variables for which levels (categories) do not have a natural ordering are called *nominal.*
- Categorical variables that do have a natural ordering of their levels are called *ordinal.*
- An *interval variable* is one that has numerical distances between any two levels of the scale.

According to the above definitions, the variables INCOME and AGE in Tables 1.1 to 1.5 and BAL_AFTER in Table 1.3 are interval-scaled variables. Because the variable RESP in Table 1.1 is categorical and has only two levels, it is called a binary variable. The variable LOSSFRQ in Table 1.2 is ordinal. (In Enterprise Miner you can change its measurement scale to interval, but I have left it as ordinal.) The variables PRIORPR and NXTPR in Table 1.5 are nominal.

In some manuals and books, interval-scaled variables are called *continuous*. Continuous variables are usually treated as interval variables as a matter of practical convenience. Therefore I use the terms *interval-scaled* and *continuous* interchangeably.

I also use the terms *ordered polychotomous variables* and *ordinal variables* interchangeably. Similarly, I use the terms *unordered polychotomous variables* and *nominal variables* interchangeably.

1.3 Defining the Target

The first step in any data mining project is to define and measure the target variable to be predicted by the model that emerges from your analysis of the data. This section presents examples of this step applied to five different business questions.

The examples shown here are intended to highlight the fact that the tasks of defining and measuring the target variable can be nontrivial.

1.3.1 Predicting Response to Direct Mail

In the example used in this book, a hypothetical auto insurance company wants to acquire customers through direct mail. The company wants to minimize mailing costs by targeting only the most responsive customers. Therefore, the company decides to use a response model. The target variable for this model will be RESP, and it is binary, taking the value of 1 for response and 0 for no response.

Table 1.1 shows a simplified version of a data set used for modeling the binary target, response (RESP).

[1] Alan Agresti, *Categorical Data Analysis,* 2nd ed. (New York, NY: John Wiley & Sons, 2002), 2.

Table 1.1

CUSTOMER	AGE	INCOME	STATUS	PC	NC	RESP
1	25	$45,000	S	1	1	0
2	45	$61,000	MC	1	2	1
3	54	.	MC	1	3	0
4	32	$24,000	MNC	0	4	0
5	43	$31,000	MC	0	5	0
6	56	$23,456	MC	1	6	1
7	78	.	W	0	7	0
8	6	$100,256	D	1	1	0
9	26	$345,678	MNC	1	2	1
10	32	$100,211	S	0	3	0
11	51	$21,312	MC	1	4	0
12	31	$83,456		0	5	1
13	23	$24,234	MNC	1	1	0
14	47	$43,566	MC	0	3	0
15	77	$12,002	MC	1	4	1
16	83	$32,454	W	1	5	0
17	25	$61,345	S	0	6	0
18	32	$76,123	MC	1	7	0
19	52	$25,324		1	8	0
20	32	$31,886	MNC	0	1	0
21	23	$78,345	S	1	8	0
22	80	$61,234	MNC	1	2	0
23	123	$76,876	S	1	4	0
24	45	$24,002		3	5	0

In Table 1.1 the variables AGE, INCOME, STATUS, PC, and NC are input variables (or explanatory variables). AGE and INCOME are numeric and, although they could theoretically be considered continuous, it is simply more practical to treat them as interval variables.

The variable STATUS is categorical and nominal-scaled. The categories of this variable are S if the customer is single and never married, MC if married with children, MNC if married without children, W if widowed, and D if divorced.

The variable PC is numeric and binary. It indicates whether the customers own a personal computer or not, taking the value 1 if they do and 0 if not. The variable NC represents the number of credit cards the customers own. You can decide whether this variable is ordinal or interval-scaled.

The target variable is RESP and takes the value 1 if the customer responded, for example, to a mailing campaign, and 0 otherwise. A binary target can be either numeric or character; I could have recorded a response as Y instead of 1, and a non-response as N instead of 0, with virtually no implications for the form of the final equation.

Note that there are some extreme values in the table. For example, one customer's age is recorded as 6. This is obviously a recording error, and the age should be corrected to show the actual value, if possible. INCOME has missing values that are shown as dots, while the nominal variable STATUS has missing values that are represented by blanks. The **Impute** node of Enterprise Miner can be used to impute such missing values. See Chapters 2, 6, and 7 for details.

1.3.2 Predicting Risk in the Auto Insurance Industry

The auto insurance company wants to examine its customer data and classify its customers into different risk groups. The objective is to align the premiums it is charging with the risk rates of its customers. If high-risk customers are charged low premiums, the loss ratios will be too high and the company will be driven out of business. If low-risk customers are charged disproportionately high rates, then the company will lose customers to its competitors. By accurately assessing the risk profiles of its customers, the company hopes to set customers' insurance premiums at an optimum level consistent with risk. A risk model is needed to assign a risk score to each existing customer.

In a risk model, *loss frequency* can be used as the target variable. Loss frequency is calculated as the number of losses due to accidents per *car-year*, where car-year is equal to the time since the auto insurance policy went into effect, expressed in years, multiplied by the number of cars covered by the policy. Loss frequency can be treated as either a continuous (interval-scaled) variable or a discrete (ordinal) variable that classifies each customer's losses into a limited number of bins. (See Chapters 5 and 7 for details about bins.) For purposes of illustration, I model loss frequency as a continuous variable in Chapter 4 and as a discrete ordinal variable in Chapters 5 and 7. The loss frequency considered here is the loss arising from an accident in which the customer was "at fault," so it could also be referred to as "at-fault accident frequency." I use *loss frequency*, *claim frequency,* and *accident frequency* interchangeably.

Table 1.2 shows what the modeling data set might look like for developing a model with loss frequency as an ordinal target.

Table 1.2

CUSTOMER	AGE	INCOME	NPRVIO	lossfrq
1	25	$45,000	0	0
2	45	$61,000	1	1
3	54	.	2	0
4	32	$24,000	3	3
5	43	$31,000	4	0
6	56	$23,456	0	0
7	78	.	1	2
8	6	$100,256	3	2
9	26	$345,678	4	1
10	32	$100,211	5	3
11	51	$21,312	3	2
12	31	.	1	1
13	23	$24,234	0	0
14	47	$43,566	1	0
15	77	$12,002	0	0
16	83	$32,454	0	2
17	25	$61,345	1	1
18	32	$76,123	1	0
19	52	$25,324	3	1
20	32	$31,886	1	0
21	23	$78,345	3	3
22	80	$61,234	2	0
23	123	$76,876	2	0
24	45	$24,002	1	1

The target variable is LOSSFRQ, which represents the accidents per car-year incurred by a customer over a period of time. This variable will be discussed in more detail in subsequent chapters in this book. For now it is sufficient to note that it is an ordinal variable that takes on values of 0, 1, 2, and 3. The input variables are AGE, INCOME, and NPRVIO. The variable NPRVIO represents the number of previous violations a customer had before he purchased the insurance policy.

1.3.3 Predicting Rate Sensitivity of Bank Deposit Products

In order to assess customers' sensitivity to an increase in the interest rate on a savings account, a bank may conduct price tests. Suppose one such test involves offering a higher rate for a fixed period of time, called the *promotion window*.

In order to assess customer sensitivity to a rate increase, it is possible to fit three types of models to the data generated by the experiment:

- a response model to predict the probability of response
- a short-term demand model to predict the expected change in deposits during the promotion period
- a long-term demand model to predict the increase in the level of deposits beyond the promotion period

The target variable for the response model is binary: response or no response. The target variable for the short-term demand model is the increase in savings deposits during the promotion period net of[2] any concomitant declines in other accounts. The target variable for the long-term demand model is the amount of the increase remaining in customers' bank accounts after the promotion period. In the case of this model, the promotion window for analysis has to be clearly defined, and only customer transactions that have occurred prior to the promotion window should be included as inputs in the modeling sample.

Table 1.3 shows what the data set looks like for modeling a continuous target.

---

[2] If a customer increased savings deposits by $100 but decreased checking deposits by $20, then the net increase is $80. Here, *net of* means *excluding*.

Table 1.3

CUSTOMER	AGE	INCOME	B_JAN	B_FEB	B_MAR	B_APR	BAL_AFTER
1	25	$45,000	$4,000	$4,230	$4,400	$4,900	$5,900
2	45	$61,000	$5,000	$4,000	$3,000	$0	$2,000
3	54	.	$1,200	$1,100	$3,000	$100	$200
4	32	$24,000	$5,234	$345	$5,678	$78	$878
5	43	$31,000	$4,000	$4,230	$4,400	$4,900	$4,950
6	56	$23,456	$2,000	$4,000	$3,000	$0	$1,000
7	78	.	$1,200	$1,100	$3,000	$100	$1,300
8	6	$100,256	$5,234	$345	$5,678	$78	$1,088
9	26	$345,678	$3,435	$4,674	$678	$80,000	$80,000
10	32	$100,211	$787	$4,230	$4,400	$4,900	$5,900
11	51	$21,312	$8,780	$7,800	$3,456	$0	$10,000
12	31	.	$5,000	$4,000	$3,000	$0	$4,000
13	23	$24,234	$4,000	$4,230	$4,400	$4,900	$5,900
14	47	$43,566	$4,674	$678	$800	$7,890	$8,890
15	77	$12,002	$5,234	$345	$5,678	$78	$1,078
16	83	$32,454	$4,000	$4,230	$4,400	$4,900	$5,900
17	25	$61,345	$2,000	$4,000	$3,000	$0	$1,000
18	32	$76,123	$1,200	$1,100	$3,000	$100	$1,100
19	52	$25,324	$5,234	$345	$5,678	$78	$1,078
20	32	$31,886	$3,435	$4,674	$678	$8,000	$9,000
21	23	$78,345	$787	$4,230	$4,400	$4,900	$5,900
22	80	$61,234	$8,780	$7,800	$3,456	$0	$100
23	123	$76,876	$5,000	$4,000	$3,000	$0	$1,034
24	45	$24,002	$4,000	$4,230	$4,400	$4,900	$7,245

The data set shown in Table 1.3 represents an attempt by a hypothetical bank to induce its customers to increase their savings deposits by increasing the interest paid to them by a predetermined number of basis points. This increased interest rate was offered (let us assume) in May 2006. Customer deposits were then recorded at the end of May 2006 and stored in the data set shown in Table 1.3 under the variable name BAL_AFTER. The bank would like to know what type of customer is likely to increase her savings balances the most in response to a future incentive of the same amount. The target variable for this is the dollar amount of change in balances from a point before the promotion period to a point after the promotion period. The target variable is continuous. The inputs, or explanatory variables, are AGE, INCOME, B_JAN, B_FEB, B_MAR, and B_APR. The variables B_JAN, B_FEB, B_MAR, and B_APR refer to customers' balances in all their accounts at the end of January, February, March, and April of 2006, respectively.

1.3.4 Predicting Customer Attrition

In banking, attrition may mean a customer closing a savings account, a checking account, or an investment account. In a model to predict attrition, the target variable can be either binary or continuous. For example, if a bank wants to identify customers who are likely to terminate their accounts at *any* time within a pre-defined interval of time in the future, it is possible to model attrition as a binary target. However, if the bank is interested in predicting the *specific* time at which the customer is likely to "attrit," then it is better to model attrition as a continuous target—time to attrition.

In the example shown below, attrition is modeled as a binary target. When you model attrition using a binary target, you must define a performance window during which you observe the occurrence or non-occurrence of the event. If a customer attrited during the performance window, the record will show 1 for the event and 0 otherwise.

Any customer transactions (deposits, withdrawals, and transfers of funds) that are used as inputs for developing the model should take place during the period prior to the performance window. The *inputs window* during which the transactions are observed, the *performance window* during which the event is observed, and the *operational lag*, which is the time delay in acquiring the inputs, are discussed in detail in Chapter 7 where an attrition model is developed.

Table 1.4 shows what the data set looks like for modeling customer attrition.

Table 1.4

CUSTOMER	AGE	INCOME	B_JAN	B_FEB	B_MAR	B_APR	ATTR
1	25	$45,000	$4,000	$4,230	$4,400	$4,900	0
2	45	$61,000	$5,000	$4,000	$3,000	$0	1
3	54	.	$1,200	$1,100	$3,000	$100	0
4	32	$24,000	$5,234	$345	$5,678	$78	0
5	43	$31,000	$4,000	$4,230	$4,400	$4,900	0
6	56	$23,456	$2,000	$4,000	$3,000	$0	1
7	78	.	$1,200	$1,100	$3,000	$100	0
8	6	$100,256	$5,234	$345	$5,678	$78	0
9	26	$345,678	$3,435	$4,674	$678	$80,000	1
10	32	$100,211	$787	$4,230	$4,400	$4,900	0
11	51	$21,312	$8,780	$7,800	$3,456	$0	1
12	31	.	$5,000	$4,000	$3,000	$0	1
13	23	$24,234	$4,000	$4,230	$4,400	$4,900	0
14	47	$43,566	$4,674	$678	$800	$7,890	0
15	77	$12,002	$5,234	$345	$5,678	$78	1
16	83	$32,454	$4,000	$4,230	$4,400	$4,900	0
17	25	$61,345	$2,000	$4,000	$3,000	$0	0
18	32	$76,123	$1,200	$1,100	$3,000	$100	0
19	52	$25,324	$5,234	$345	$5,678	$78	0
20	32	$31,886	$3,435	$4,674	$678	$80,000	0
21	23	$78,345	$787	$4,230	$4,400	$4,900	0
22	80	$61,234	$8,780	$7,800	$3,456	$0	0
23	123	$76,876	$5,000	$4,000	$3,000	$0	0
24	45	$24,002	$4,000	$4,230	$4,400	$4,900	0

In the data set shown in Table 1.4, the variable ATTR represents the customer attrition observed during the performance window, consisting of the months of June, July, and August of 2006. The target variable takes the value of 1 if a customer attrits during the performance window and 0 otherwise. Table1.4 shows the input variables for the model. They are AGE, INCOME, B_JAN, B_FEB, B_MAR, and B_APR. The variables B_JAN, B_FEB, B_MAR, and B_APR refer to customers' balances for all of their accounts at the end of January, February, March, and April of 2006, respectively.

1.3.5 Predicting a Nominal Categorical (Unordered Polychotomous) Target

Assume that a hypothetical bank wants to predict, based on the products a customer currently owns and other characteristics, which product the customer is likely to purchase next. For example, a customer may currently have a savings account and a checking account, and the bank would like to know if the customer is likely to open an investment account, or open an IRA, or take out a mortgage. The target variable for this situation is nominal. Models with nominal targets are also used by market researchers who need to understand consumer preferences for different products or brands. Chapter 6 shows some examples of models with nominal targets.

Table 1.5 shows what a data set might look like for modeling a nominal categorical target.

Table 1.5

CUSTOMER	AGE	INCOME	PRIORPR	NXTPR
1	25	$45,000	A	X
2	45	$61,000	B	Z
3	54	.	C	Y
4	32	$24,000	A	X
5	43	$31,000	B	Z
6	56	$23,456	C	Z
7	78	.	C	Z
8	6	$100,256	A	X
9	26	$345,678	AB	X
10	32	$100,211	CD	Z
11	51	$21,312	AC	Y
12	31	.	AB	X
13	23	$24,234	CD	Z
14	47	$43,566	D	Z
15	77	$12,002	E	Z
16	83	$32,454	A	X
17	25	$61,345	B	X
18	32	$76,123	A	Z
19	52	$25,324	A	Y
20	32	$31,886	C	X
21	23	$78,345	D	Z
22	80	$61,234	A	Z
23	123	$76,876	B	Z
24	45	$24,002	D	X

In Table 1.5, the input data includes the variable PRIORPR which indicates the product or products owned by the customer of a hypothetical bank at the beginning of the performance window. The *performance window*, defined in the same way as in Section 1.3.4, is the time period during which a customer's purchases are observed. Given that a customer owned certain products at the beginning of the performance window, we observe the next product that the customer purchased during the performance window and indicate it by the variable NXTPR.

For each customer, the value for the variable PRIORPR indicates the product that was owned by the customer at the beginning of the performance window. The letter A might stand for a savings account, B might stand for a certificate of deposit, etc. Similarly, the value for the variable NXTPR indicates the first product purchased by a customer during the performance window. For example, if the customer owned product B at the beginning of the performance window and purchased products X and Z, in that order, during the performance window, then the variable NXTPR takes the value X. If the customer purchased Z and X, in that order, the variable NXTPR takes the value Z, and the variable PRIORPR takes the value B on the customer's record.

1.4 Sources of Modeling Data

It is important to distinguish between two different scenarios by which data becomes available for modeling. For example, consider a marketing campaign. In the first scenario, the data is based on an experiment carried out by conducting a marketing campaign on a well-designed sample of customers drawn from the target population. In the second scenario, the data is a sample drawn from the results of a past marketing campaign and not from the target population. While the latter scenario is clearly less desirable, it is often necessary to make do with whatever data is available. In such cases, you can make some adjustments through observation weights to compensate for the lack of perfect compatibility between the modeling sample and the target population.

In either case, for modeling purposes, the file with the marketing campaign results is appended to data on customer characteristics and customer transactions. Although transaction data is not always available, these tend to be key drivers for predicting the attrition event.

1.4.1 Comparability between the Sample and the Target Universe

Before launching a modeling project it is necessary to verify that the sample is a good representation of the target universe. This can be done by comparing the distributions of some key variables in the sample and the target universe. For example, if the key characteristics are age and income, then you should compare the age and income distribution between the sample and the target universe.

1.4.2 Observation Weights

If the distributions of key characteristics in the sample and the target population are different, sometimes observation weights are used to correct for any bias. In order to detect the difference between the target population and the sample, it is necessary to have some prior knowledge of the target population. Assuming that age and income are the key characteristics, you can derive the weights as follows: Divide income into, let's say, four groups and age into, say, three groups. Suppose that the target universe has N_{ij} people in the ith age group and j^{th} income group, and assume that the sample has n_{ij} people in the same age-income group. In addition, suppose the total number of people in the target population is *N*, and the total number of people in the sample is n. In this case, the appropriate observation weight is $(N_{ij} / N)/(n_{ij} / n)$ for the individual in the ith age group and j^{th} income group in the sample. These observation weights should be constructed and included for each record in the modeling sample prior to launching Enterprise Miner, in effect creating an additional variable in your data set. In Enterprise Miner, you assign the role of **Frequency** to this variable in order for the modeling tools to consider these weights in estimating the models. The situation described here inevitably arises when you do not have a scientific sample drawn from the target population, which is very often the case.

However, there is another source of bias that is often deliberately introduced. This bias is due to over-sampling of rare events. For example, in response modeling, if the response rate is very low, it is necessary to include all the responders available and only a random fraction of non-responders. The bias introduced by such over-sampling is corrected by adjusting the predicted probabilities with prior probabilities. This technique will be discussed in Section 4.7.2.

1.5 Pre-Processing the Data

Pre-processing has several purposes:

- eliminate obviously irrelevant data elements, e.g., name, social security number, street address, etc., that clearly have no effect on the target variable
- convert the data to an appropriate measurement scale, especially converting categorical (nominal-scaled) data to interval-scaled when appropriate
- eliminate variables with highly skewed distributions
- eliminate inputs which are really target variables disguised as inputs
- impute missing values

Although many cleaning tasks can be done within Enterprise Miner, as the following examples show, there are some that should be done prior to launching Enterprise Miner.

1.5.1 Data Cleaning Before Launching SAS Enterprise Miner

Data vendors sometimes treat interval-scaled variables, such as birth date or income, as character variables. If a variable such as birth date is entered as a character variable, it would be treated by Enterprise Miner as a categorical variable with many categories. To avoid such a situation, it would be better to derive a numeric variable from the character variable and then drop the original character variable from your data set. For example, in one modeling project, I first converted the birth date to SAS format, derived age from the birth date, and then used age as an input variable.

Similarly, income is sometimes represented as a character variable. The character A may stand for $20K ($20,000), B for $30K, etc. To convert the income variable to an ordinal or interval scale, it is best to create a new version of the income variable in which all the values are numeric, and then eliminate the character version of income.

Another situation which requires data cleaning that cannot be done within Enterprise Miner arises when the target variable is disguised as an input variable. For example, a financial institution would like to model customer attrition in its brokerage accounts. A model is to be developed to predict the probability of attrition during a time interval of three months in the future. The institution decides to develop the model based on actual attrition during a performance window of three months. The objective is to predict attritions based on customers' demographic and income profiles and balance activity in their brokerage accounts prior to the window. The binary target variable takes the value of 1 if the customer attrits and 0 otherwise. If a customer's balance in his brokerage account is 0 for two consecutive months, then he is considered an attritor, and the target value is set to 1. If the data set includes both the target variable (attrition/no attrition) and the balances during the performance window, then the account balances may be inadvertently treated as input variables. To prevent this, inputs which are really target variables disguised as input variables should be removed before launching Enterprise Miner.

These examples demonstrate that while there is no uniformly applicable solution to data cleaning, every effort should be made to ensure that instances such as those illustrated above are resolved before launching Enterprise Miner.

1.5.2 Data Cleaning After Launching SAS Enterprise Miner

Display 1.1 shows an example of a variable that is highly skewed. The variable is MS, which indicates the marital status of a customer. The variable RESP represents customer response to mail. It takes the value of 1 if a customer responds, and 0 otherwise. In this hypothetical sample, there are only 100 customers with marital status M (married), and 2900 with S (single). None of the married customers are responders. An unusual situation such as this may cause the marital status variable to play a much more significant role in the predictive model than is really warranted, because the model tends to infer that all the married customers were non-responders because they were married. The real reason there were no responders among them is simply that there were so few married customers in the sample.

Display 1.1

The SAS System

The FREQ Procedure

Frequency
Percent
Row Pct
Col Pct

Table of MS by RESP			
	RESP(RESPONSE)		
MS(MARITAL STATUS)	0	1	Total
M	100 3.33 100.00 3.42	0 0.00 0.00 0.00	100 3.33
S	2826 94.20 97.45 96.58	74 2.47 2.55 100.00	2900 96.67
Total	2926 97.53	74 2.47	3000 100.00

Variables such as the one shown in Display 1.1 can produce spurious results if used in the model. You can identify these variables using the **StatExplore** node, set their roles to **Rejected** in the **Input Data** node, and drop them from the table using the **Drop** node.

The **Filter** node can be used for eliminating observations with extreme values, although I do not recommend elimination of observations. Correcting them or capping them instead may be a better alternative, in order to avoid introducing any bias into the model parameters. The **Impute** node offers a variety of methods for imputing missing values. These nodes will be discussed in detail in the next chapter. Imputing missing values is necessary when you use the **Regression** or **Neural Network** nodes.

1.6 Alternative Modeling Strategies

The choice of modeling strategy depends on the modeling tool and the number of inputs under consideration for modeling. Here are examples of two possible strategies to consider in using the **Regression** node.

1.6.1 Regression with a Moderate Number of Input Variables

Pre-process the data:

- Eliminate obviously irrelevant variables.
- Convert nominal-scaled inputs with too many levels to numeric interval-scaled inputs, if appropriate.
- Create composite variables (such as average balance in a savings account during the six months prior to a promotion campaign) from the original variables if necessary. This can also be done with Enterprise Miner using the **SAS Code** node.

Next, use Enterprise Miner to perform these tasks:

- Impute missing values.
- Transform the input variables.
- Partition the modeling data set into train, validate, and test (when the available data is large enough) samples. Partitioning can be done prior to imputation and transformation, because Enterprise Miner automatically applies these to all parts of the data.
- Run the **Regression** node with the Stepwise option.

1.6.2 Regression with a Large Number of Input Variables

Pre-process the data:

- Eliminate obviously irrelevant variables.
- Convert nominal-scaled inputs with too many levels to numeric interval-scaled inputs, if appropriate.
- Combine variables if necessary.

Next, use Enterprise Miner to perform these tasks:

- Impute missing values.
- Make a preliminary variable selection. (Note: This step is not included in Section 1.6.1.)
- Group categorical variables (collapse levels).

- Transform interval-scaled inputs.
- Partition the data set into train, validate, and test samples.
- Run the **Regression** node with the Stepwise option.

The main difference between the steps outlined in Sections 1.6.1 and 1.6.2 is that Enterprise Miner is used to make a preliminary variable selection in Section 1.6.2 (because the number of inputs is large) and not in Section 1.6.1.

The steps given in Sections 1.6.1 and 1.6.2 are only two of many possibilities. For example, one can use the **Decision Tree** node to make a variable selection and create dummy variables to then use in the **Regression** node.

Getting Started with Predictive Modeling

2.1 Introduction

This chapter introduces you to SAS Enterprise Miner 5.2 and some of the preprocessing and data cleaning tools (nodes) needed for data mining and predictive modeling projects. The following topics are covered:

- opening Enterprise Miner 5.2
- starting a new project:
  - naming the project
  - providing a directory where Enterprise Miner stores the data sets and programs that it generates for the project
  - specifying the directory where the user's data sets are located.
- creating a data source for each data set used in the project:
  - providing the name and location of the data set
  - setting the measurement scales and roles of the input variables
  - creating a profit matrix for decision-making.

- using some of the nodes for data cleaning, data exploration, variable selection, and variable transformation:
  - **Input Data** node (for specifying the data set for the project)
  - **StatExplore** node (for univariate analysis of the inputs and exploring the relationship of inputs to the target)
  - **MultiPlot** node (also for univariate analysis of the inputs and exploring the relationship of inputs to the target)
  - **Impute** node (for imputing missing values)
  - **Data Partition** node (for partitioning the data set into subsets for use in training, validating, and testing models)
  - **Filter** node (for examining and filtering outliers)
  - **Variable Selection** node (for selecting important variables for modeling)
  - **Transform Variables** node (for performing various transformations to improve model accuracy)
  - **SAS Code** node (for writing custom code)
  - **Drop** node (for dropping variables).

Note: The order in which the nodes are presented is not necessarily the order in which they should generally be used.

Enterprise Miner's modeling tools—**Decision Tree** node, **Neural Network** node, and **Regression** node—are not included in this chapter as they are covered extensively in Chapters 4, 5, and 6.

2.2 Opening SAS Enterprise Miner 5.2

To start Enterprise Miner 5.2, click on the Enterprise Miner icon on your desktop[1]. The logon screen will appear as shown in Display 2.1.

[1] All the illustrations presented here are from sessions in which Enterprise Miner 5.2 is installed on a personal computer.

Display 2.1

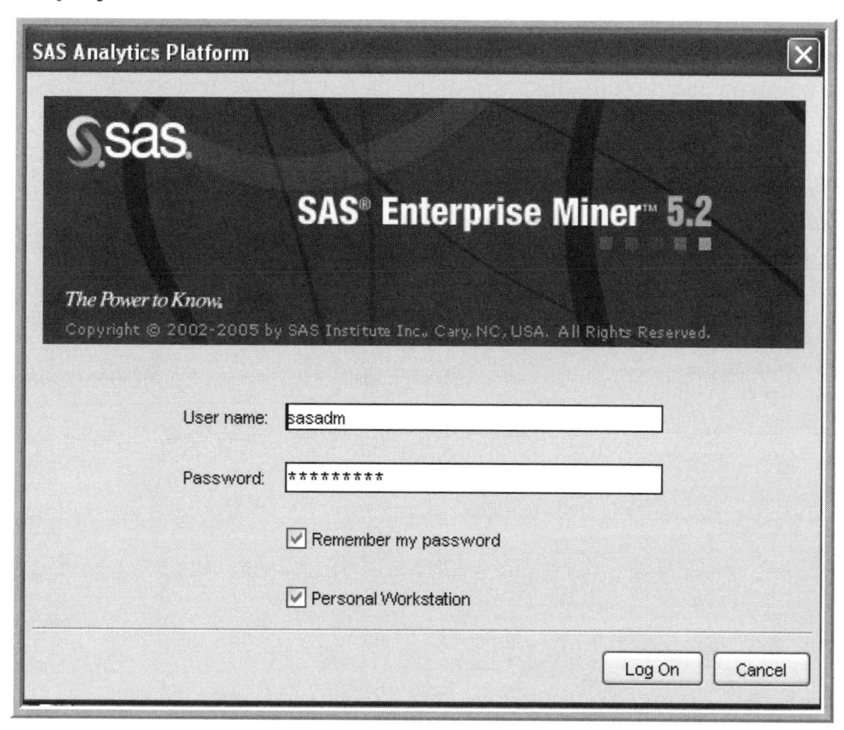

Enter your user name and password in the logon screen. Check the **Personal Workstation** box if you are using Enterprise Miner 5.2 on a personal computer. When you click the **Log On** button, the **Enterprise Miner** window opens, as shown in Display 2.2.

Display 2.2

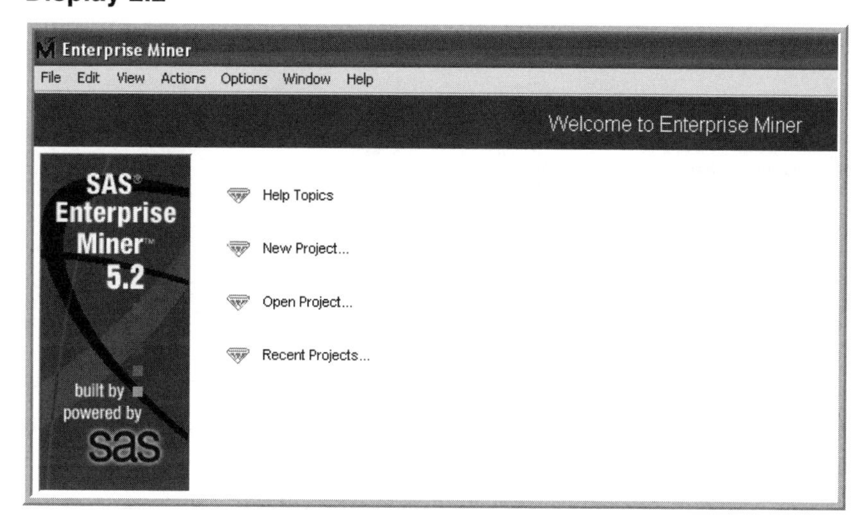

2.3 Creating a New Project in SAS Enterprise Miner 5.2

Creating a new project involves naming the project, as well as indicating the directory where you want the project to be saved and the directory where the data for the project exists. These steps are shown below.

When you select **New Project** from the **Enterprise Miner** window shown in Display 2.2, the **Create New Project** window opens, as shown in Display 2.3.

Display 2.3

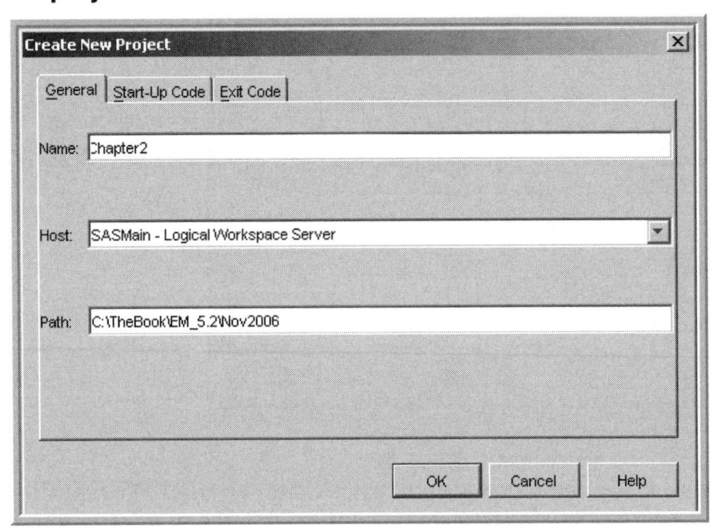

In this window enter the **Name** of the project and the **Path** where you want to save the project. I typed `Chapter2` in the **Name** box, and typed `C:\TheBook\EM_5.2\Nov2006` in the **Path** box. This is the directory where the project will be stored. Next, I selected the **Start-Up Code** tab. On the **Start-Up Code** tab, I typed the LIBNAME statement as shown in Display 2.4.

Display 2.4

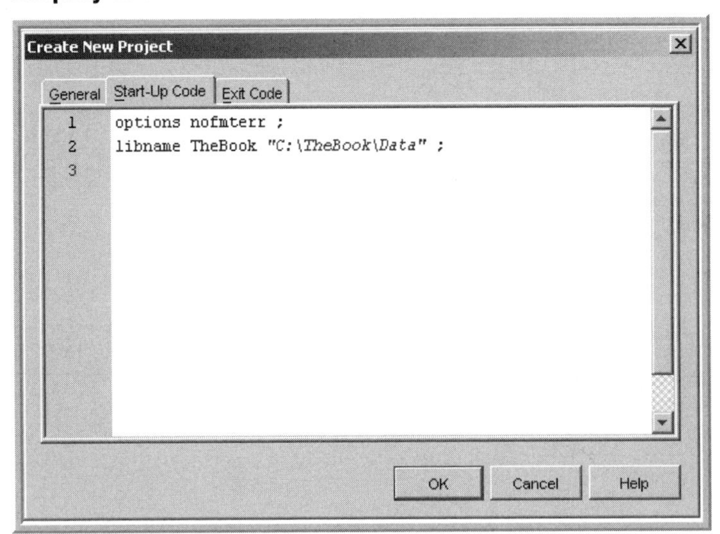

The data for this project is located in the folder `C:\TheBook\Data`. This will be indicated by the libref `TheBook`. When you select **OK**, the Enterprise Miner 5.2 interface window opens, showing the new project. This window is shown in Display 2.5.

2.4 The SAS Enterprise Miner Window

This is the window where you create the process flow diagram for your data mining project. In Display 2.5, I have numbered the components of the **Enterprise Miner** window for easy reference.

Display 2.5

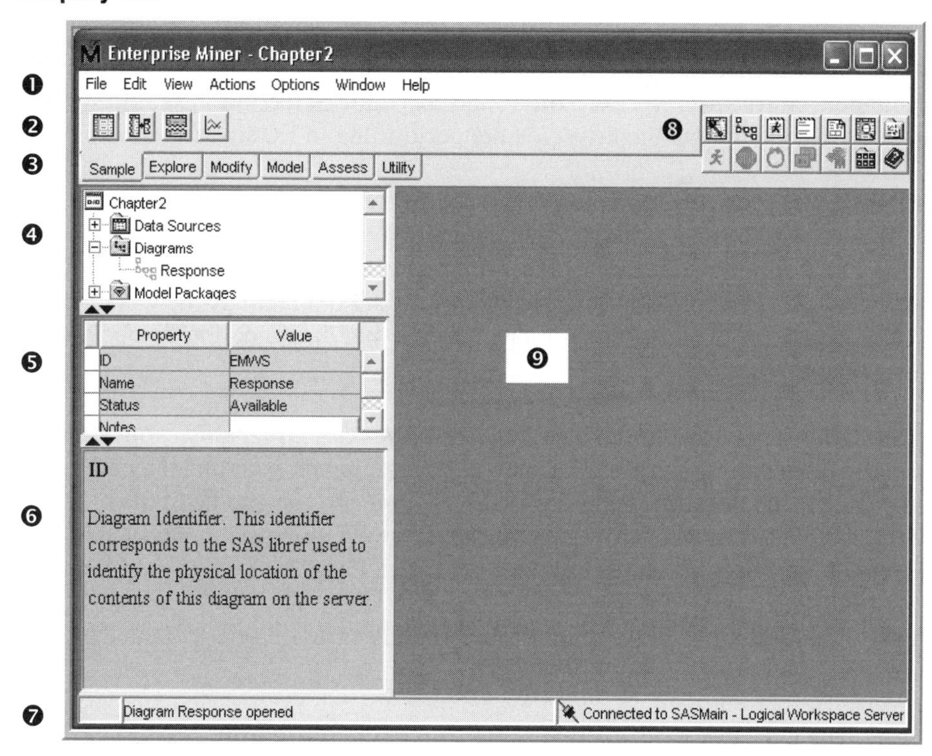

A brief description of these components follows:

➊ **Menu Bar**: The menu bar has seven items. These are **File**, **Edit**, **View**, **Actions**, **Options**, **Window**, and **Help**. By clicking on any one of these buttons, you will see a drop-down menu. For example, by clicking on **View→Property Sheet→Advanced**, you can reset the properties list from **Basic** to **Advanced**. This is illustrated in Display 2.6.

Display 2.6

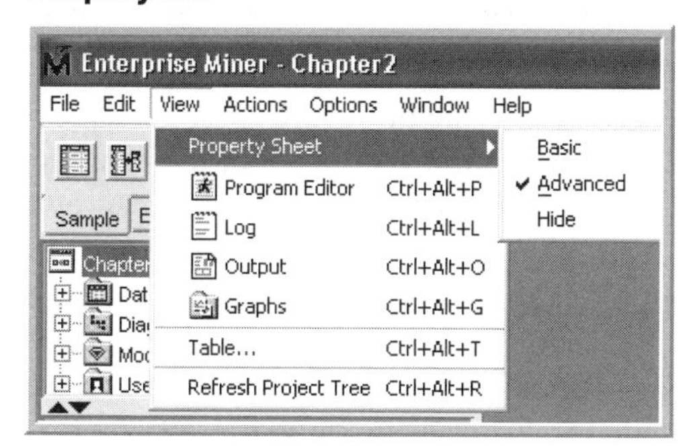

❷ **Toolbar**: The toolbar contains the Enterprise Miner node (tool) icons. The icons displayed on the **toolbar** change according the **Node (Tool) Group** tab you select in the next line indicated by ❸ in Display 2.5.

❸ **Node (Tool) Groups**: These tabs are for selecting different groups of nodes. The node groups are **Sample**, **Explore**, **Modify**, **Model**, **Assess**, and **Utility**. The toolbar ❷ changes according to the node group selected. If you select the **Sample** tab on this line, you will see the icons for **Input Data**, **Sample**, **Data Partition**, and **Time Series** in ❷. If you select the **Explore** tab, you will see the icons for **Association**, **Cluster**, **Multiplot**, **Path Analysis**, **SOM/Kohonen**, **StatExplore**, **Text Miner**, and **Variable Selection** in ❷. Similarly, other tabs can be used according to the tasks you are performing in Enterprise Miner.

❹ **Project Panel**: The project panel is for viewing, creating, deleting, and modifying the **Data Sources**, **Diagrams**, **Model Packages**, and **Users**. For example, if you want to create a data source (that is, tell Enterprise Miner where your data is and give information about the variables, etc.), you click on **Data Sources** and proceed. For creating a new diagram, you right-click on **Diagrams** and proceed. To open an existing diagram, double-click on the diagram you want.

❺ **Properties Panel**: The properties of **Project**, **Data Sources**, **Diagrams**, **Nodes**, **Model Packages**, and **Users** are shown in this panel. Note that, in our example, the nodes are not yet created; hence, you do not see them in Display 2.5. You can view and edit the properties of any object selected. If you want to specify or change any options in a node such as **Decision Tree** or **Neural Network**, you must use the **Properties Panel**.

❻ **Help Panel**: This displays a description of the property that you select in the **Properties Panel**.

❼ **Status Bar**: This indicates the execution status of the Enterprise Miner task.

❽ **Toolbar Shortcut Buttons**: These are shortcut buttons for **Create Data Source**, **Create Diagram**, **Run**, etc. To display the text name of these buttons, position the mouse pointer over the button.

❾ **Diagram Workspace**: This is used for building and running the process flow diagram for the project with various nodes (tools) of Enterprise Miner.

2.5 Creating a SAS Data Source

You need to create a data source for each data set you use in the project. Creating a data source means providing Enterprise Miner with the relevant information about the data set being imported—the data set's name, location, library path, the role assignments and measurement levels of its variables, and the decision variables attributed to the target such as the profit matrix, etc. The profit matrix, also referred to as **Decision Weights** in Enterprise Miner 5.2, is used in decisions such as assigning a target class to an observation and assessing the models. (The use of profit matrices is discussed in detail in Chapter 4.) Enterprise Miner saves all of this information, or *metadata*, as different data sets in a folder called **DataSources** in the **Project Directory**.

To create a data source, click on the toolbar shortcut button or right-click on **Data Sources** in the **Project Panel**, as shown in Display 2.7.

Display 2.7

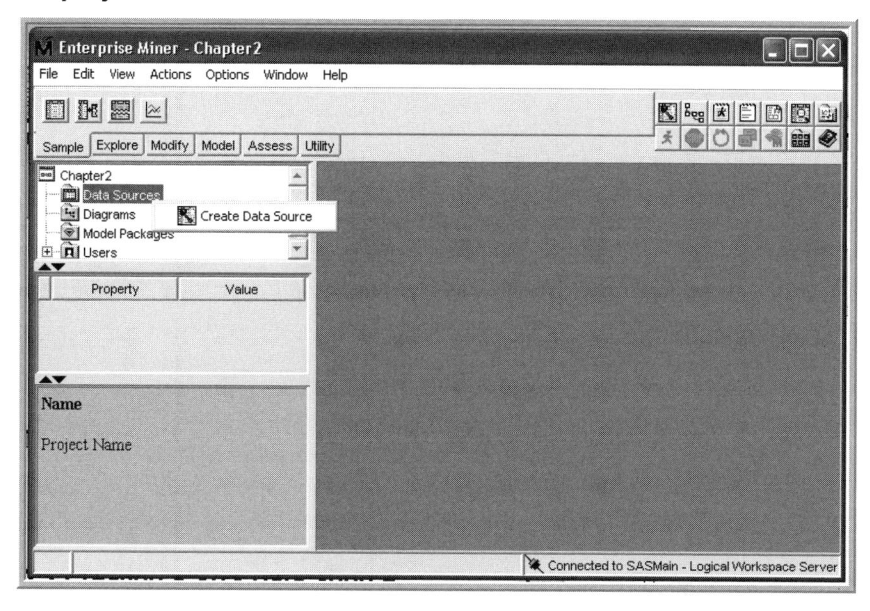

When you click on **Create Data Source**, the **Data Source Wizard** window opens, and Enterprise Miner prompts you to enter the data source. This window is shown in Display 2.8.

Display 2.8

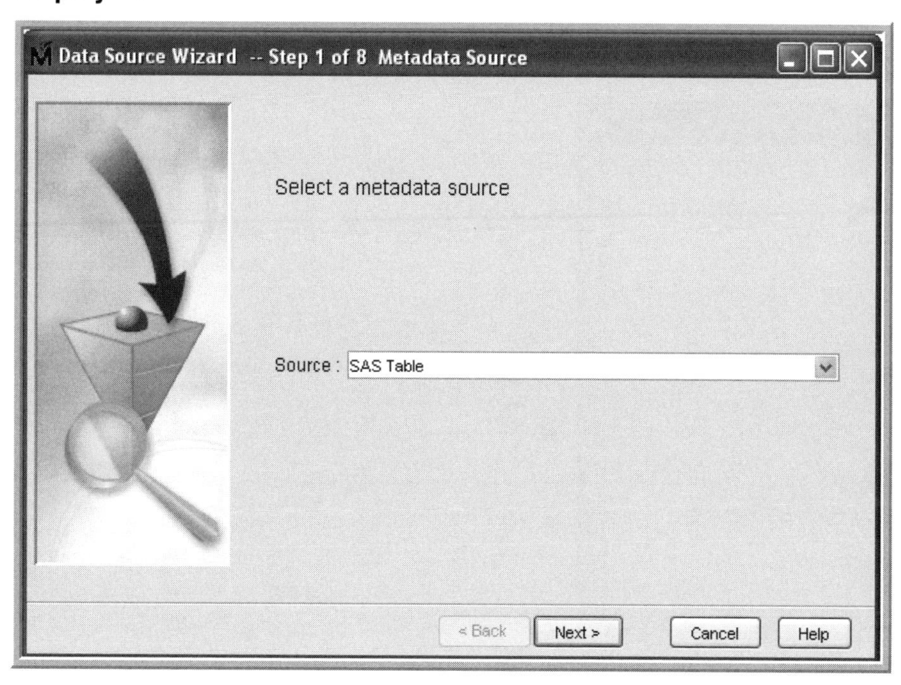

Is the source a **SAS Table** or is it a **Metadata Repository**? If, as in my example, you are using a SAS data set in your project, enter **SAS Table** in the **Source** box and click on **Next>**. **SAS Table** is, by default, already selected in the **Source** box. So, you can simply click on **Next>**. Then another window opens, prompting you to give the location of the SAS data set. This window is shown in Display 2.9.

Display 2.9

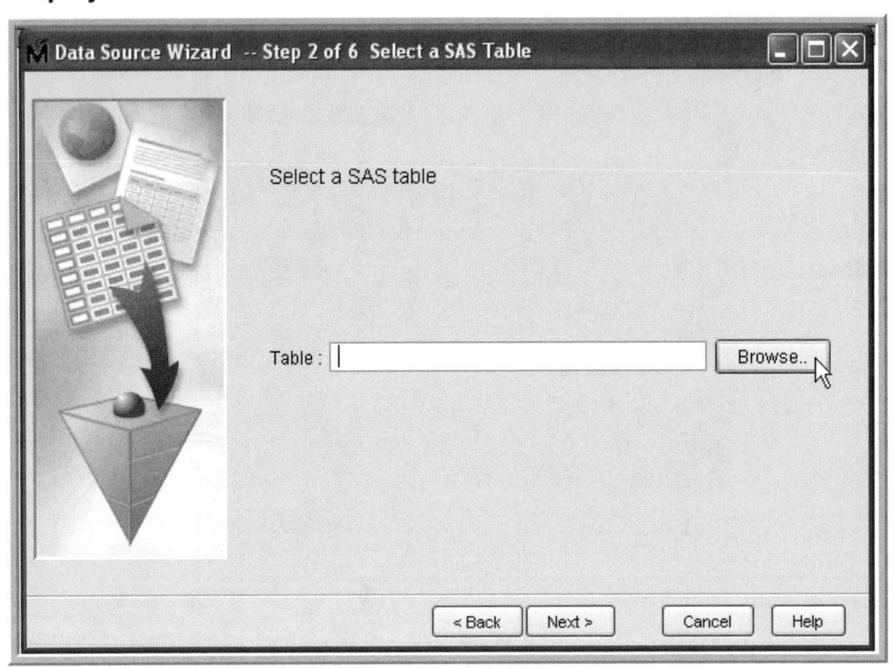

When you click **Browse..**, a window opens that shows the list of library references. This window is shown in Display 2.10.

Display 2.10

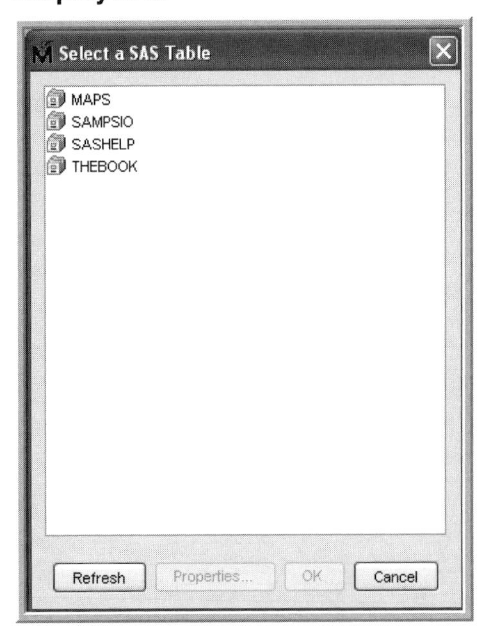

Since the data for my example project is in the library **TheBook**, when I double-click on **THEBOOK**, the window opens with a list of all the data sets in that library. This window is shown in Display 2.11.

Display 2.11

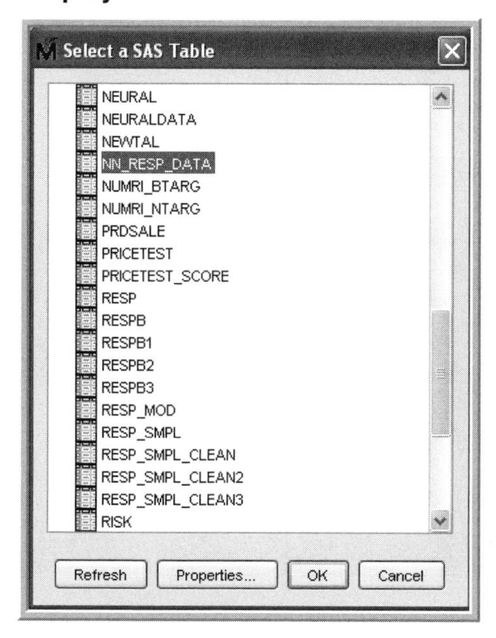

I select the data set named NN_RESP_DATA. When I click **OK**, the **Data Source Wizard** window opens as shown in Display 2.12.

Display 2.12

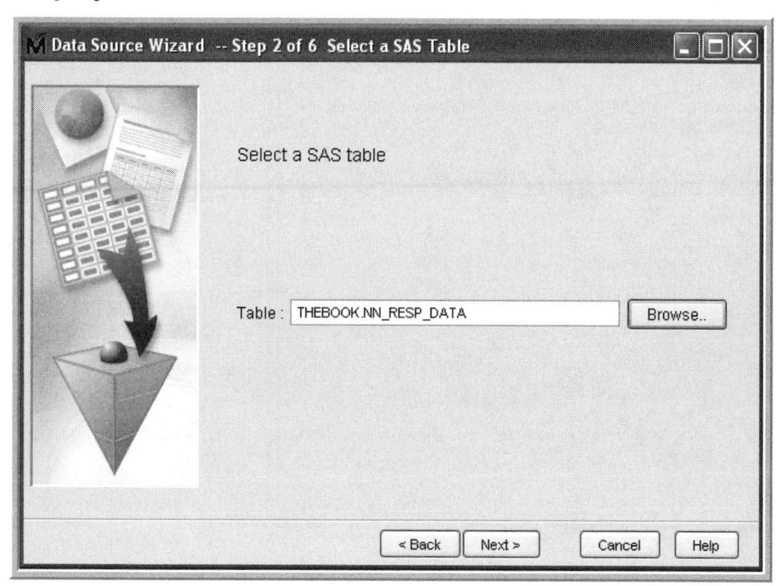

This display shows the libref and the data set name, which I entered. When I press **Next>**, another window opens displaying the **Table Properties**. This is shown in Display 2.13.

Display 2.13

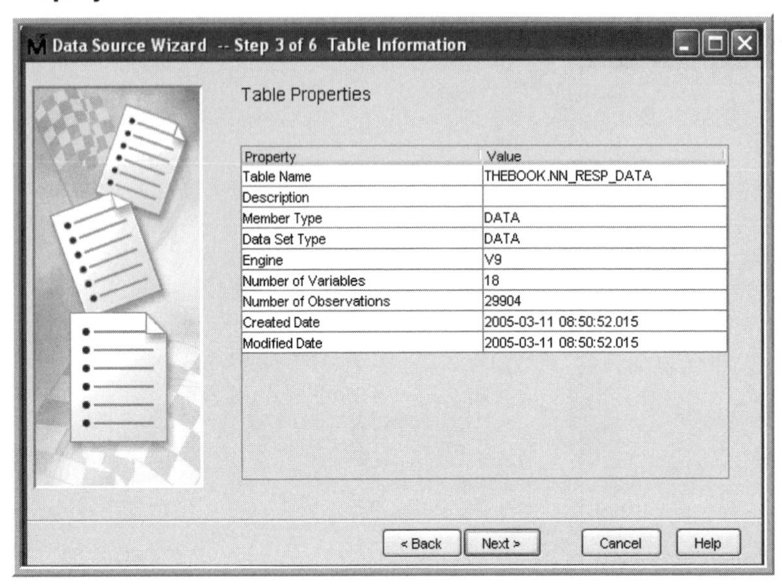

Clicking on **Next>** opens another window that shows **Metadata Advisor Options**. This is shown in Display 2.14A.

Display 2.14A

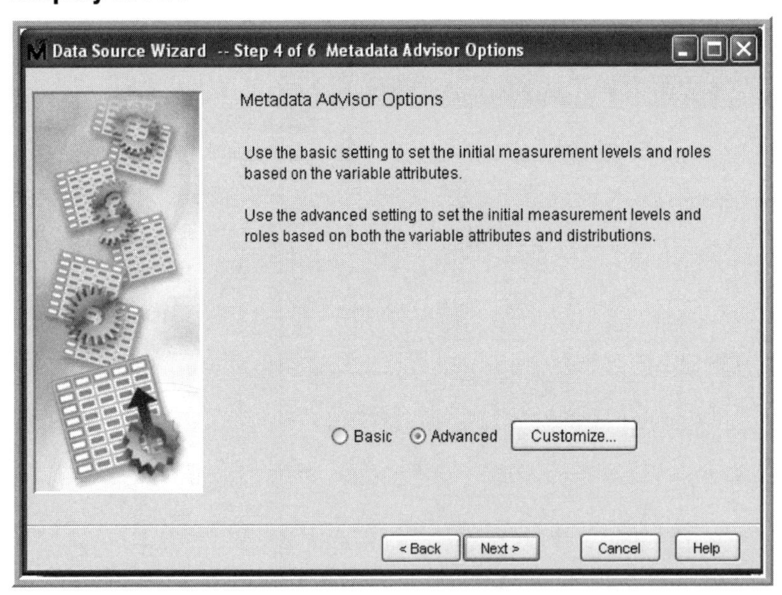

Display 2.14A provides a snapshot of the **Metadata Advisor Options** window, which is used for defining the metadata. Metadata is data about data sets. It specifies how each variable is used in the modeling process. The metadata contains information about the role of each variable, its measurement scale, etc.

If you select the **Basic** option, the initial measurement levels and roles are based on the variable attributes. This means that if a variable is numeric, its measurement scale is designated to be interval, irrespective of how many distinct levels the variable may have. For example, a numeric binary variable will initially be given the interval scale. If your target variable is binary in numeric form, it will be treated as an interval-scaled variable, and it will be treated as such in the subsequent nodes. If the subsequent node is a **Regression** node, Enterprise Miner automatically uses ordinary least squares regression, instead of the logistic regression, which is usually appropriate with a binary target variable.

By selecting the **Basic** option, all character variables will be considered nominal. You may reset the measurement scales and roles as appropriate.

By selecting the **Advanced** option, Enterprise Miner applies a bit more logic as it automatically sets the variable roles and measurement scales. If a variable is numeric and has more than 20 distinct values, Enterprise Miner sets its measurement scale to interval. In addition, if you select the **Advanced** option, you can customize the measurement scales. For example, by default the **Advanced** option sets the measurement scale of any numeric variable to nominal if it takes 20 or fewer unique values, but you can change this number by clicking the **Customize** button.

Display 2.14B shows the default settings for the **Advanced Advisor Options**.

Display 2.14B

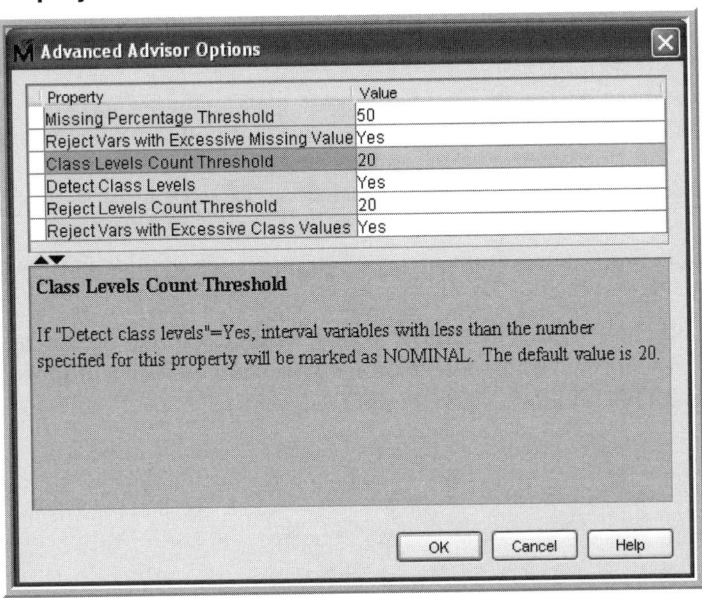

One advantage of selecting the **Advanced** option is that Enterprise Miner will automatically set the role if each unary variable to **Rejected**. If any of the settings are not appropriate, you can change them later in the window shown in Display 2.15.

In this example, I changed the **Class Levels Count Threshold** property to 5 and closed the **Advanced Advisor Options** window by clicking on **OK** and then clicking on **Next>**. This brings up the window shown in Display 2.15.

Display 2.15

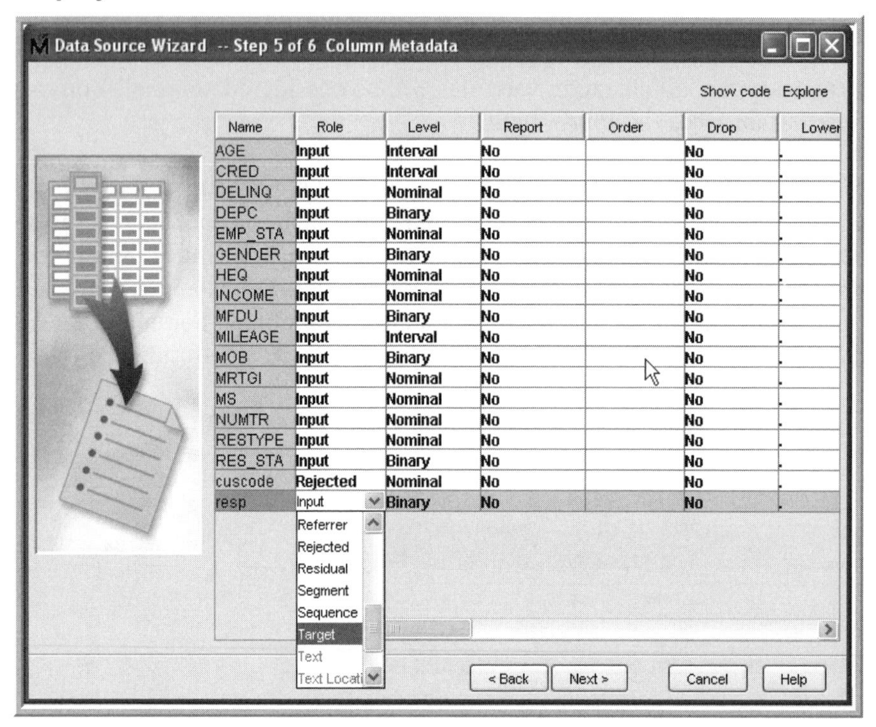

In this window, a table is shown with the variable names, model roles, and measurement levels of the variables in the data set. For the purpose of this illustration, I specify the model role of the variable RESP as the target, and click **Next>**. Enterprise Miner recognizes that the target is binary. Hence, it opens another window with the question **Do you want to build models based on the values of the decisions?** This window is shown in Display 2.16.

Display 2.16

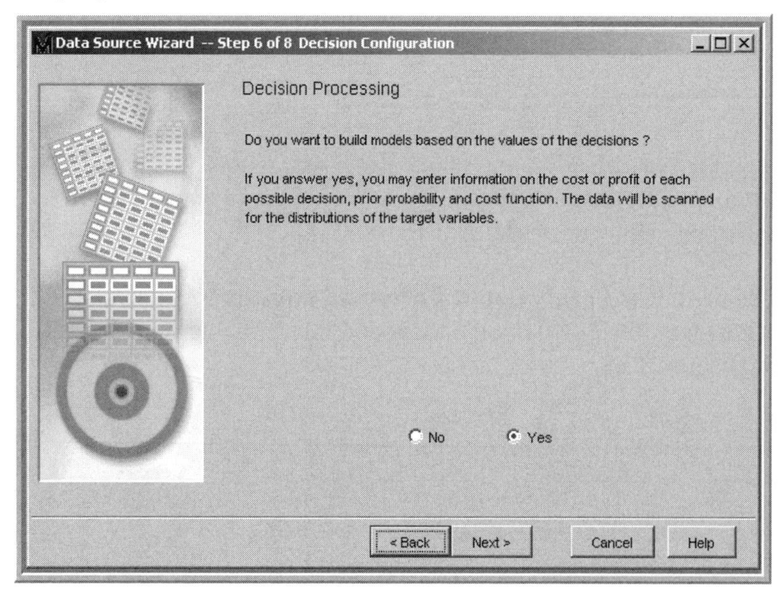

If you are using a profit matrix (decision weights), cost variables, and posterior probabilities, you should select **Yes**, and click **Next >** to enter these values. (These matrices can also be entered and modified at later stages.)

A window opens with tabs for **Targets**, **Prior Probabilities**, **Decisions**, and **Decision Weights**.

The window for the **Targets** tab is shown in Display 2.17.

Display 2.17

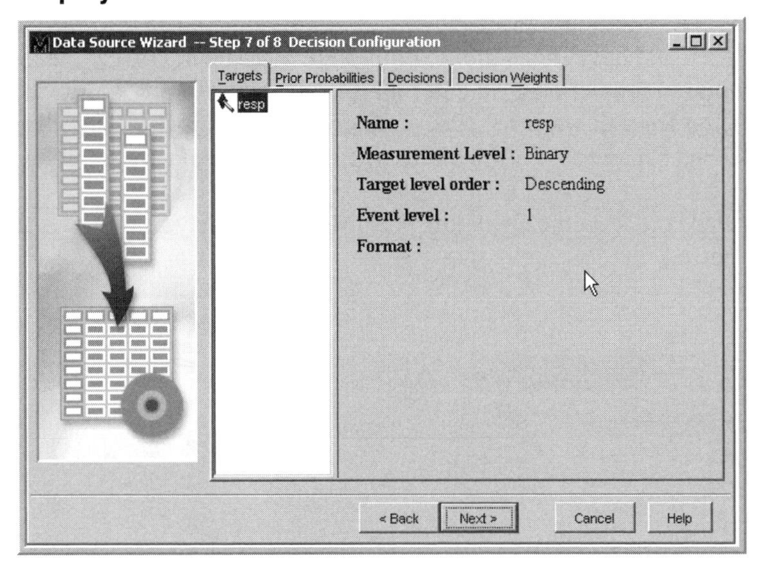

The **Targets** tab displays the name of the target variable and its measurement level. It also gives the target levels of interest. In this example, the variable RESP is the target and is binary, which means that it has two levels, namely response, indicated by 1, and non-response, indicated by 0. The event of interest is response. That is, the model is set up to estimate the probability of response. If the target has more than two levels, this window will show all of its levels. In later chapters, I will model an ordinal target that has more than two levels, each level indicating frequency of losses or accidents, with 0 indicating no accidents, 1 indicating one accident, and 2 indicating two accidents, and so on.

Display 2.18 shows the **Prior Probabilities** tab.

Display 2.18

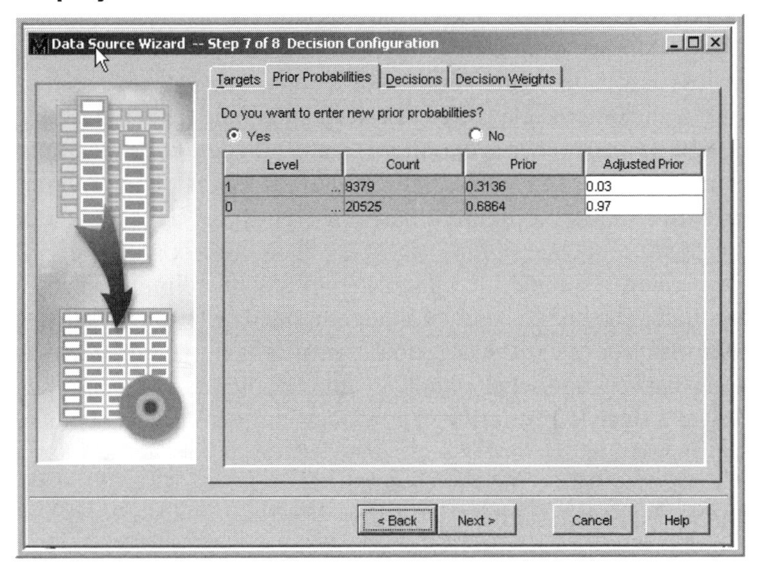

This tab shows (in the column labeled **Prior**) the probabilities of response and non-response calculated by Enterprise Miner for the sample used for model development. In the modeling sample I used in this example, the responders are over-represented. In the sample there are 31.36% responders and 68.64% non-responders. These are called *prior probabilities*. So the models developed from the modeling sample at hand will be biased unless a correction is made for the bias caused by over-representation of the responders. If you enter these prior probabilities in the **Adjusted Priors** column as I have done above, Enterprise Miner will correct the models for the bias and produce unbiased predictions. To enter these adjusted prior probabilities, select **Yes** in response to the question **Do you want to enter new prior probabilities?** Then enter the probabilities that you calculated separately for the entire population (.03 and .97 in my example).

In order to enter a profit matrix, click on the **Decision Weights** tab. Display 2.19 shows the **Decision Weights** (profit matrix) window.

Display 2.19

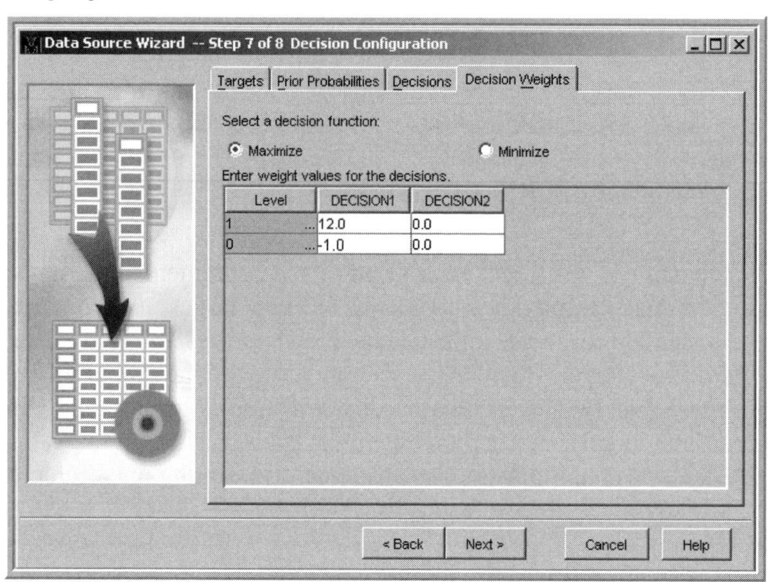

The columns of this matrix refer to the different decisions that need to be made based on the model's predictions. In this example, **Decision1** means classifying or labeling a customer as a responder, while **Decision2** means classifying a customer as a non-responder. The entries in the matrix indicate the profit or loss associated with a correct or incorrect assignment, so the matrix in this example implies that if a customer is classified as a responder, and he is in fact a responder, then the profit is $12. If a customer is classified as a responder, but she is in fact a non-responder, then there will be a loss of $1. Other cells of the matrix can be interpreted similarly. In developing predictive models, Enterprise Miner assigns target levels to the records in a data set. In the case of a response model, assigning target levels to the records means classifying each customer as a responder or non-responder. In a later step of any modeling project, Enterprise Miner also compares different models, on the basis of a user-supplied criterion, to select the best model. In order to have Enterprise Miner use the criterion of profit maximization when assigning target levels to the records in a data set and when choosing among competing models, select **Maximize** for the option **Select a decision function**. The values in the matrix shown here are arbitrary, and given only for illustration.

Display 2.20 shows the window for cost variables.

Display 2.20

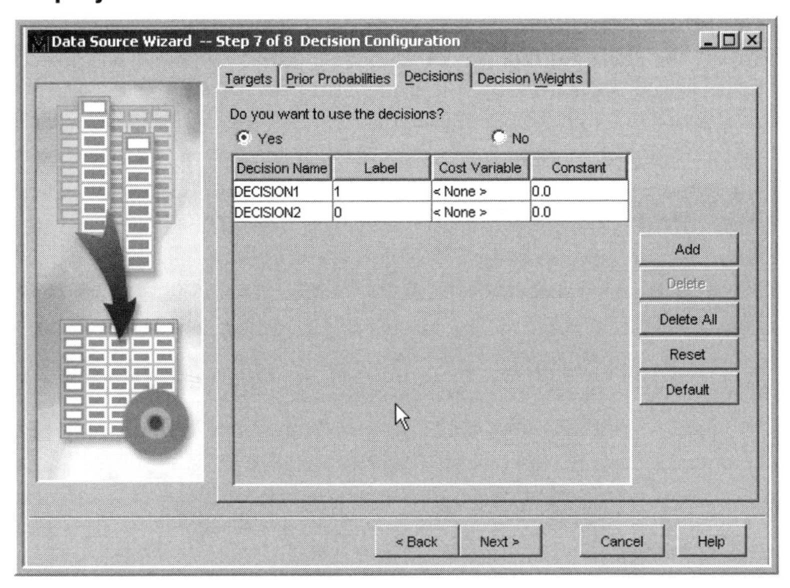

If you want to maximize profit instead of minimizing cost, then there is no need to enter cost variables. Costs are already taken into account in profits. Therefore, in this example, cost variables are not entered. When you click on **Next>**, another window opens showing the data set and its **Role**. Display 2.21 shows this window. In the display, the data set NN_RESP_DATA is being assigned a role of **Raw**.

Display 2.21

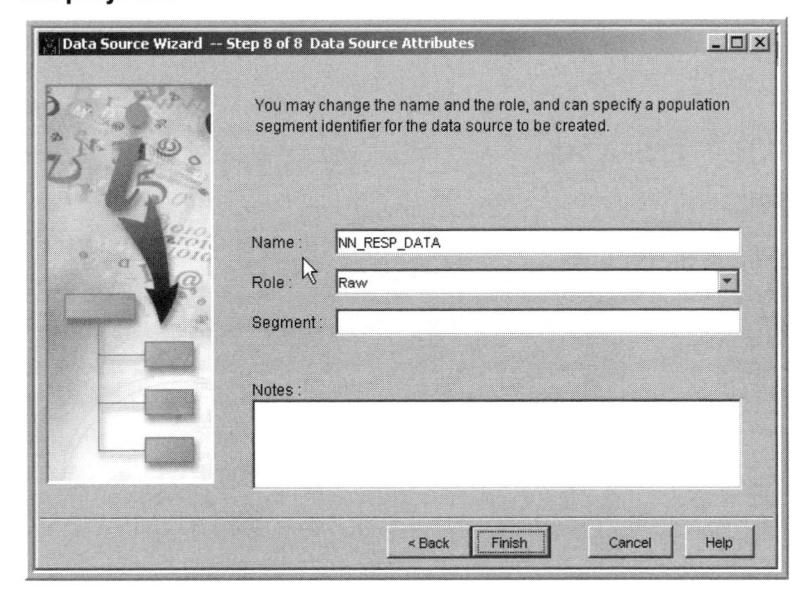

Other options for **Role** are **Train**, **Validate**, **Test**, **Score**, **Document**, and **Transaction**. Since I plan to create the **Train**, **Validate**, and **Test** data sets from the sample data set, I leave its role as **Raw**. When I click on **Finish**, the **Data Source Wizard** closes, and the project window opens as shown in Display 2.22.

Display 2.22

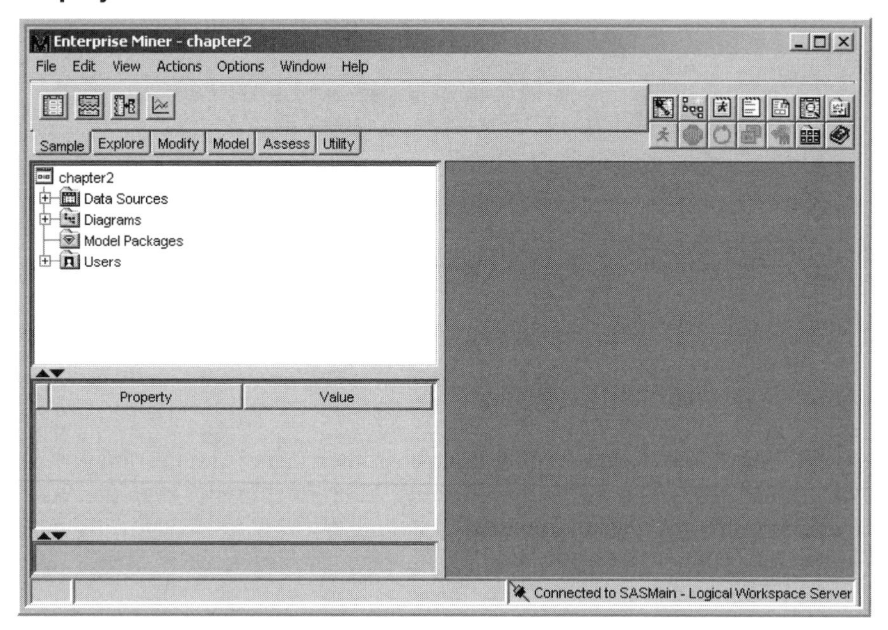

I have upgraded the **Property Sheet** from **Basic** to **Advanced** by clicking on **View→Property Sheet→Advanced**, as shown in Display 2.23.

Display 2.23

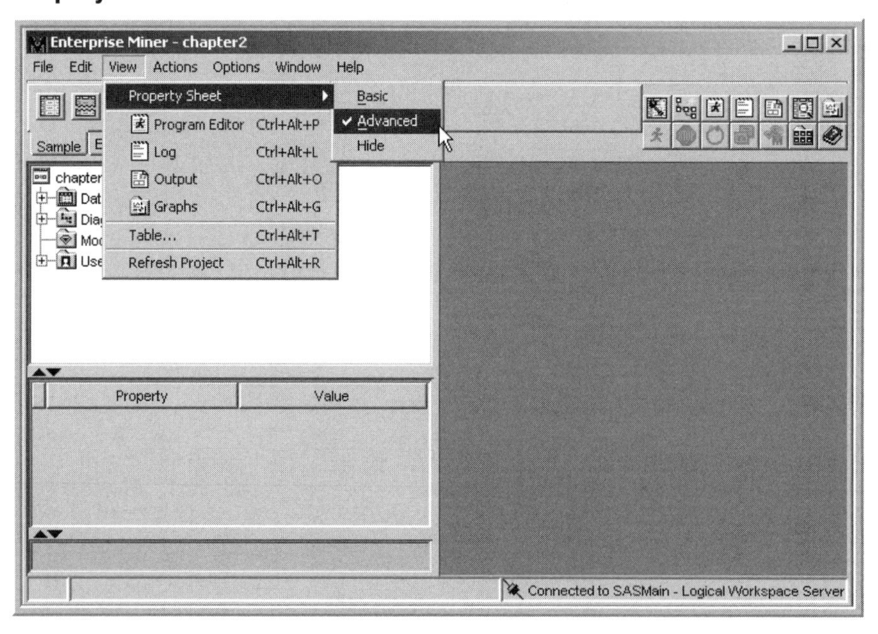

Display 2.24 shows **Advanced** properties of the project named Chapter2.

Display 2.24

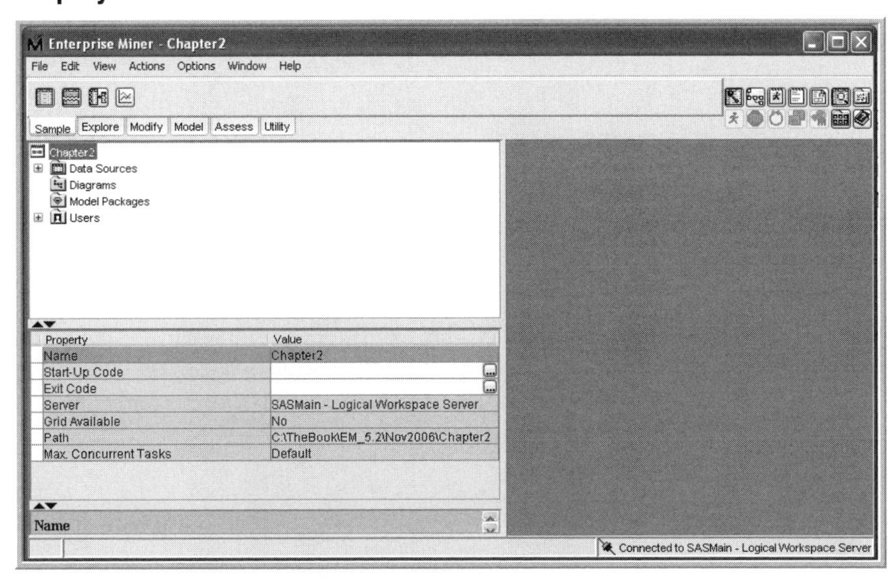

In order to see the properties of the data set, expand **Data Sources** in the project panel by clicking on the plus sign located to the left of **Data Sources**. Then when you select the data set NN_RESP_DATA, the properties panel shows the properties of the data source as shown in Display 2.25.

You can view and edit the properties. You can also see a list of variables by clicking on [...] located on the right of the **Variables** property in the **Value** column. The arrow in Display 2.25 shows where to click the mouse pointer, and Display 2.26 shows the window that opens as a result.

Display 2.25

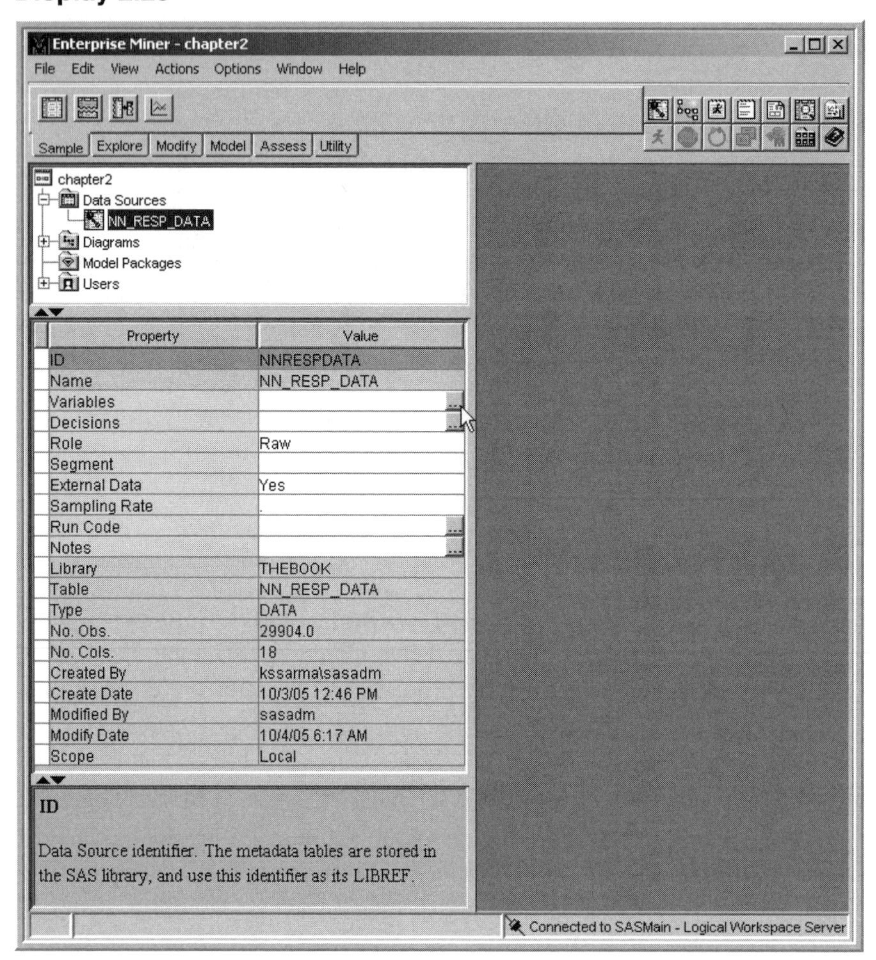

Display 2.26

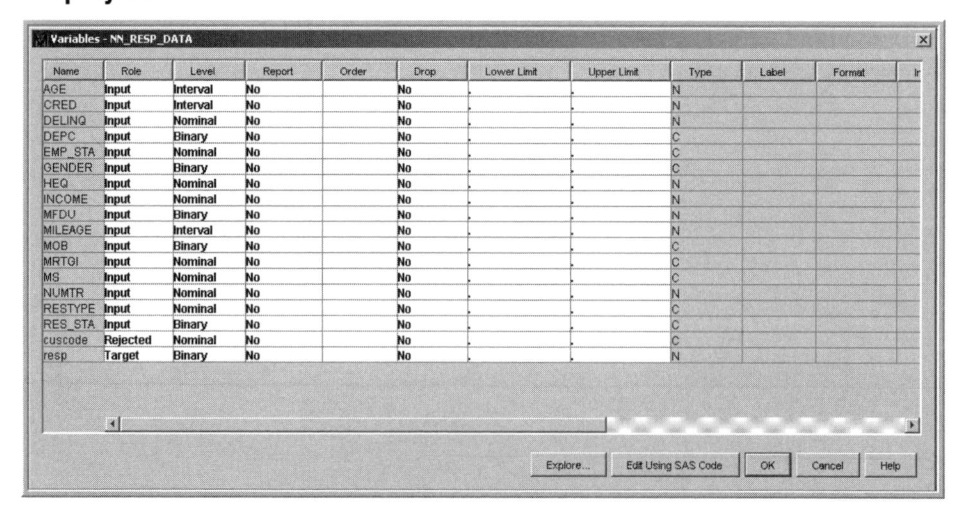

Name	Role	Level	Report	Order	Drop	Lower Limit	Upper Limit	Type	Label	Format
AGE	Input	Interval	No		No	.	.	N		
CRED	Input	Interval	No		No	.	.	N		
DELINQ	Input	Nominal	No		No	.	.	N		
DEPC	Input	Binary	No		No	.	.	C		
EMP_STA	Input	Nominal	No		No	.	.	C		
GENDER	Input	Binary	No		No	.	.	C		
HEQ	Input	Nominal	No		No	.	.	N		
INCOME	Input	Nominal	No		No	.	.	N		
MFDU	Input	Binary	No		No	.	.	N		
MILEAGE	Input	Interval	No		No	.	.	N		
MOB	Input	Binary	No		No	.	.	C		
MRTGI	Input	Nominal	No		No	.	.	C		
MS	Input	Nominal	No		No	.	.	C		
NUMTR	Input	Nominal	No		No	.	.	N		
RESTYPE	Input	Nominal	No		No	.	.	C		
RES_STA	Input	Binary	No		No	.	.	C		
cuscode	Rejected	Nominal	No		No	.	.	C		
resp	Target	Binary	No		No	.	.	N		

This window shows the name, role, measurement scale, etc., for each variable in the data set. You can change the role of any variable, change its measurement scale, or drop a variable from the data set. If a variable is dropped from the data set, it will not be available in any of the subsequent nodes.

2.6 Creating a Process Flow Diagram

To create a process flow diagram, right-click on **Diagrams** in the project panel (shown in Display 2.25) and click on **Create Diagram**. You will be prompted to enter the name of the diagram in a text box labeled **Diagram Name**. After entering a name for your diagram, click **OK**. Then a blank workspace opens, as shown in Display 2.27A, where you create your process flow diagram.

Display 2.27A

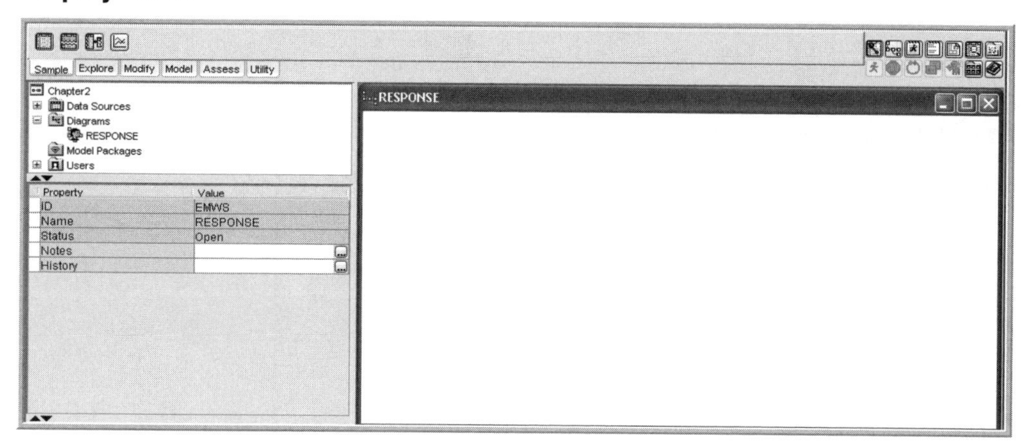

2.6.1 Input Data Node

This is the first node in any diagram (unless you start with the SAS Code node). In this node, you specify the data set that you want to use in the diagram. There may be several data sources that you may have already created for this project, as discussed in Section 2.5. From these data sources, you need to select one for this diagram. A data set can be assigned to an input in one of two ways:

- When you expand **Data Sources** in the project panel by clicking on the [+] on the left of **Data Sources**, all the data sources will appear. Then click on the icon to the left of the data set you want, and drag it to the Diagram Workspace. This will create the **Input Data** node with the desired data set assigned to it.
- Alternatively, first drag the **Input Data** node from the toolbar into the Diagram Workspace. Then set the **Data Source** property of the **Input Data** node to the name of the data set. To do this, select the **Input Data** node, then click on [...] located to the right of the **Data Source** property as shown in Display 2.27B. The **Select Data Source** window opens, as shown in Display 2.28. Then click on the data set you want to use in the diagram. Then click **OK**.

When you follow either procedure above, the **Input Data** node is created as shown in Display 2.27B.

Display 2.27B

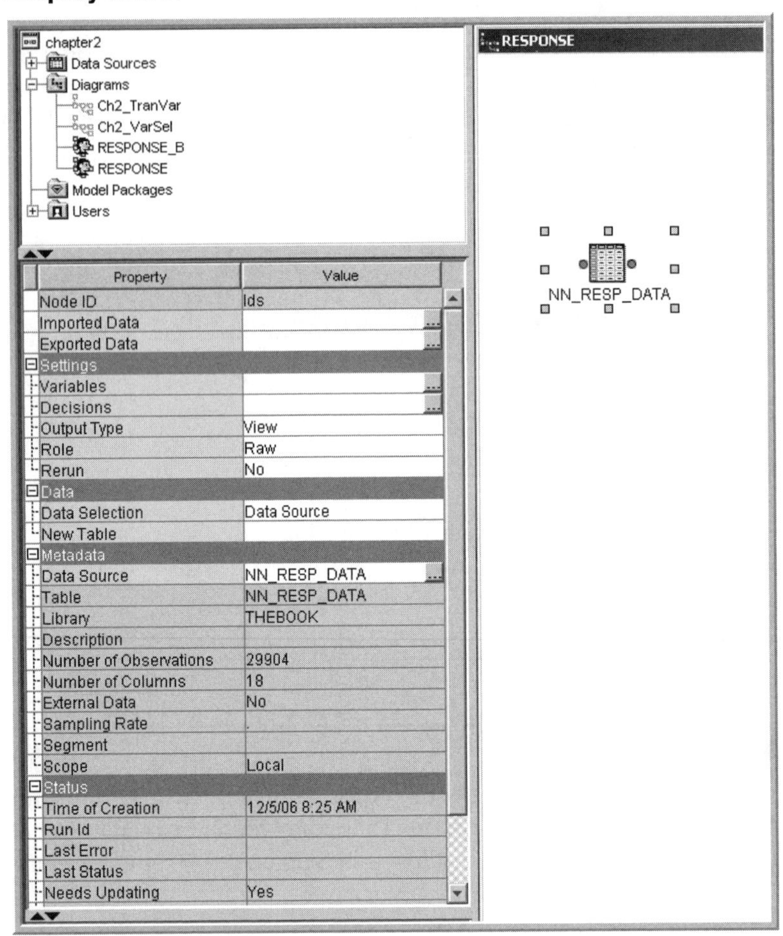

Display 2.28

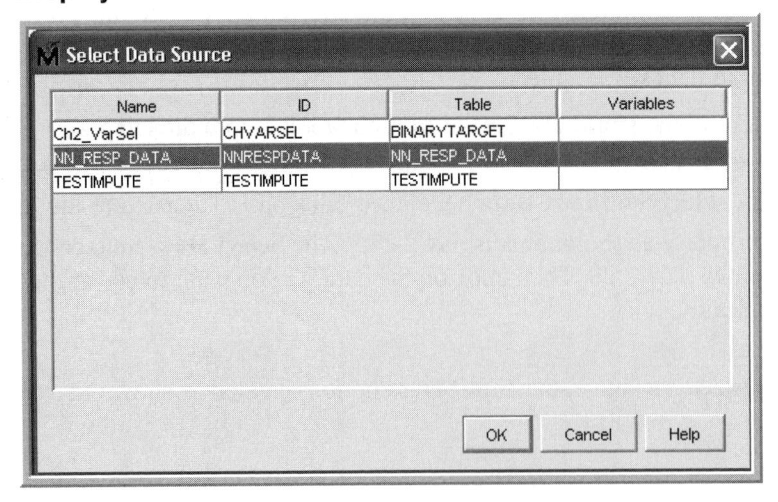

2.6.2 Tools for Initial Data Exploration

To explore the data, you can use the Enterprise Miner nodes **MultiPlot** and **StatExplore**. These nodes enable you to examine the distributions of the input variables and their relationships with the target variable. In order to use these nodes you must create a process flow diagram, as shown in Display 2.29.

To include **MultiPlot** and **StatExplore** nodes in the process flow diagram, click on the **Explore** tab in the node groups (see Display 2.5). The exploration tools are now visible on the toolbar. When you place your mouse pointer over a specific exploration tool, the tool's name becomes visible. Drag the **StatExplore** and **MultiPlot** tools onto the Diagram Workspace and connect the nodes as shown in Display 2.29.

Display 2.29

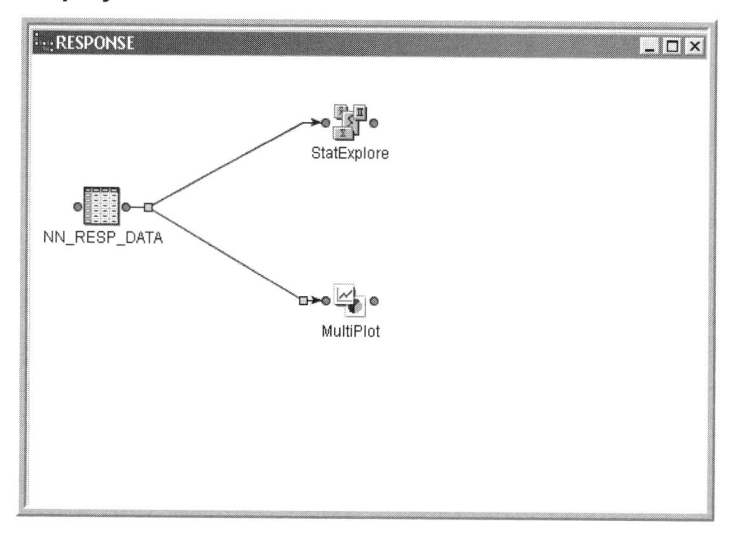

2.6.2.1 StatExplore Node

The **StatExplore** node can be used to find out which input variables are closely related to the target.

When you run the **StatExplore** node and open the **Results** window, you will see a graph window and an output window. The graph window, shown in Display 2.30A, is called **Chi-Square Plot**. This plot shows the statistic known as Cramer's V on the vertical axis and the inputs on the horizontal axis. A Chi-Square plot is shown for categorical targets.

Display 2.30A

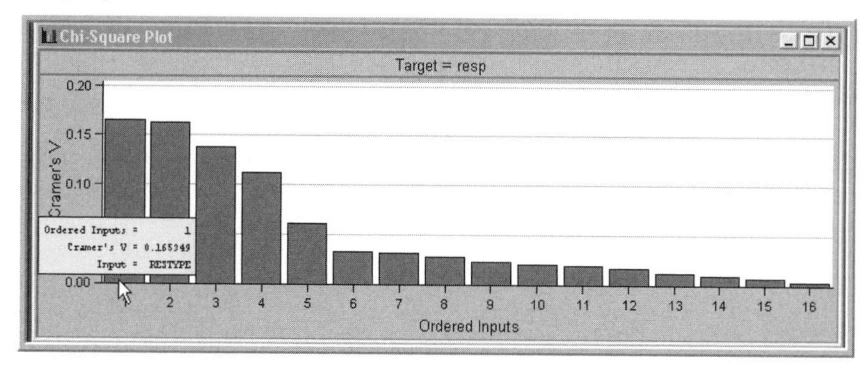

When you place the mouse pointer over a bar, you can see the variable name and the Cramer's V value for that variable. In this example, the top variable is RESTYPE. This variable indicates the customer's residence type (homeowner or renter). Cramer's V for RESTYPE is 0.165347.

In general, Cramer's V measures the strength of relationship between two categorical variables. A detailed explanation of Cramer's V is given in the appendix to this chapter.

Cramer's V is automatically generated for all categorical inputs when the target is also categorical. To calculate Cramer's V for continuous inputs such as AGE, you must convert the variable from continuous to categorical by creating intervals, or bins, with each bin representing a category. To accomplish this, set the **Interval Variables** property of the **StatExplore** node to **Yes**, and specify the **Number of Bins** property to the desired number of bins, or leave it at its default value of 5, as shown in Display 2.30B. To get a Chi-Square plot, make sure that the **Chi-Square** property of the **StatExplore** node is set to **Yes**.

Display 2.30B

Property	Value
Node ID	Stat
Imported Data	...
Exported Data	...
Variables	...
Use Segment Variables	No
Variable Selection	
Hide Rejected Variables	Yes
Number of Selected Variables	1000
Chi-Square Statistics	
Chi-Square	Yes
Interval Variables	Yes
Number of Bins	5
Correlation Statistics	
Correlations	Yes
Pearson Correlations	Yes
Spearman Correlations	No
Status	
Time of Creation	12/2/06 2:18 PM
Run Id	
Last Error	
Last Status	
Needs Updating	Yes
Needs to Run	Yes
Time of Last Run	
Run Duration	
Grid Host	

For example, the interval-scaled variable AGE is grouped into five bins, which are 18–32.4, 32.4–46.8, 46.8–61.2, 61.2–75.6, and 75.6–90. In the appendix to this chapter, a step-by-step illustration of the calculation of the Chi-Square statistic and Cramer's V is shown for the variable AGE.

The Chi-Square value shows the strength of relationship between the target and the binned variable. More specifically, it shows if the distribution of the target levels differs significantly from bin to bin among the bins constructed from the interval variable.

From the output sub-window in the **Results** window of the **StatExplore** node, you can see the modal value of each input for each target level. For the input RESTYPE, the modal values are shown in Output 2.1.

Output 2.1

Variable=RESTYPE

Target	Target Value	Numcat	NMiss	Mode	Mode Pct	Mode2	Mode2Pct
OVERALL		4	0	HOME	54.73	RENTER	40.42
resp	0	4	0	HOME	60.19	RENTER	35.11
resp	1	4	0	RENTER	52.05	HOME	42.77

This output reports the modal values of the inputs by the target levels. In this example, the target has two levels: 0 and 1. The columns labeled **Mode Pct** and **Mode2Pct** exhibit the first modal value and the second modal value, respectively. The first row of the output is labeled _OVERALL_. The _OVERALL_ row values for **Mode** and **Mode Pct** indicate that the most predominant category in the sample is homeowners, indicated by HOME. The second row indicates the modal values for non-responders, and the third row shows the modal values for the responders. The first modal value for the responders is RENTER, suggesting that the renters in general are more likely to respond than home owners in this marketing campaign. These numbers can be verified by running PROC FREQ from the Program Editor, as shown in Display 2.30C.

Display 2.30C

```
1  proc freq data=TheBook.NN_RESP_DATA;
2    table RESTYPE*RESP/nopercent norow missing ;
3  run ;
```

The results of PROC FREQ are shown in Output 2.2.

Output 2.2

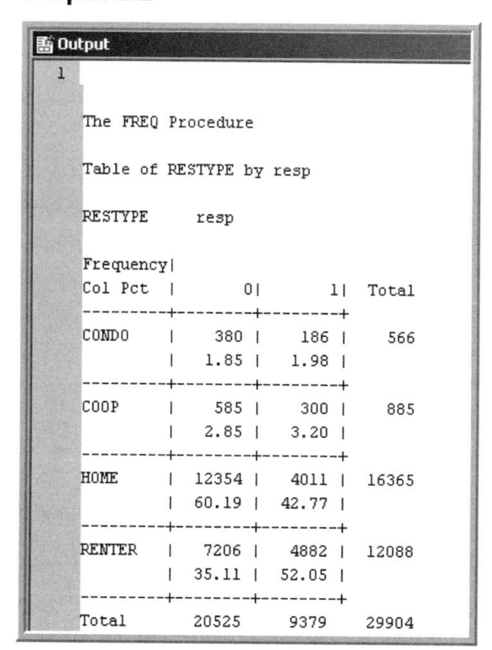


```
1

The FREQ Procedure

Table of RESTYPE by resp

RESTYPE     resp

Frequency|
Col Pct  |       0|       1|  Total
---------+--------+--------+
CONDO    |    380 |    186 |    566
         |   1.85 |   1.98 |
---------+--------+--------+
COOP     |    585 |    300 |    885
         |   2.85 |   3.20 |
---------+--------+--------+
HOME     |  12354 |   4011 |  16365
         |  60.19 |  42.77 |
---------+--------+--------+
RENTER   |   7206 |   4882 |  12088
         |  35.11 |  52.05 |
---------+--------+--------+
Total       20525     9379    29904
```

2.6.2.2 MultiPlot Node

The **MultiPlot** node can be used for visualizing the data. You can examine the distributions of the variables and relationships among variables. You must make sure that the **Property Sheet** is set to **Advanced**. The property sheet can be set to **Advanced** by clicking **View→Property Sheet→Advanced**.

To explore the variables, click on [...] located to the right of the **Variables** property. This opens a window with all the variables, as shown in Display 2.31.

Display 2.31

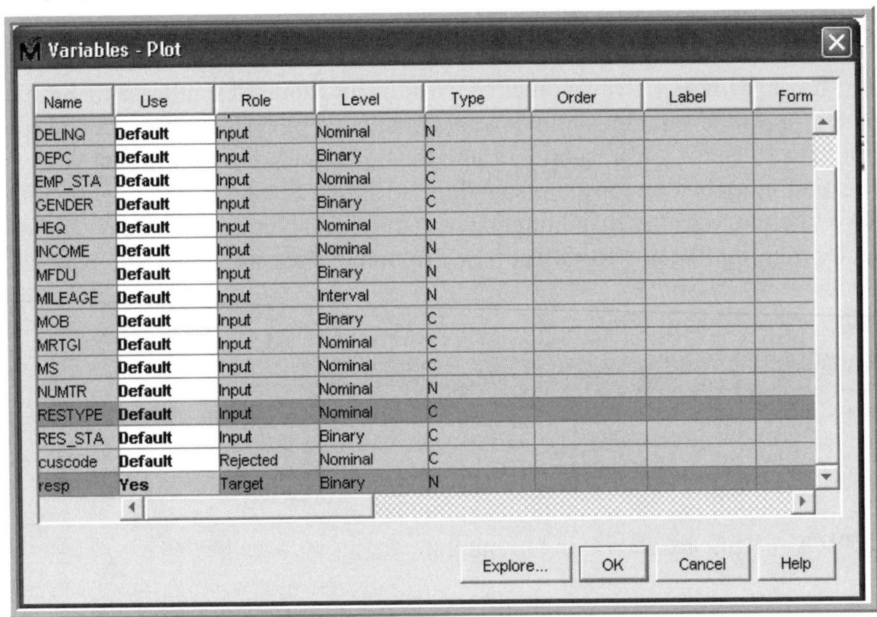

Name	Use	Role	Level	Type	Order	Label	Form
DELINQ	Default	Input	Nominal	N			
DEPC	Default	Input	Binary	C			
EMP_STA	Default	Input	Nominal	C			
GENDER	Default	Input	Binary	C			
HEQ	Default	Input	Nominal	N			
INCOME	Default	Input	Nominal	N			
MFDU	Default	Input	Binary	N			
MILEAGE	Default	Input	Interval	N			
MOB	Default	Input	Binary	C			
MRTGI	Default	Input	Nominal	C			
MS	Default	Input	Nominal	C			
NUMTR	Default	Input	Nominal	N			
RESTYPE	Default	Input	Nominal	C			
RES_STA	Default	Input	Binary	C			
cuscode	Default	Rejected	Nominal	C			
resp	Yes	Target	Binary	N			

To examine the histograms, select the variable of interest and click on **Explore**. Display 2.32 shows the histograms generated for the variables RESTYPE and RESP.

Display 2.32

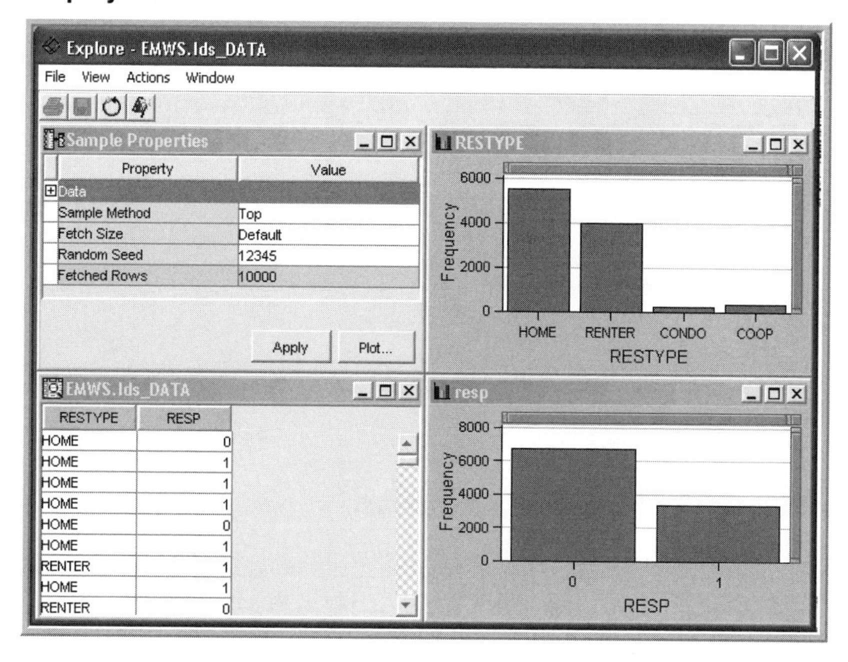

To examine the relationship between a categorical input such as RESTYPE and the target variable RESP, click the **Plot** button located at the bottom of the **Sample Properties** pane in the **Explore** window shown in Display 2.32. Then the **Select a Chart Type** window appears, as shown in Display 2.33.

Display 2.33

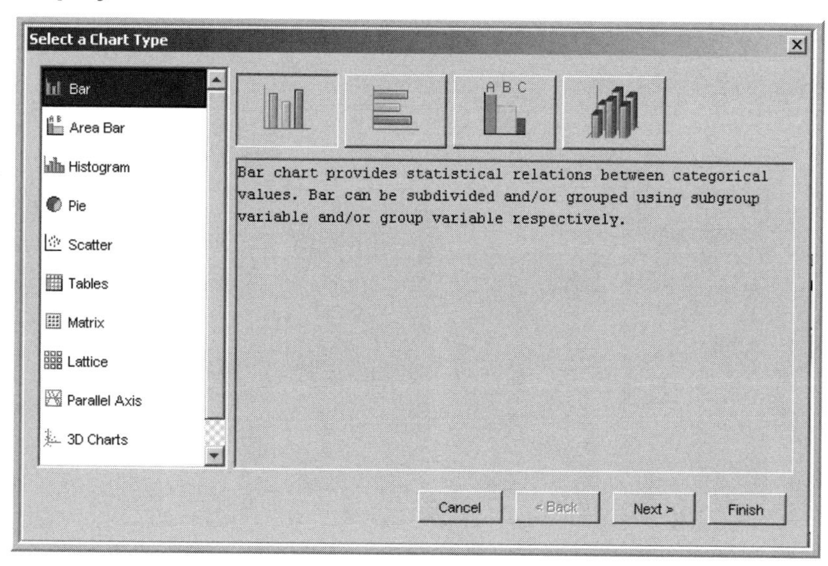

Select the bar chart and click on **Next>**. The **Select Chart Roles** window will open, as shown in Display 2.34.

Display 2.34

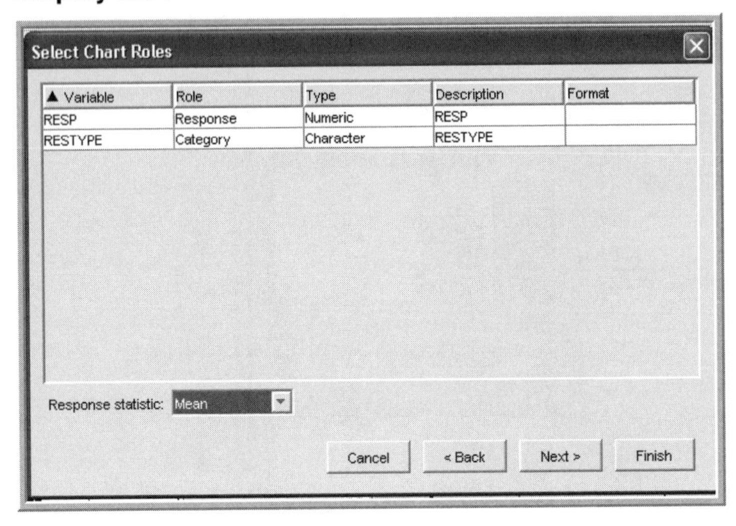

By clicking in the box under **Role**, you get a list of roles that can be assigned to the variables listed in the first column of the window shown in Display 2.34. In this example, I set the role of the variables RESP and RESTYPE to **Response** and **Category**, respectively. I set the **Response statistic** to **Mean**, clicked **Next>** four times, and then clicked on **Finish**.

The resulting chart is shown in Display 2.35.

Display 2.35

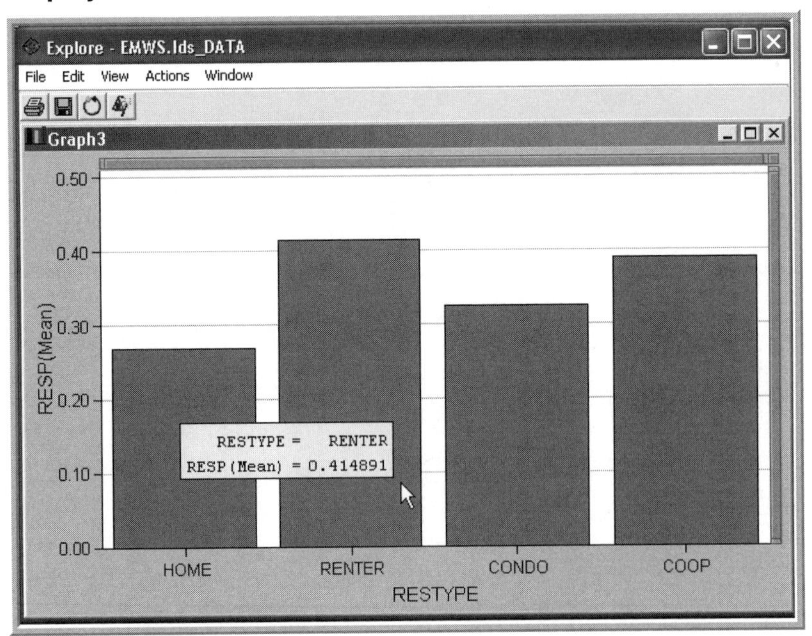

By passing the mouse pointer over the bars, you can see the response rate for a given level of the categorical input. From Display 2.35, you can see that the response rate is 41% among renters.

You can also examine the relationship of the inputs to the target variables by setting the **Type of Chart** property of the **MultiPlot** node to **Bar Charts**, and then running the **MultiPlot** node. By opening the results, you can see charts such as the one shown in Display 2.36.

Display 2.36

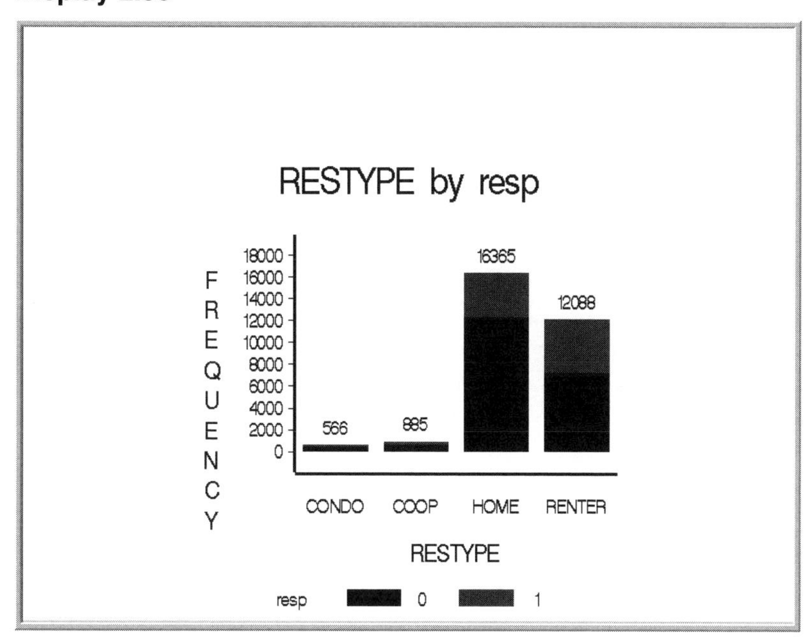

The examples given above demonstrate some of the capabilities of **MultiPlot** and **StatExplore**. For more details, see the Enterprise Miner help.

2.6.3 Impute Node

The **Impute** node is used for imputing missing values of inputs. This node icon, along with other icons of the Modify group, appear on the toolbar when you select the **Modify** tab of the **Enterprise Miner** window (see Display 2.5).

Display 2.37 shows the **Impute** node selected.

Display 2.37

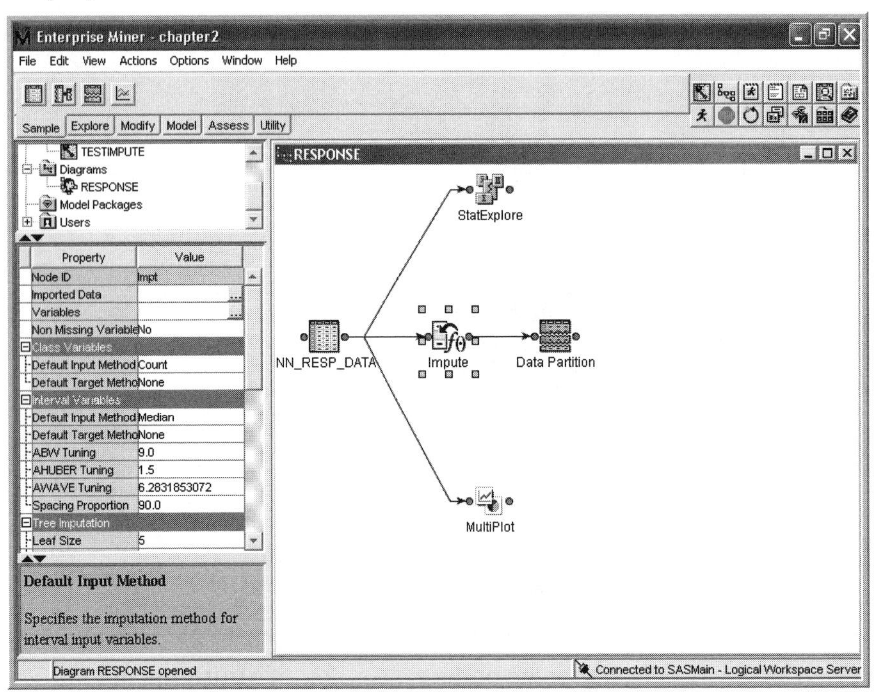

The **Impute** node has a number of techniques for imputation. To select a particular technique, you must first select the **Impute** node on the Diagram Workspace and set the **Default Input Method** property to the desired technique. For interval variables, the options available are Mean, Median, Mid-Range, Distribution, Tree, Tree Surrogate, Mid-Minimum space, Tukey's Biweight, Huber, Andrew's wave, Default constant, and None. For class variables, the options are Count, Default constant value, Distribution, Tree, Tree surrogate, and None.

2.6.4 Data Partition Node

In model building it is necessary to partition the sample into **Training**, **Validation**, and **Test** subsamples. The training data is used for developing the model using tools such as **Regression**, **Decision Tree**, and **Neural Network**. During the training process, these tools generate a number of models. The validation data set is used to evaluate these models, and then to select the best one. The process of selecting the best model is often referred to as *fine tuning*. The test data set is used for an independent assessment of the final model.

The **Data Partition** node is shown in Display 2.38.

Display 2.38

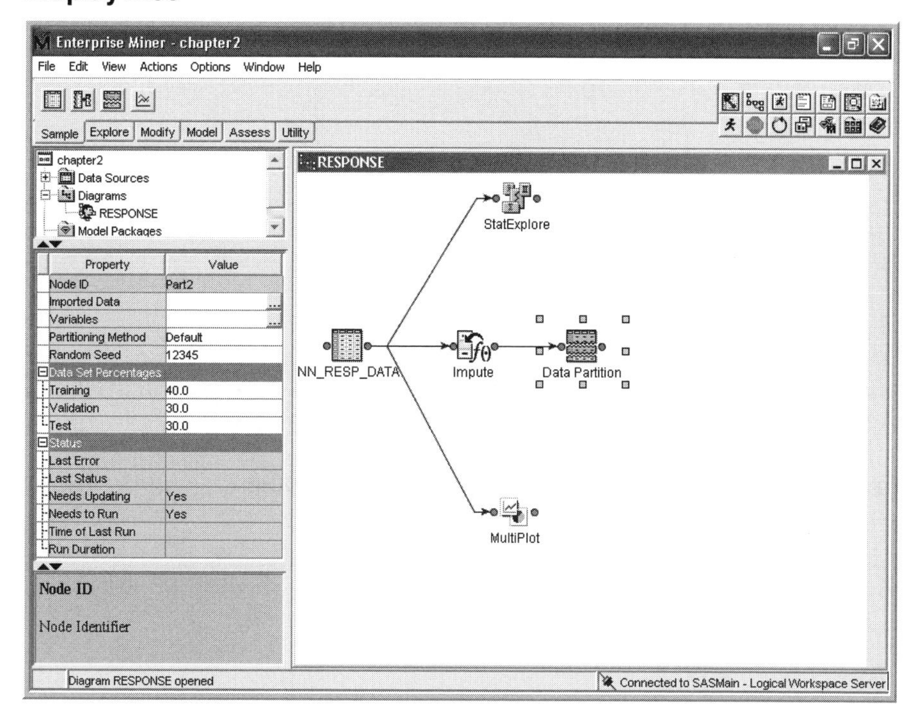

In the **Data Partition** node, you can specify the method of partitioning by setting the **Partitioning Method** property to one of four values: **Default**, **Simple random**, **Cluster** and **Stratified**. In the case of a binary target such as response, the stratified sampling method results in uniform proportions of responders in each of the partitioned data sets. Hence, I set the **Partitioning Method** property to **Stratified**, which is the default for binary targets. The default proportion of records allocated to these three data sets are 40%, 30%, and 30%, respectively. These proportions can be changed by resetting the **Training**, **Validation**, and **Test** properties under the **Data Set Percentage** property. In order to reset the properties, you should first drag the **Data Partition** node into the **Diagram Workspace**, and select it, so that the **Property Panel** will show the properties of the **Data Partition** node as shown in Display 2.38.

2.6.5 Filter Node

The **Filter** node can be used for eliminating observations with extreme values (outliers) in the variables.

You should not use this node routinely to eliminate outliers. While it may be reasonable to eliminate some outliers for very large data sets for predictive models, the outliers often have interesting information that leads to insights about the data and customer behavior.

Before using this node, you should first find out the source of any extreme value. If the extreme value is due to an error, the error should be corrected. If there is no error, you can truncate the value so that the extreme value does not have an undue influence on the model.

Display 2.39 shows the flow diagram with the **Filter** node.

Display 2.39

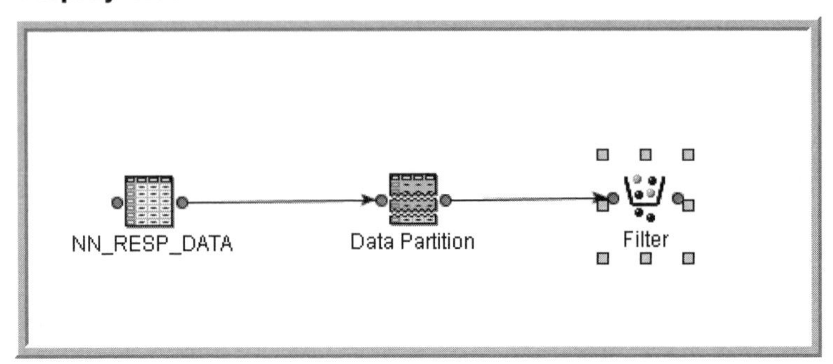

In the process flow shown in Display 2.39, the **Filter** node follows the **Data Partition** node. Alternatively, you can use the **Filter** node before the **Data Partition** node.

To use the **Filter** node, select it, and set the **Default Filtering Method** property for **Interval variables** to one of the values listed there as shown in Display 2.40A.

Display 2.40A

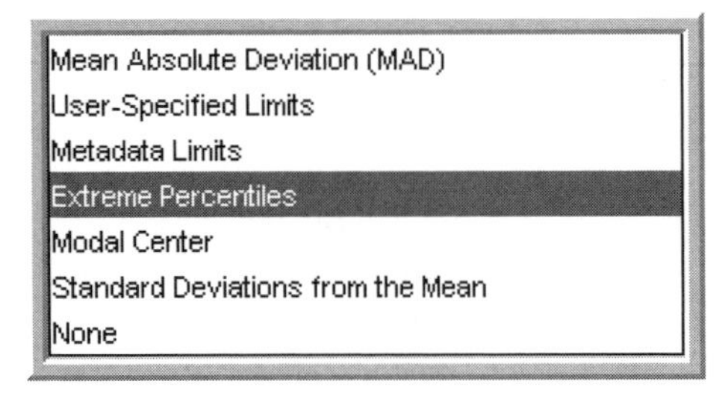

For class variables, the **Default Filtering Method** property can be set to one of the values as shown in Display 2.40B.

Display 2.40B

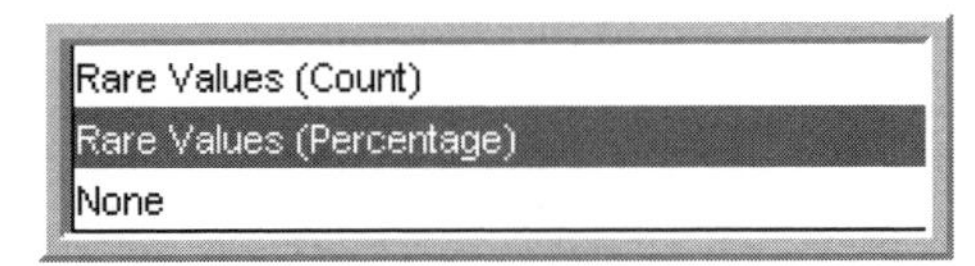

Display 2.40C shows the properties panel of the **Filter** node.

Display 2.40C

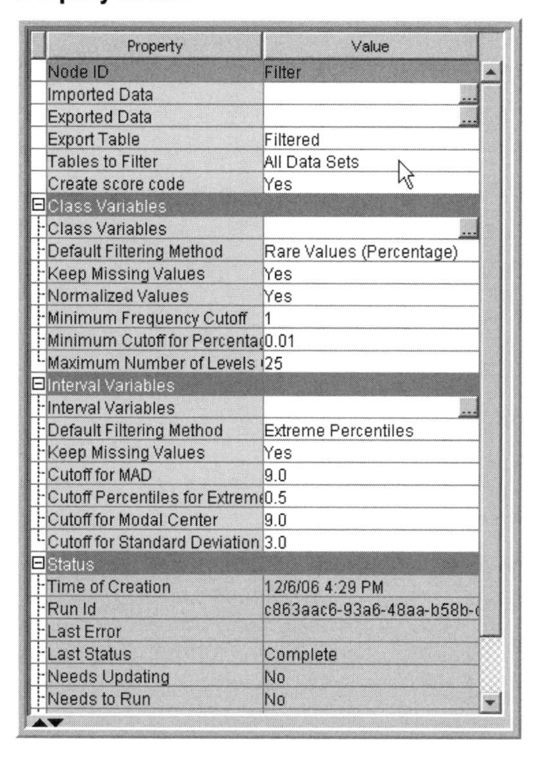

I set the **Tables to Filter** property to **All Data Sets** so that outliers are filtered in all three data sets—training, validation, and test—and then ran the **Filter** node.

The **Results** window displays the number of observations eliminated due to outliers of the variables. Output 2.3A (from the output of the **Results** window) shows the number of records excluded from the **Train, Validate**, and **Test** data sets due to outliers.

Output 2.3A

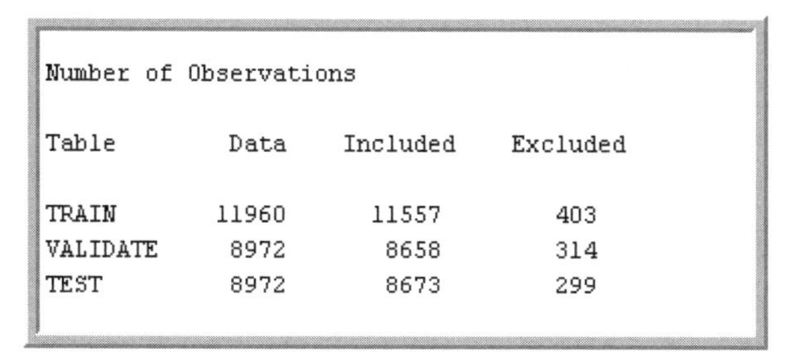

Number of Observations

Table	Data	Included	Excluded
TRAIN	11960	11557	403
VALIDATE	8972	8658	314
TEST	8972	8673	299

Outputs 2.3 to 2.5 show summary statistics for original and filtered data for the variable AGE.

Output 2.3

Data Role=TRAIN Variable=AGE

Statistics	Original	Filtered
Number Missing	0.0000	0.0000
Minimum	18.0000	22.0000
Maximum	90.0000	88.0000
Mean	49.6672	49.6075
Std Deviation	15.7092	15.3838
Skewness	0.1867	0.1598
Kurtosis	-0.7058	-0.7781

Output 2.4

Data Role=VALIDATE Variable=AGE

Statistics	Original	Filtered
Number Missing	0.0000	0.0000
Minimum	18.0000	22.0000
Maximum	90.0000	88.0000
Mean	49.1291	49.0687
Std Deviation	15.4426	15.1197
Skewness	0.2047	0.1799
Kurtosis	-0.6851	-0.7676

Output 2.5

Data Role=TEST Variable=AGE

Statistics	Original	Filtered
Number Missing	0.0000	0.0000
Minimum	18.0000	22.0000
Maximum	90.0000	88.0000
Mean	48.9826	49.0687
Std Deviation	15.4225	15.1197
Skewness	0.2069	0.1799
Kurtosis	-0.7138	-0.7676

2.6.6 Variable Selection Node

The **Variable Selection** node can be used to make a preliminary selection of the variables to be included in predictive modeling. There are a number of alternative methods with various options for selecting variables. The methods of variable selection depend on the measurement scales of the inputs and the targets.

These options are discussed in detail in Chapter 3. In this chapter, I present a brief introduction to the techniques used by the **Variable Selection** node for different types of targets, and I show how to set the properties of the **Variable Selection** node for choosing an appropriate technique.

There are two basic techniques used by the **Variable Selection** node. They are the **R-Square** selection method and the **Chi-Square** selection method. Both of these techniques select variables based on the strength of their relationship with the target variable. For interval targets, only the **R-Square** selection method is available. For binary targets, both the **R-Square** and **Chi-Square** selection methods are available.

2.6.6.1 R-Square Selection Method

To use the R-Square selection method, you must set the **Target Model** property to **R-Square**, as shown in the Display 2.41.

Display 2.41

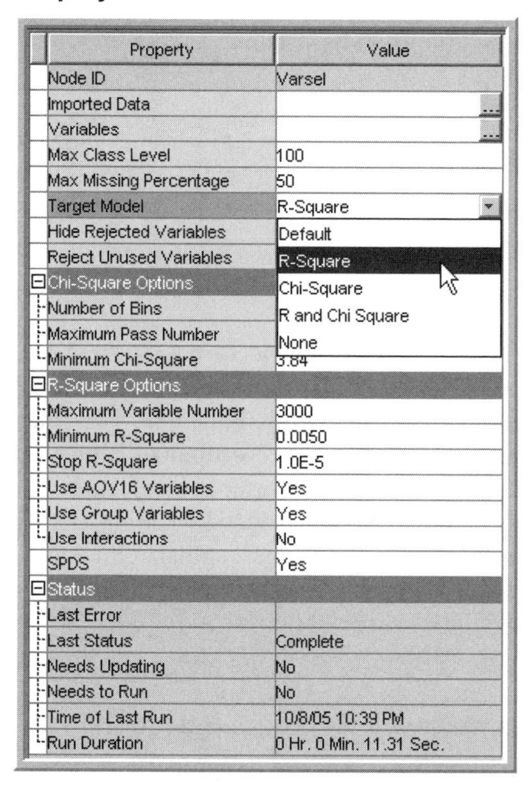

When the R-Square selection method is used, the variable selection is done in two steps. In Step 1, the **Variable Selection** node computes an R-Square value (the squared correlation coefficient) based on the relationship between the target and each input variable, and then assigns the **Rejected** role to those variables that have a value less than the minimum R-Square. The default minimum R-Square cut-off is set to 0.005. You can change the value of the minimum R-Square by setting the **Minimum R-Square** property to a value other than the default.

The R-Square value (or the squared correlation coefficient) is the proportion of variation in the target variable explained by a single input variable, ignoring the effect of other input variables.

In Step 2, the **Variable Selection** node performs a forward stepwise regression to evaluate the variables chosen in the first step. Those variables that have a stepwise R-Square improvement less than the cut-off criterion have the role of **rejected**. The default cut-off for R-Square improvement is set to 0.0005. This can be changed by setting the **Stop R-Square** property to a different value.

For interval variables, the R-Square value used in Step 1 is calculated directly by means of a linear regression of the target variable on the interval variable, assessing only the linear relation between the interval variable and the target. To detect nonlinear relations, the **Variable Selection** node creates binned variables from each interval variable. The binned variables are called **AOV16** variables. Each **AOV16** variable has a maximum of 16 intervals of equal width. The **AOV16** variable is treated as a class variable. A one-way analysis of variance is performed to calculate the R-Square between an **AOV16** variable and the target. These **AOV16** variables are included in Step 1 and Step 2 of the selection process.

Some of the interval variables may be selected both in their original form and in their binned (**AOV16**) form. If you set the **Use AOV16 Variables** property to **Yes**, then only **AOV16** variables appear in the **Variable Selection** pane of the **Results** window, and these will be passed on to the next node, with their **Role** set to **Input**.

For variable selection, the number of categories of a nominal categorical variable can be reduced by combining categories that have similar distribution of the target levels. To use this option, set the **Use Group Variables** property to **Yes**.

2.6.6.2 Chi-Square Selection Method

When you select the Chi-Square selection method, the **Variable Selection** node creates a tree based on Chi-Square maximization. The **Variable Selection** node first bins interval variables and then uses the binned variable rather than the original inputs in building the tree. The default number of bins is 50, but the number can be changed by setting the **Number of Bins** property to a different value. Any split with a Chi-Square below the specified threshold is rejected. The default value for the Chi-Square threshold is 3.84. This number can be changed by setting the **Minimum Chi-Square** property to the desired level. The inputs that give the best splits are included in the final tree, and passed to the next node with the role of **Input**. All other inputs are given the role of **Rejected**.

2.6.6.3 Variable Selection Node: An Example with R-Square Selection

As a means of illustrating how our results depend on the settings of the properties of the **Variable Selection** node, I experiment with a small set of variables and show you the results below.

Display 2.42 shows the process flow with the **Variable Selection** node.

Display 2.42

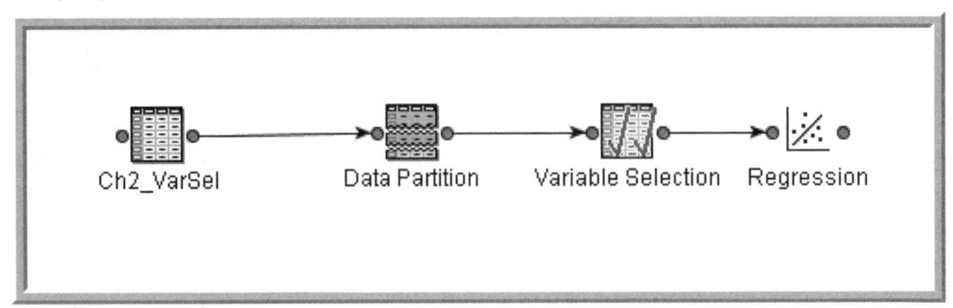

Display 2.43 shows the **Variables** window of the **Variable Selection** node. The data set is partitioned such that 50% is allocated to training, 30% to validation, and 20% to test.

Display 2.43

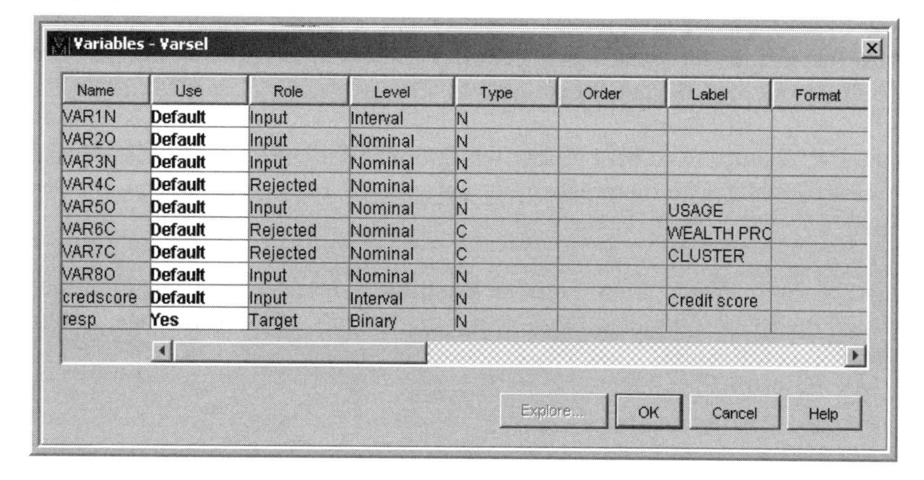

Name	Use	Role	Level	Type	Order	Label	Format
VAR1N	Default	Input	Interval	N			
VAR2O	Default	Input	Nominal	N			
VAR3N	Default	Input	Nominal	N			
VAR4C	Default	Rejected	Nominal	C			
VAR5O	Default	Input	Nominal	N		USAGE	
VAR6C	Default	Rejected	Nominal	C		WEALTH PRC	
VAR7C	Default	Rejected	Nominal	C		CLUSTER	
VAR8O	Default	Input	Nominal	N			
credscore	Default	Input	Interval	N		Credit score	
resp	Yes	Target	Binary	N			

Since the target variable is binary in this example, any of the above options shown in Display 2.44 could be used. I chose the R-Square selection method by setting the **Target Model** property to **R-Square**.

Display 2.44

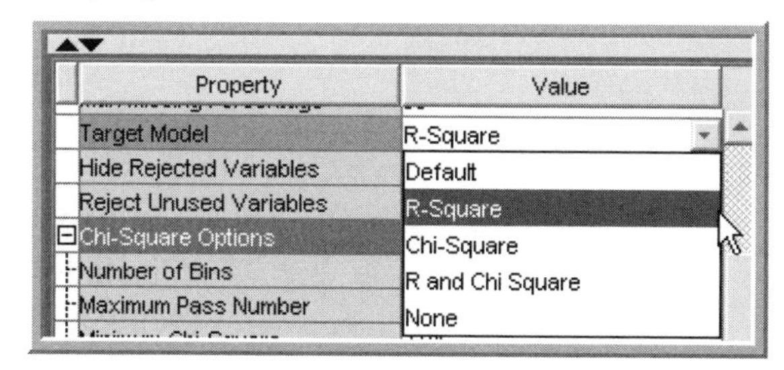

Next, I select the threshold values for the **Minimum R-Square** and **Stop R-Square** properties. The **Minimum R-Square** property is used in Step 1, as described in Section 2.6.6.1. The **Stop R-Square** property is used in Step 2, which is also described in Section 2.6.6.1. In this example, the **Minimum R-Square** property is set to the default value, and the **Stop R-Square** property to 0.00001. The **Stop R-Square** property is deliberately set at a very low level so that most of the variables selected in the first step will also be selected in the second step. In other words, I effectively by-passed the second step so that all or most of the variables selected in Step 1 are passed to the **Tree** node or **Regression** node that follows. I did this because both the **Decision Tree** and **Regression** nodes make their own selection of variables, and I wanted the second step of variable selection to be done in these nodes rather than in the **Variable Selection** node.

Display 2.45

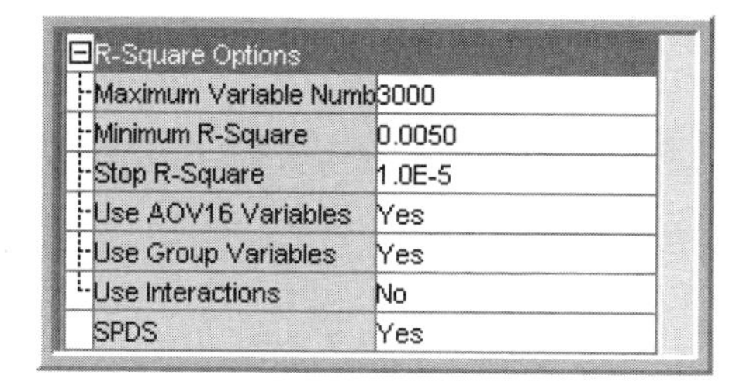

R-Square Options	
Maximum Variable Numb	3000
Minimum R-Square	0.0050
Stop R-Square	1.0E-5
Use AOV16 Variables	Yes
Use Group Variables	Yes
Use Interactions	No
SPDS	Yes

From Display 2.45, you can see that I set the **Use AOV16 Variables** property to **Yes**. If I chose **No**, the **Variable Selection** node (i.e., the underlying PROC DMINE) would still create the AOV16 variables and include them in Step 1, but not in Step 2. In addition, if you select **No**, these variables will not be passed to the next node.

When you set the **Use Group Variables** property to **Yes**, the categories of the class variables are combined, and new variables with the prefix of G_ are created. (This is explained in more detail in the next chapter.)

With the settings shown in Display 2.45, I ran the **Variable Selection** node, and after the run was completed I opened the **Results** window by selecting the **Variable Selection** node, right-clicking on it, and then clicking on **Results**.

Display 2.46 shows the **Results** window.

Display 2.46

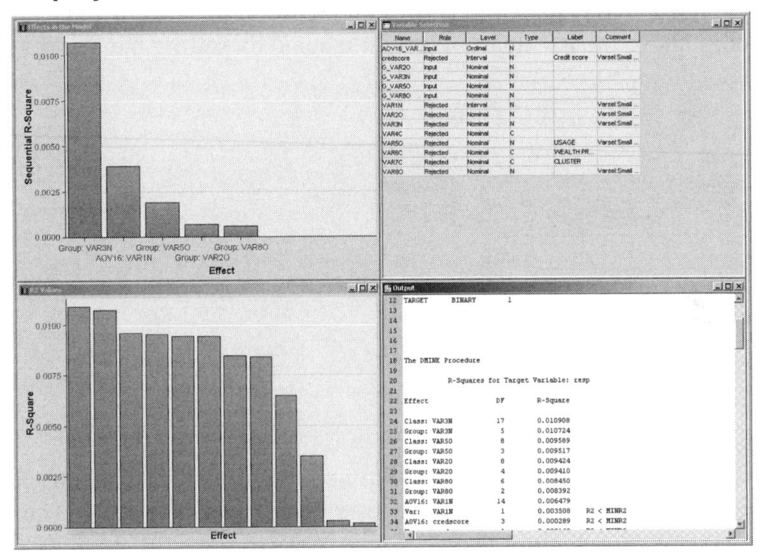

The bar chart in the top left quadrant of Display 2.46 shows the variables selected in Step 2. The bottom left quadrant shows the variables selected in Step 1. The output shown in the top right quadrant shows the variables selected in Step 2. The output pane in the bottom right quadrant of the **Results** window shows the variables selected in Step 1 and Step 2. By scrolling down the output pane, you first see the variables selected in Step 1. These also appear in Output 2.6.

Output 2.6

The DMINE Procedure

R-Squares for Target Variable: resp

Effect	DF	R-Square	
Class: VAR3N	17	0.010908	
Group: VAR3N	5	0.010724	
Class: VAR50	8	0.009589	
Group: VAR50	3	0.009517	
Class: VAR20	8	0.009424	
Group: VAR20	4	0.009410	
Class: VAR80	6	0.008450	
Group: VAR80	2	0.008392	
AOV16: VAR1N	14	0.006479	
Var: VAR1N	1	0.003508	R2 < MINR2
AOV16: credscore	3	0.000289	R2 < MINR2
Var: credscore	1	0.000162	R2 < MINR2

Scrolling down further in the output pane, you can see the variables selected in Step 2. These are shown in Output 2.7.

Output 2.7

The DMINE Procedure

Effects Chosen for Target: resp

Effect	DF	R-Square	F Value	p-Value	Sum of Squares	Error Mean Square
Group: VAR3N	5	0.010724	19.369933	<.0001	8.628783	0.089095
AOV16: VAR1N	14	0.003912	2.529776	0.0013	3.147905	0.088882
Group: VAR50	3	0.001891	5.715847	0.0007	1.521686	0.088741
Group: VAR20	4	0.000673	1.525318	0.1918	0.541304	0.088720
Group: VAR80	2	0.000567	2.574017	0.0763	0.456572	0.088689

The variables shown in Output 2.7 were selected in both Step 1 and Step 2.

The variables selected by the **Variable Selection** node are given the role of **Input** and passed to the next node. The **Variable Selection** area in the top right quadrant of the **Results** window (shown in Display 2.46) shows the variables with their assigned roles.

The next node in the process flow is the **Regression** node. By opening the **Variables** window of the **Regression** node, you can see the variables passed to it by the **Variable Selection** node. These variables are shown in Display 2.47.

Display 2.47

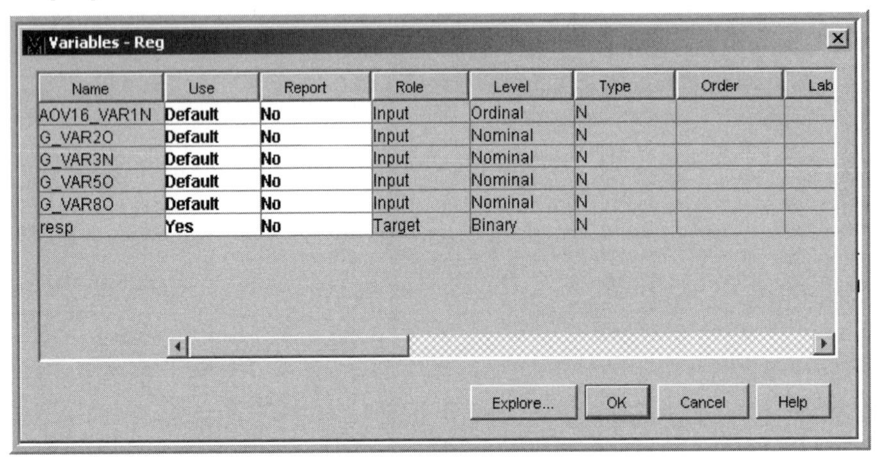

Name	Use	Report	Role	Level	Type	Order	Lab
AOV16_VAR1N	Default	No	Input	Ordinal	N		
G_VAR2O	Default	No	Input	Nominal	N		
G_VAR3N	Default	No	Input	Nominal	N		
G_VAR5O	Default	No	Input	Nominal	N		
G_VAR8O	Default	No	Input	Nominal	N		
resp	Yes	No	Target	Binary	N		

Display 2.48 (taken from the output area of the **Results** window) shows that VAR1N is rejected in Step 1, but AOV16_VAR1N is selected in both Step 1 and Step 2, indicting a nonlinear relationship between VAR1N and the target.

The definitions of the AOV16 variables and the grouped class variables (G variables) are included in the SAS code generated by the **Variable Selection** node. This code can be accessed by selecting the **Variable Selection** node, right-clicking on it, and clicking on **Results→View→Scoring→SAS Code**. Display 2.48 shows a segment of the SAS code.

Display 2.48

```
1   ******************************************;
2   *** Begin Scoring Code from PROC DMINE ***;
3   ******************************************;
4
5   length _WARN_ $ 4;
6   label _WARN_ = "Warnings";
7   _WARN_ = ' ';
8
9   if (VAR1N = .) then AOV16_VAR1N = 3;
10  else
11  if (VAR1N <= 5.8125) then AOV16_VAR1N = 1;
12  else
13  if (VAR1N <= 10.625) then AOV16_VAR1N = 2;
14  else
15  if (VAR1N <= 15.4375) then AOV16_VAR1N = 3;
16  else
17  if (VAR1N <= 20.25) then AOV16_VAR1N = 4;
18  else
19  if (VAR1N <= 25.0625) then AOV16_VAR1N = 5;
20  else
21  if (VAR1N <= 29.875) then AOV16_VAR1N = 6;
22  else
23  if (VAR1N <= 34.6875) then AOV16_VAR1N = 7;
24  else
25  if (VAR1N <= 39.5) then AOV16_VAR1N = 8;
26  else
27  if (VAR1N <= 44.3125) then AOV16_VAR1N = 9;
28  else
29  if (VAR1N <= 49.125) then AOV16_VAR1N = 10;
30  else
31  if (VAR1N <= 53.9375) then AOV16_VAR1N = 11;
32  else
33  if (VAR1N <= 58.75) then AOV16_VAR1N = 12;
34  else
35  if (VAR1N <= 63.5625) then AOV16_VAR1N = 13;
36  else
37  if (VAR1N <= 68.375) then AOV16_VAR1N = 14;
38  else
39  if (VAR1N <= 73.1875) then AOV16_VAR1N = 15;
```

Display 2.48 (*continued*)

```
length _NORM12 $ 12;
_NORM12 = put( VAR3N , BEST12. );
%DMNORMIP( _NORM12 )
drop _NORM12;
select(_NORM12);
  when(".           " ) G_VAR3N = 5;
  when("1           " ) G_VAR3N = 4;
  when("2           " ) G_VAR3N = 3;
  when("3           " ) G_VAR3N = 2;
  when("4           " ) G_VAR3N = 2;
  when("5           " ) G_VAR3N = 1;
  when("6           " ) G_VAR3N = 1;
  when("7           " ) G_VAR3N = 1;
  when("8           " ) G_VAR3N = 2;
  when("9           " ) G_VAR3N = 2;
  when("10          " ) G_VAR3N = 0;
  when("11          " ) G_VAR3N = 2;
  when("12          " ) G_VAR3N = 5;
  when("13          " ) G_VAR3N = 5;
  when("14          " ) G_VAR3N = 5;
  when("15          " ) G_VAR3N = 5;
  when("17          " ) G_VAR3N = 5;
  when("19          " ) G_VAR3N = 5;
  otherwise _WARN_ = 'U';
end;
```

2.6.6.4 Variable Selection Node: An Example with Chi-Square Selection

To perform variable selection using the Chi-Square criterion, set the **Target Model** property to **Chi-Square**. Enterprise Miner constructs a CHAID type of tree, and the variables used in the tree become the selected variables. The relative importance of the variables is shown in Display 2.49.

Display 2.49

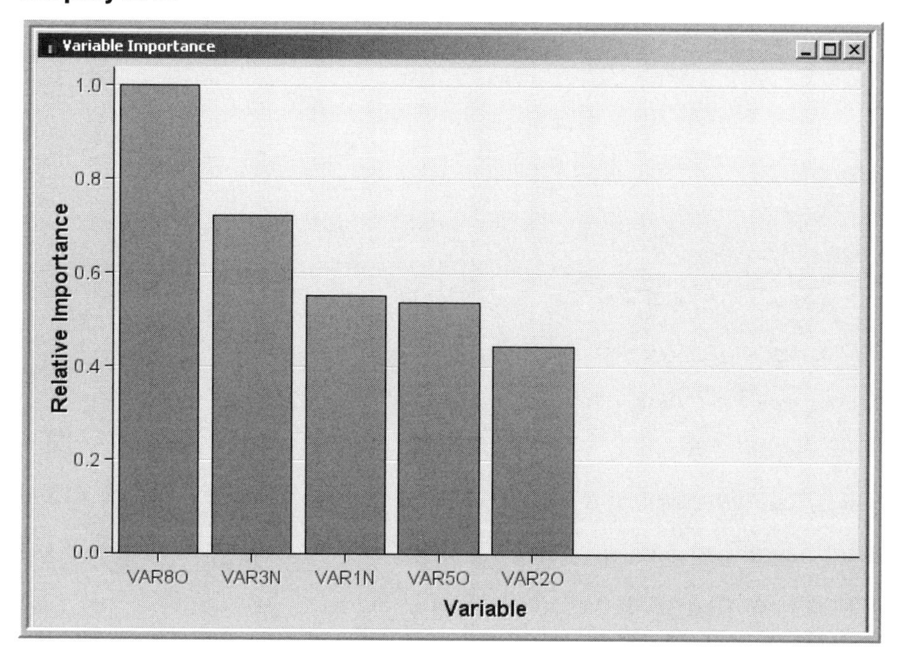

Display 2.50 shows the variables selected by the Chi-Square selection process.

Display 2.50

Variable Selection

Name	Role	Level	Type	Label	Comment
credscore	Rejected	Interval	N	Credit score	Varsel:Small ...
VAR1N	Input	Interval	N		
VAR2O	Input	Nominal	N		
VAR3N	Input	Nominal	N		
VAR4C	Rejected	Nominal	C		
VAR5O	Input	Nominal	N	USAGE	
VAR6C	Rejected	Nominal	C	WEALTH PR...	
VAR7C	Rejected	Nominal	C	CLUSTER	
VAR8O	Input	Nominal	N		

2.6.6.5 Saving the SAS Code Generated by the Variable Selection Node

If the **Target Model** property is set to **R-Square**, the **Use AOV16 Variables** property is set to **Yes**, and the **Use Group Variables** property is set to **Yes**, then the **Variable Selection** node will generate SAS code with definitions of the AOV16 and Group variables. When you right-click on the **Variable Selection** node and then click on **Results**, the **Results** window opens. If you click **View → SAS Results→Flow Code** in the **Results** window you can view the SAS code. A partial listing of this code is shown in Display 2.51.

Display 2.51

```
*******************************************;
*** Begin Scoring Code from PROC DMINE ***;
*******************************************;

length _WARN_ $ 4;
label _WARN_ = "Warnings";
_WARN_ = ' ';

if (VAR1N = .) then AOV16_VAR1N = 3;
else
if (VAR1N <= 5.8125) then AOV16_VAR1N = 1;
else
if (VAR1N <= 10.625) then AOV16_VAR1N = 2;
else
if (VAR1N <= 15.4375) then AOV16_VAR1N = 3;
else
if (VAR1N <= 20.25) then AOV16_VAR1N = 4;
else
if (VAR1N <= 25.0625) then AOV16_VAR1N = 5;
else
if (VAR1N <= 29.875) then AOV16_VAR1N = 6;
else
if (VAR1N <= 34.6875) then AOV16_VAR1N = 7;
else
if (VAR1N <= 39.5) then AOV16_VAR1N = 8;
else
if (VAR1N <= 44.3125) then AOV16_VAR1N = 9;
else
if (VAR1N <= 49.125) then AOV16_VAR1N = 10;
else
if (VAR1N <= 53.9375) then AOV16_VAR1N = 11;
else
if (VAR1N <= 58.75) then AOV16_VAR1N = 12;
else
if (VAR1N <= 63.5625) then AOV16_VAR1N = 13;
else
if (VAR1N <= 68.375) then AOV16_VAR1N = 14;
else
if (VAR1N <= 73.1875) then AOV16_VAR1N = 15;
else AOV16_VAR1N = 16;
```

This can be saved by clicking **File→Save as** and giving a name to the file where the SAS code is being saved and pointing to the directory where the file is being saved. The code can also be saved by clicking **Results → View → Scoring → SAS Code**, followed by **File→Save as**.

2.6.6.6 The Procedures Behind the Variable Selection Node

Enterprise Miner uses **PROC DMINE** in variable selection. Prior to running this procedure, it creates a data mining database (DMDB) and a data mining database catalog (DMDBCAT) using **PROC DMDB**. In the **Results** window of the **Variable Selection** node, you can see the **log** by clicking on **View** on the menu bar and selecting **log**.

Display 2.52 shows the **log** from **PROC DMDB**.

Display 2.52

```
Run proc dmdb with the specified maxlevel criterion.
*------------------------------------------------------------* ;
* EM: DMDBClass Macro ;
*------------------------------------------------------------* ;
%macro DMDBClass;
    VAR20(ASC) VAR3N(ASC) VAR50(ASC) VAR80(ASC) resp(DESC)
%mend DMDBClass;
*------------------------------------------------------------* ;
* EM: DMDBVar Macro ;
*------------------------------------------------------------* ;
%macro DMDBVar;
    VAR1N credscore
%mend DMDBVar;
*------------------------------------------------------------*;
* EM: Create DMDB;
*------------------------------------------------------------*;
libname _spdslib SPDE "C:\DOCUME~1\sasadm\LOCALS~1\Temp\SAS Temporary Files\_TD4908";
Libref _SPDSLIB was successfully assigned as follows:
Engine:        SPDE
Physical Name: C:\DOCUME~1\sasadm\LOCALS~1\Temp\SAS Temporary Files\_TD4908\
proc dmdb batch data=EMWS1.Part_TRAIN dmdbcat=WORK.EM_DMDB
maxlevel = 101
out=_spdslib.EM_DMDB
;
class %DMDBClass;
var %DMDBVar;
target
resp
;
run;
```

Display 2.53 shows the **log** from **PROC DMINE**.

Display 2.53

```
*------------------------------------------------------------* ;
* Varsel: Input Variables Macro ;
*------------------------------------------------------------* ;
%macro INPUTS;
    VAR1N VAR20 VAR3N VAR50 VAR80 CREDSCORE
%mend INPUTS;
proc dmine data=_spdslib.EM_DMDB dmdbcat=WORK.EM_DMDB
minr2=0.005 maxrows=3000 stopr2=0.00001 NOINTER USEGROUPS outest=WORK._Varsel_OUTESTDMINE
NOMONITOR
PSHORT
;
var %INPUTS;
target resp;
code file="C:\TheBook\EM_5.2\Nov2006\Chapter2B\Workspaces\EMWS1\Varsel\EMFLOWSCORE.sas";
code file="C:\TheBook\EM_5.2\Nov2006\Chapter2B\Workspaces\EMWS1\Varsel\EMPUBLISHSCORE.sas";
run;
```

2.6.7 Transform Variables Node

Sections 2.6.7.1 and 2.6.7.2 describe the different transformation methods available in the Enterprise Miner **Transform Variables** node. Sections 2.6.7.3 to 2.6.7.6 show some examples of how to make the transformations, save the code, and pass the transformed variables to the next node.

2.6.7.1 Transformations for Interval Inputs

- Simple Transformations

  The available simple transformations are Log, Square Root, Inverse, Square, Exponential, and Standardize. They can be applied to any interval-scaled input. These simple transformations can be used irrespective of whether the target is categorical or continuous.

- Binning Transformations

  In Enterprise Miner, there are three ways of binning an interval-scaled variable. To use these as default transformations, select the **Transform Variables** node, and set the value of the **Interval Inputs** property to **Bucket**, **Quantile**, or **Optimal** in the **Default Methods** section.

  - **Bucket**: The Bucket option creates buckets by dividing the input into n equal-sized intervals and grouping the observations into the n buckets. The resulting number of observations in each bucket may differ from bucket to bucket. For example if AGE is divided into the four intervals 0–25, 25–50, 50–75, and 75–100 then the number of observations in the interval 0–25 (bin 1) may be 100, the number of observations in the interval 25–50 (bin 2) may be 2000, the number of observations in the interval 50–75 (bin 3) may be 1000, and the number of observations in the interval 75–100 (bin 4) may be 200.
  - **Quantile**: The Quantile option groups the observations into quantiles (bins) with an equal number of observations in each. If there are 20 quantiles, then each quantile consists of 5% of the observations.
  - **Optimal Binning for Relationship to Target**: This transformation is available for binary targets only. The input is split into a number of bins, and the splits are placed so as to make the distribution of the target levels (for example, response and non-response) in each bin significantly different from the distribution in the other bins. To help you understand optimal binning on the basis of relationship to target, consider two possible ways of binning. The first method involves binning the input variable recursively. Suppose you have an interval input X that takes on values ranging from 0 to 320. A new variable with 64 levels can be created very simply by dividing this range into 64 sub-intervals as follows:

    $$0 \le X < 5,\ 5 \le X < 10, \ldots\ldots, 310 \le X < 315,\ 315 \le X \le 320.$$

    The values the new variable can take are 5, 10, 15... 315, which can be called *splitting values*. You can split the data into two parts at each splitting value and determine the Chi-Square value from a contingency table, with columns representing the splitting value of the input X and rows representing the levels of the target variable. Suppose the splitting value tested is 105. At that point, the contingency table will look like the following:

	$X < 105$	$X \ge 105$
Target level		
1	n_{11}	n_{12}
0	n_{01}	n_{02}

where, n_{11}, n_{12}, n_{01}, and n_{02} denote the number of records in each cell of the contingency table. The computation of the Chi-Square statistic is detailed in Chapter 4.

The splitting value that gives the maximum Chi-Square determines the first split of the data into two parts. Then each part is split further into two parts using the same procedure, giving two more splitting values and four partitions. The process continues until no more partitioning is possible. That is, there is no splitting value that gives a Chi-Square above the specified threshold. At the end of the process, the splitting values chosen define the optimal bins. This is illustrated in Display 2.54.

Display 2.54

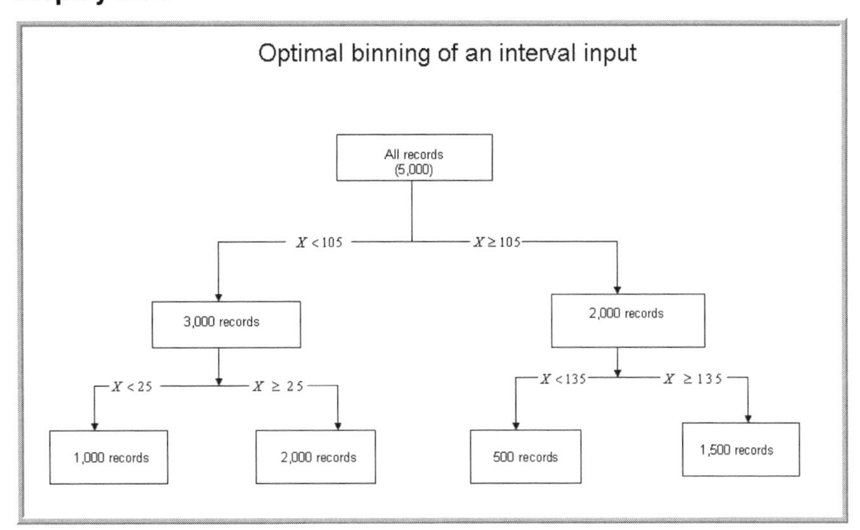

From Display 2.54, the optimal bins created for the input X are $X < 25,\ 25 \le X < 105,\ 105 \le X < 135$ and $135 \le X$. Four optimal bins are created from the initial 64 bins of equal size. The optimal bins are defined by selecting the best split value at each partition.

The second method involves starting with the 64 or so initial bins and collapsing them to a smaller number of bins by combining adjacent bins that have similar distribution of the target levels. Enterprise Miner does not follow the steps outlined above exactly, but Enterprise Miner's method of finding optimal bins is essentially what is described in the second method above. There are other ways of finding the optimal bins. Note that in the **Transform Variables** node the optimal bins are created for one interval-scaled variable at a time in contrast to the **Decision Tree** node where the bins are created with *many inputs of different measurement scales*, interval as well as categorical.

- **Best Power Transformations**

  The **Transform Variables** node selects the best power transformations from among $X, \log(X), sqrt(X),\ e^X,\ X^{1/4},\ X^2,$ and X^4, where X is the input. There are four criteria of "best" available:

  - **Maximum Normal**: To find the transformation that maximizes normality, sample quantiles from each of the transformations listed above are compared with the theoretical quantiles of a normal distribution. The transformation that yields quantiles that are closest to the normal distribution is chosen.

Suppose Y is obtained by applying one of the above transformations to X. For example, the 0.75-sample quantile of the transformed variable Y is that value of Y at or below which 75% of the observations in the data set fall. The 0.75-quantile for a standard normal distribution is 0.6745 given by $P(Z \leq 0.6745) = 0.75$, where Z is a normal random variable with mean 0 and standard deviation 1. The 0.75-sample quantile for Y is compared with 0.6745, and similarly the other quantiles are compared with the corresponding quantiles of the standard normal distribution.

- **Maximum Correlation**: This is available only for continuous targets. The transformation that yields the highest linear correlation with the target is chosen.
- **Equalize Spread with Target Levels**: This method requires a class target. The method first calculates variance of a given transformed variable within each target class. Then for each transformation it calculates the variances of these variances. It chooses the transformation that yields the smallest variance of the variances.
- **Optimal Maximum Equalize Spread with Target Level**: This method requires a class target. It chooses the method that equalizes spread with the target.

2.6.7.2 Transformations of Class Inputs

For class inputs, two types of transformations are available.

- **Group Rare Levels transformation**: This transformation combines the rare levels into a separate group, _OTHER_. To define a rare level, you define a cutoff value using the **Cutoff Value** property.
- **Dummy Indicators Transformation**: This transformation creates a dummy indicator variable (0 or 1) for each level of the class variable.

To choose one of these available transformations, select the **Transform Variables** node and set the value of the **Class Inputs** property to the desired transformation.

By setting the **Class Inputs** property to **Dummy Indicators Transformation**, as mentioned before, you will get a dummy variable for each category of the class input. You can test the significance of these dummy variables using the **Regression** node. The groups that correspond to the dummy variables, which are not statistically significant, can be combined, or omitted from the regression. If they are omitted from the regression, it implies their effects are captured in the intercept term.

2.6.7.3 Transformations of Targets

In the **Transform Variables** node of Enterprise Miner 5.2, the target variables can also be transformed by setting the **Interval Targets** and **Class Targets** properties. If you click the **Value** column in the **Interval Targets** property or **Class Targets** property, a list of all available transformations drops down and lets you select the desired property.

2.6.7.4 Selecting Default Methods

To select a default method for transforming interval inputs, select the **Transform Variables** node in the **Diagram Workspace** and set the **Interval Inputs** property (under the **Default Methods** section in the properties panel) to one of the transformations described in 2.6.7.1. The selected method will be applied to all interval variables. Similarly, set the **Class Inputs** property to one of the methods described in 2.6.7.2. Enterprise Miner applies the chosen default method to all inputs of the same type.

2.6.7.5 Overriding the Default Methods

If you do not want to use the default method of transformation for a particular input, you can override it by right-clicking on [...] located on the right side of the **Variables** property of the **Transform Variables** node, as shown in Display 2.55. This opens the **Variables** window as shown in Display 2.56.

Display 2.55

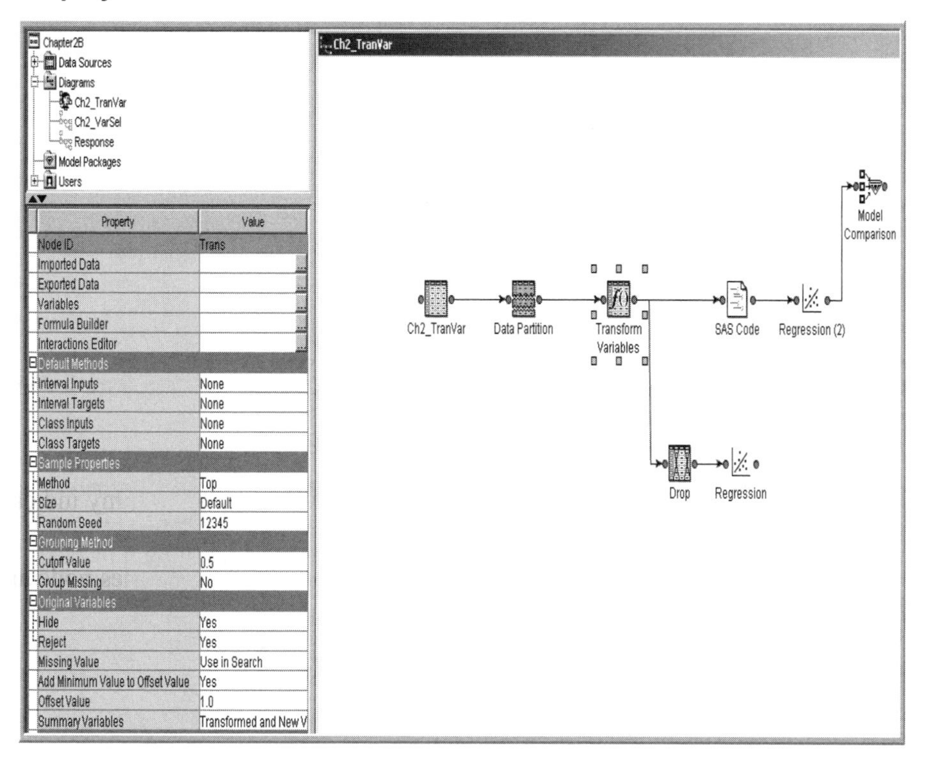

Display 2.56

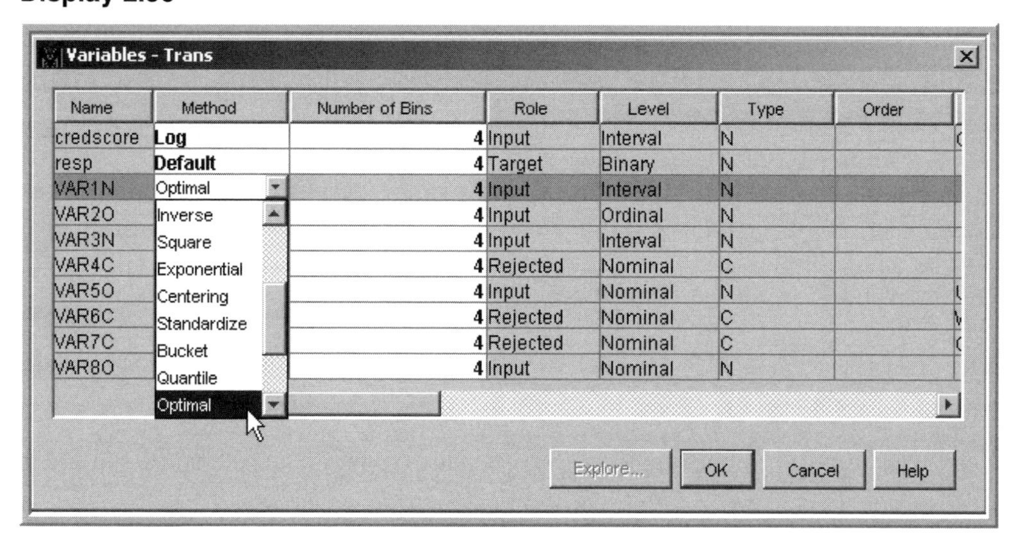

In the **Variables** window, click on the **Method** column for the variable VAR1N, and select **Optimal**. Similarly, select the **log** transformation for the variable CREDSCORE (Credit score). After these changes are made, the resulting **Variables** window is shown in Display 2.63, which also indicates that **Optimal** binning is used for the variable VAR1N and **log** is used for the variable CREDSCORE. Transformations are not performed for other variables since the default methods are set to **NONE** in the properties panel (see Display 2.57).

Display 2.57

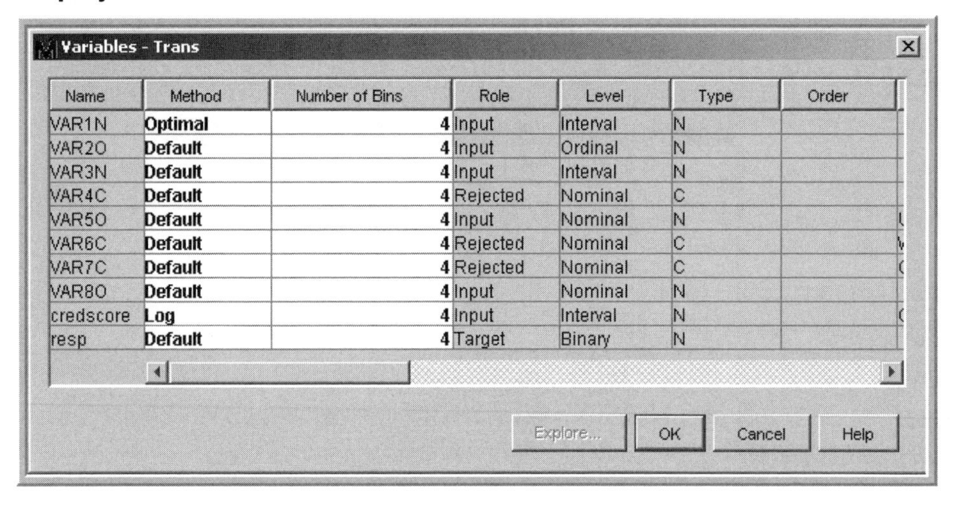

Variables - Trans

Name	Method	Number of Bins	Role	Level	Type	Order
VAR1N	Optimal	4	Input	Interval	N	
VAR2O	Default	4	Input	Ordinal	N	
VAR3N	Default	4	Input	Interval	N	
VAR4C	Default	4	Rejected	Nominal	C	
VAR5O	Default	4	Input	Nominal	N	
VAR6C	Default	4	Rejected	Nominal	C	
VAR7C	Default	4	Rejected	Nominal	C	
VAR8O	Default	4	Input	Nominal	N	
credscore	Log	4	Input	Interval	N	
resp	Default	4	Target	Binary	N	

Explore... OK Cancel Help

2.6.7.6 Saving the SAS Code Generated by the Transform Variables Node

Right-click on the **Transform Variables** node and run it. Then click on **Results**. In the **Results** window, click **View→SAS Results→Flow Code** to see the SAS code, as shown in Display 2.58.

Display 2.58

Flow Code

```
1   *------------------------------------------------------------*;
2   * Computed Code;
3   *------------------------------------------------------------*;
4   *------------------------------------------------------------*;
5   * TRANSFORM: credscore , log(credscore + 1);
6   *------------------------------------------------------------*;
7   label LOG_credscore = 'Transformed: Credit score';
8   if credscore + 1 > 0 then LOG_credscore = log(credscore + 1);
9   else LOG_credscore = .;
10  *------------------------------------------------------------*;
11  * TRANSFORM: VAR1N , Optimal Binning(4);
12  *------------------------------------------------------------*;
13  label OPT_VAR1N = 'Transformed VAR1N';
14  length OPT_VAR1N $21;
15  if (VAR1N < 13.5) then
16  OPT_VAR1N = "01:low -13.5, MISSING";
17  else
18  if (VAR1N >= 13.5) then
19  OPT_VAR1N = "02:13.5-high";
```

This code can be saved by clicking **File→Save as**, and typing in the path of the file and the filename.

If you want to pass both the original and transformed variables to the next node, you must change the **Hide** and **Reject** properties to **No** in the **Original Variables** area in the properties panel and then run the **Transform Variables** node. If you open the **Variables** property in the next node, which is the **SAS Code** node in this case, you will see both the original and transformed variables passed to it from the **Transform Variables** node, as shown in Display 2.59.

Display 2.59

Variables - EMCODE

Name	Use	Report	Role	Level	Type	Order	
credscore	Yes	No	Input	Interval	N		Cre
LOG_credscore	Yes	No	Input	Interval	N		Tra
OPT_VAR1N	Yes	No	Input	Nominal	C		Tra
resp	Yes	No	Target	Binary	N		
VAR1N	Yes	No	Input	Interval	N		
VAR2O	Yes	No	Input	Ordinal	N		
VAR3N	Yes	No	Input	Interval	N		
VAR4C	No	No	Rejected	Nominal	C		
VAR5O	Yes	No	Input	Nominal	N		US
VAR6C	No	No	Rejected	Nominal	C		WE

Explore... OK Cancel Help

2.6.8 SAS Code Node

The **SAS Code** node is used to incorporate SAS procedures and external SAS code into the process flow of a project. In addition, you can perform DATA step programming in this node. The **SAS Code** node is the starting point for creating custom nodes and extending the functionality of Enterprise Miner 5.2.

The **SAS Code** node can be used at any position in the sequence of nodes in the process flow. For example, in Chapters 6 and 7, custom programming is done in the **SAS Code** node to create special lift charts.

The following sections explore the **SAS Code** node so you can become familiar with the macro variables and macros used internally. The macro variables refer to imported and exported data sets, libraries, etc.

First, look at a simple process flow with the **SAS Code** node, as shown in Display 2.60.

Display 2.60

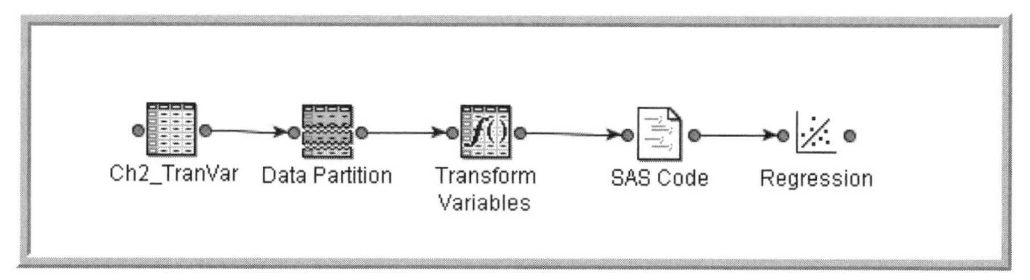

Display 2.61 shows the variables contained in the data sets passed by the **Transform Variables** node to the **SAS Code** node.

Display 2.61

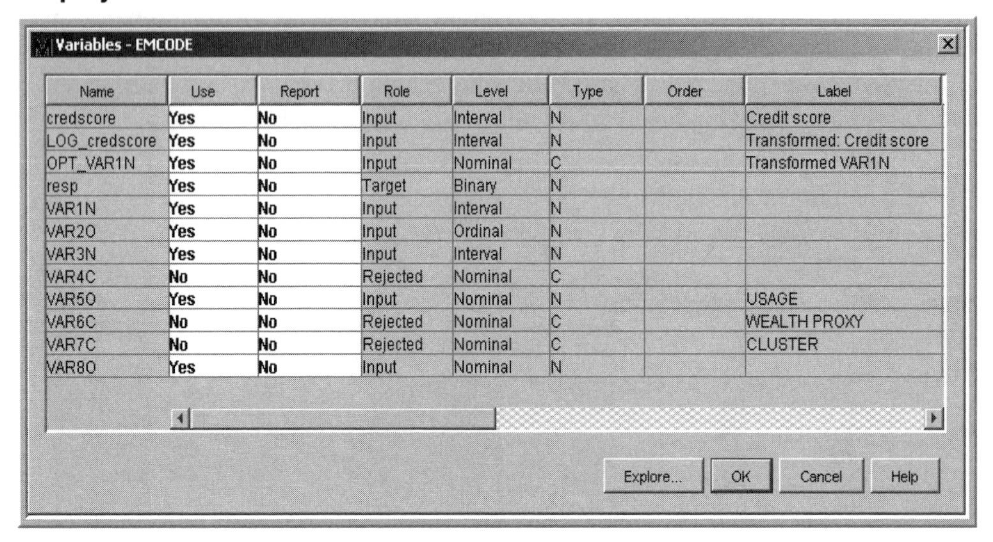

Name	Use	Report	Role	Level	Type	Order	Label
credscore	Yes	No	Input	Interval	N		Credit score
LOG_credscore	Yes	No	Input	Interval	N		Transformed: Credit score
OPT_VAR1N	Yes	No	Input	Nominal	C		Transformed VAR1N
resp	Yes	No	Target	Binary	N		
VAR1N	Yes	No	Input	Interval	N		
VAR2O	Yes	No	Input	Ordinal	N		
VAR3N	Yes	No	Input	Interval	N		
VAR4C	No	No	Rejected	Nominal	C		
VAR5O	Yes	No	Input	Nominal	N		USAGE
VAR6C	No	No	Rejected	Nominal	C		WEALTH PROXY
VAR7C	No	No	Rejected	Nominal	C		CLUSTER
VAR8O	Yes	No	Input	Nominal	N		

To open the **SAS Code** window, select the **SAS Code** node in the **Diagram Workspace** and click on [...] located to the right of the **SAS Code** property in the properties panel, as shown in Display 2.62.

Display 2.62

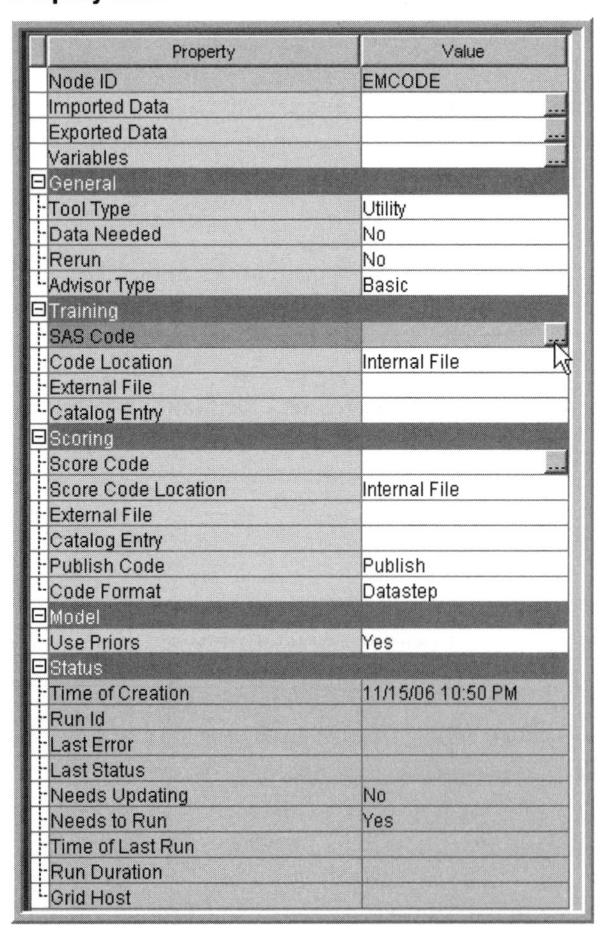

Property	Value
Node ID	EMCODE
Imported Data	...
Exported Data	...
Variables	...
General	
Tool Type	Utility
Data Needed	No
Rerun	No
Advisor Type	Basic
Training	
SAS Code	...
Code Location	Internal File
External File	
Catalog Entry	
Scoring	
Score Code	...
Score Code Location	Internal File
External File	
Catalog Entry	
Publish Code	Publish
Code Format	Datastep
Model	
Use Priors	Yes
Status	
Time of Creation	11/15/06 10:50 PM
Run Id	
Last Error	
Last Status	
Needs Updating	No
Needs to Run	Yes
Time of Last Run	
Run Duration	
Grid Host	

The **SAS Code** window is shown in Display 2.63.

Display 2.63

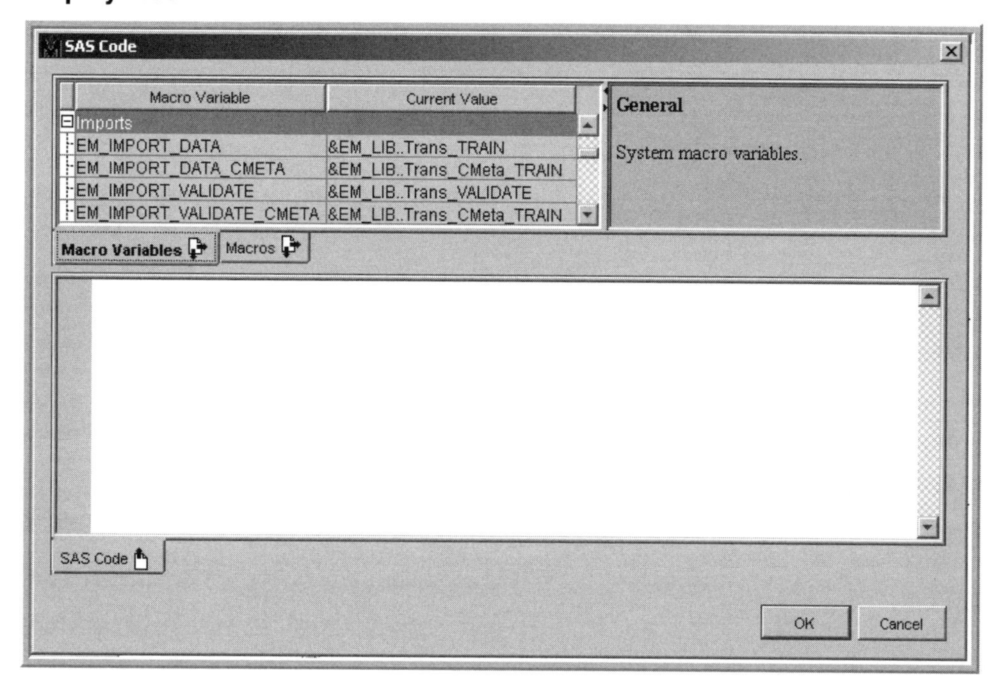

The **SAS Code** window has three tabs: **SAS Code**, **Macro Variables**, and **Macros**. These tabs can be used to place the macro variables, macros, and SAS code in different locations of the **SAS Code** window.

Display 2.64 shows the **Macro Variables** and the **Macros** tabs placed at the top and bottom of the **SAS Code** window.

Display 2.64

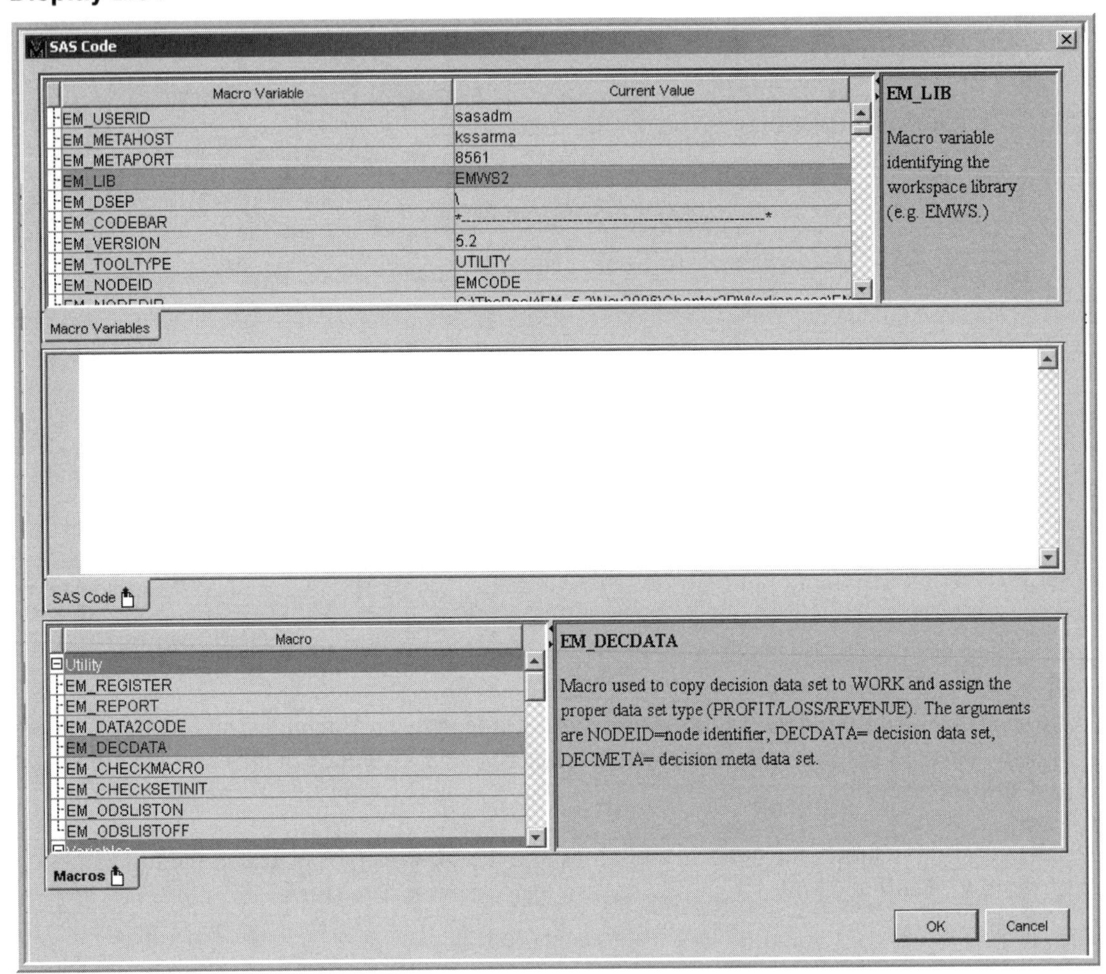

In Display 2.64, if you highlight the macro variable EM_LIB, its description appears in the right panel of the window. By scrolling down inside the **Macro Variables** tab, you can see the macro variables identifying the data sets imported from the predecessor nodes, as shown in Display 2.65.

Display 2.65

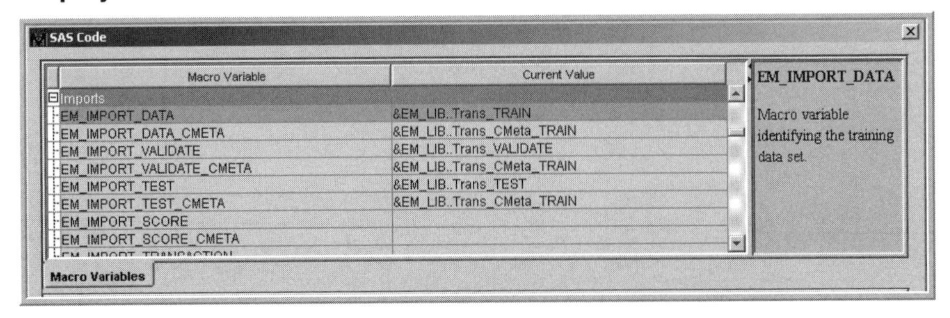

In Display 2.65, the macro variable EM_IMPORT_DATA is highlighted. Its current value is &EM_LIB.Trans_TRAIN. The first part of the current value is the macro variable &EM_LIB whose value is EMWS2 (see Display 2.64), which is the *libref.* The second part is **Trans**, which indicates that the data set is created by the **Transform Variables** node. The third part is **TRAIN**, which means this data set is the training data set. From the description given in the right panel of

Display 2.65, it is clear that the highlighted macro variable refers to the training data set. By scrolling down farther, you can see the macro names identifying exported data sets, files, number of variables, and code statements.

Display 2.66 shows some sample SAS code. In this code, I created a composite variable and included it in the exported data set.

Display 2.66

```
data &em_export_train;
     set &em_import_data ;
     total_tran = var3n + var1n ;
     label total_tran = "Total Transactions Created in SAS Code node" ;
run ;
```

SAS Code

In the sample code shown in Display 2.67, I create a new variable called total_tran, which is the sum of Var3N and Var1N. Var1N and Var3N represent the number of withdrawals and deposits from a bank per month, respectively. The composite variable total_tran represents total transactions, and it captures the customer's overall activity.

The next node after the **SAS Code** node in the process flow is the **Regression** node. If you select the **Regression** node, and click on [...], located to the right of the **Variables** property, all the variables including the newly created variable total_tran appear as shown in Display 2.67. Before you do this, you must update the path.

Display 2.67

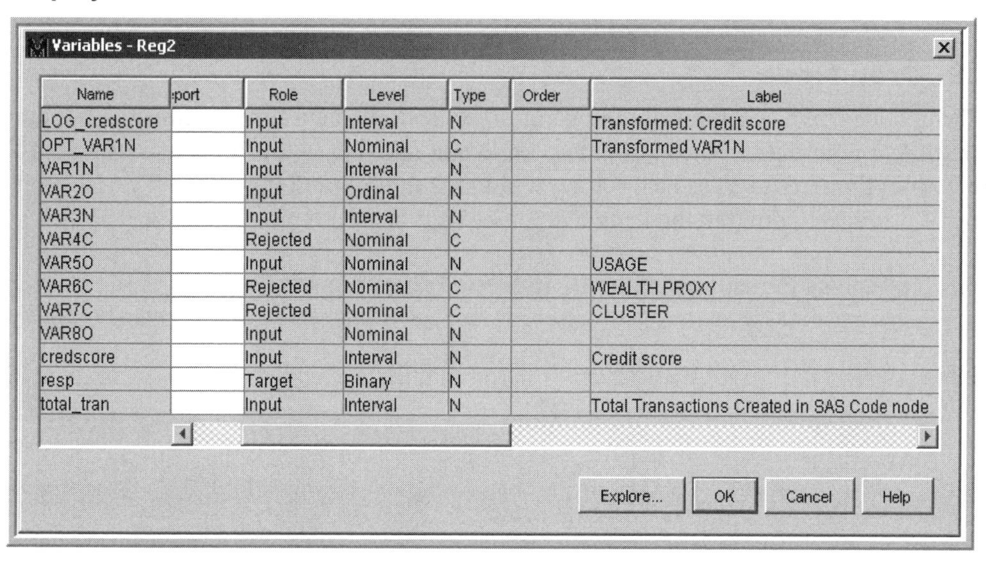

Variables - Reg2

Name	port	Role	Level	Type	Order	Label
LOG_credscore		Input	Interval	N		Transformed: Credit score
OPT_VAR1N		Input	Nominal	C		Transformed VAR1N
VAR1N		Input	Interval	N		
VAR2O		Input	Ordinal	N		
VAR3N		Input	Interval	N		
VAR4C		Rejected	Nominal	C		
VAR5O		Input	Nominal	N		USAGE
VAR6C		Rejected	Nominal	C		WEALTH PROXY
VAR7C		Rejected	Nominal	C		CLUSTER
VAR8O		Input	Nominal	N		
credscore		Input	Interval	N		Credit score
resp		Target	Binary	N		
total_tran		Input	Interval	N		Total Transactions Created in SAS Code node

Explore... OK Cancel Help

The **SAS Code** node can be used with code located in an external file. To use an external file, select the **SAS Code** node and set the **Code Location** property to **External File**, and set the **External File** property to the full path of the file and the filename. This is shown in Display 2.68.

Display 2.68

Property	Value
Node ID	EMCODE2
Imported Data	...
Exported Data	...
Variables	...
⊟General	
Tool Type	Utility
Data Needed	No
Rerun	No
Advisor Type	Basic
⊟Training	
SAS Code	...
Code Location	External File
External File	C:\TheBook\EM_5.2\Programs\LTV.SAS
Catalog Entry	
⊟Scoring	
Score Code	...
Score Code Location	Internal File
External File	
Catalog Entry	
Publish Code	Publish
Code Format	Datastep
⊟Model	
Use Priors	Yes
⊟Status	
Time of Creation	11/16/06 3:01 PM
Run Id	
Last Error	
Last Status	
Needs Updating	Yes
Needs to Run	Yes
Time of Last Run	
Run Duration	
Grid Host	

2.6.9 Drop Node

The **Drop** node can be used for dropping variables from the data set or metadata. After an initial exploration of your data, you can drop any irrelevant variable in this node by setting the **Drop from Tables** property to **Yes**. This setting will result in dropping the selected variables from the table that is to be exported to the next node.

Display 2.69 shows the settings of the **Drop from Tables** property.

Display 2.69

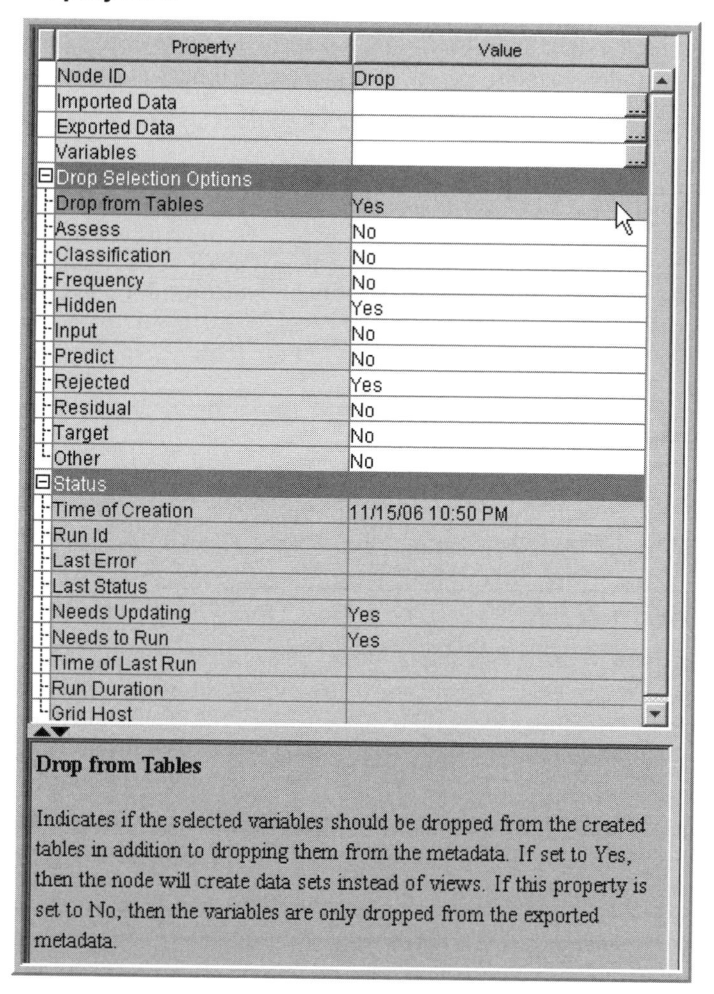

To select the variables to be dropped, click on [...], located to the right of the **Variables** property of the **Drop** node. If you want to drop any variable, click in the **Drop** column next to that variable and select **Yes**, as shown in Display 2.70.

Display 2.70

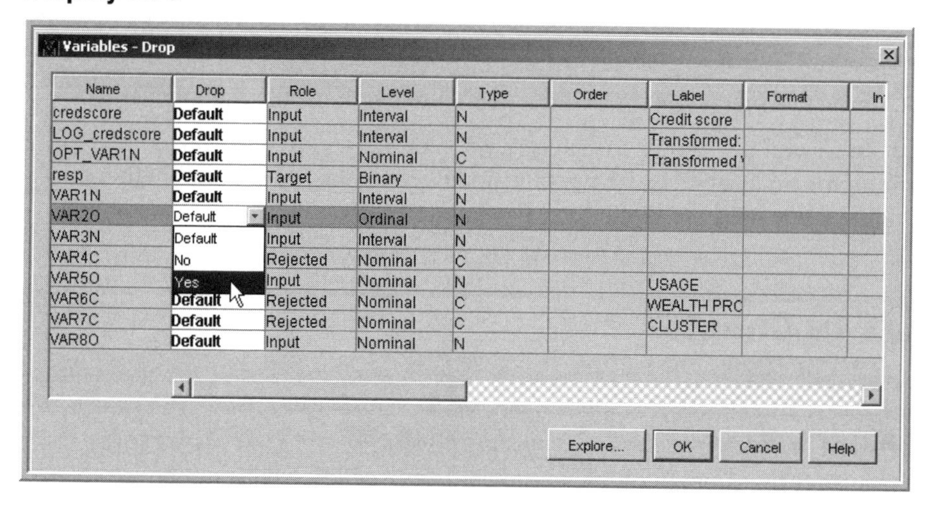

2.6.10 Decision Tree Node

The **Decision Tree** node is discussed extensively in Chapters 4 and 7 using several examples. In these chapters, the **Decision Tree** node is used to develop models with binary and continuous targets.

2.6.11 Neural Network Node

This node is covered in Chapters 5 and 7, where I develop models for predicting response and risk for an automobile insurance company, plus price sensitivity and attrition for a hypothetical bank.

2.6.12 Regression Node

This node is discussed in Chapters 6 and 7. Models are developed for binary and continuous targets.

2.6.13 Model Comparison Node

You can use the **Model Comparison** node to evaluate a model or to compare models using your test data set. Alternative models to predict risk and response are evaluated in Chapter 7 using the **Model Comparison** node. Chapter 7 includes a detailed discussion of lift charts and capture rates in the training, validation, and test data sets for each model compared.

If you want to evaluate a model or compare models using the test data, you should use the **Model Comparison** node.

2.7 Summary

- This chapter introduced you to Enterprise Miner 5.2.
- After reading this chapter, you should be able to open Enterprise Miner 5.2, define a new project, and create data sources for the project.
- The chapter gave an overview of the tools (nodes) that are available for structuring and customizing data mining tasks. It provided information to help decide which tools to use and which values to assign to the various properties of the tools in order to perform data cleaning and exploration tasks for the data mining project.
- Some of the tools mentioned here are covered in depth in subsequent chapters.
- There are several other tools in Enterprise Miner 5.2 such as **Cluster** node, **Principal Components** node, and **User Interface** node, which are not covered in this book. You can find information on these in the SAS Enterprise Miner help and documentation.

2.8 Appendix to Chapter 2

2.8.1 Cramer's V

Cramer's V measures the strength of the relationship between two categorical variables. Suppose you have a categorical target Y with K distinct categories $Y_1, Y_2, \ldots Y_K$, and a categorical input

X with M distinct categories $X_1, X_2, \ldots. X_M$. If there are n_{ij} observations in the ith category in X, and j^{th} category in Y, then you can make a contingency table of the following type:

$X \setminus Y$	Y_1	Y_2	....	Y_K	Row Total
X_1	n_{11}	n_{12}	....	n_{1K}	$n_{1.}$
X_2	n_{11}	n_{11}	....	n_{2K}	$n_{2.}$
..					
X_M	n_{M1}	n_{M2}	....	n_{MK}	$n_{M.}$
Column Total	$n_{.1}$	$n_{.2}$	....	$n_{.K}$	N

To calculate Cramer's V, you must first calculate the Chi-Square statistics under the null hypothesis that there is no association between X and Y.

$$\chi^2 = \sum_{i=1}^{M}\sum_{j=1}^{K}\frac{(n_{ij} - E_{ij})^2}{E_{ij}}, \text{ where } E_{ij} = n_{.j}\left(\frac{n_{i.}}{N}\right), \; i = 1, \ldots M \text{ and } j = 1, ..K.$$

Cramer's V = $\sqrt{\dfrac{\chi^2}{N \min(M-1, K-1)}}$, which takes a value between 0 and 1 for all tables larger than 2x2. For a 2x2 table, Cramer's V takes a value between -1 and $+1$.

Suppose that Y is a binary variable with two levels, and X is a 16-level categorical variable. Then Cramer's V = $\sqrt{\dfrac{\chi^2}{N}}$.

2.8.2 Calculation of Chi-Square Statistic and Cramer's V for a Continuous Input

As pointed out in Section 2.6.2.1, the AGE variable is divided into five groups (bins). Table 2.1 shows the number of responders (RESP = 1) and non-responders (RESP = 0).

Table 2.1

RESP	0	1	Total
Age Group			
18 - 32.4	3830	1959	5789
32.4 - 46.8	4952	2251	7203
46.8 - 61.2	6915	3181	10096
61.2 - 75.6	3882	1597	5479
75.6 - 90	946	391	1337
	20525	9379	29904
Overall resp rate	0.686363	0.313637	

Table 2.2 shows the first step in calculating the Chi-Square value. It shows the expected number of observations in each cell under the null hypothesis of independence between rows and columns.

Table 2.2

RESP	0	1	Total
Age Group			
18 - 32.4	3973	1816	5789
32.4 - 46.8	4944	2259	7203
46.8 - 61.2	6930	3166	10096
61.2 - 75.6	3761	1718	5479
75.6 - 90	918	419	1337

The expected number of observations is obtained by applying the overall response rate to each row total given in Table 2.1.

The Chi-Square for each cell is calculated as $\frac{(O-E)^2}{E}$, where O is the observed frequency as shown in Table 2.1, and E is the expected frequency (under the null hypothesis) as given in Table 2.1. The Chi-Square values for each cell are shown in Table 2.3.

Table 2.3

RESP	0.0000	1.0000
Age Group		
18 - 32.4	5.1722	11.3187
32.4 - 46.8	0.0134	0.0292
46.8 - 61.2	0.0304	0.0666
61.2 - 75.6	3.9202	8.5789
75.6 - 90	0.8748	1.9143

These cell-specific Chi-Square values can be retrieved from the **Results** window of the **StatExplore** node by clicking **View→Summary Statistics→Cell** in the menu bar. These cell-specific Chi-Square values are shown in Display 2.71.

Display 2.71

Target	Input	Target: For...	Input: Forma...	Frequency ...	Target: Num...	Input: Nume...	Chi Square	Log Chi Squ...
resp	AGE	0	18:32.4	3830	0	18	5.172157	1.64329
resp	AGE	0	32.4:46.8	4952	0	33	0.01336	-4.31549
resp	AGE	0	46.8:61.2	6915	0	47	0.03043	-3.49234
resp	AGE	0	61.2:75.6	3882	0	62	3.920158	1.366132
resp	AGE	0	75.6:90	946	0	76	0.874759	-0.13381
resp	AGE	1	18:32.4	1959	1	18	11.31875	2.42646
resp	AGE	1	32.4:46.8	2251	1	33	0.029237	-3.53232
resp	AGE	1	46.8:61.2	3181	1	47	0.066592	-2.70917
resp	AGE	1	61.2:75.6	1597	1	62	8.578872	2.149302
resp	AGE	1	75.6:90	391	1	76	1.914323	0.649364

By summing the cell Chi-Squares, you get the overall Chi-Square (χ^2) of 31.9864 for the AGE variable. Since there are 29904 observations in the sample, Cramer's V can be calculated $\sqrt{\frac{\chi^2}{29904}}$ which is equal to 0.03267.

By selecting the **Chi-Square Plot** (or clicking in it), and selecting the **Table** in the **Results** window, you can see the Chi-Square and Cramer's V for all variables. A partial view of this table is shown in Display 2.72.

Display 2.72

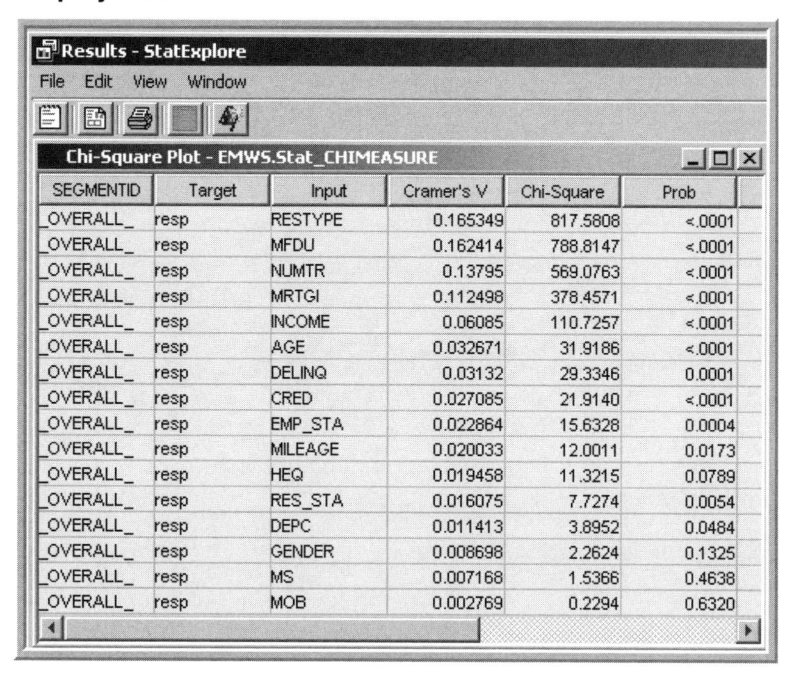

Results - StatExplore

File Edit View Window

Chi-Square Plot - EMWS.Stat_CHIMEASURE

SEGMENTID	Target	Input	Cramer's V	Chi-Square	Prob
OVERALL	resp	RESTYPE	0.165349	817.5808	<.0001
OVERALL	resp	MFDU	0.162414	788.8147	<.0001
OVERALL	resp	NUMTR	0.13795	569.0763	<.0001
OVERALL	resp	MRTGI	0.112498	378.4571	<.0001
OVERALL	resp	INCOME	0.06085	110.7257	<.0001
OVERALL	resp	AGE	0.032671	31.9186	<.0001
OVERALL	resp	DELINQ	0.03132	29.3346	0.0001
OVERALL	resp	CRED	0.027085	21.9140	<.0001
OVERALL	resp	EMP_STA	0.022864	15.6328	0.0004
OVERALL	resp	MILEAGE	0.020033	12.0011	0.0173
OVERALL	resp	HEQ	0.019458	11.3215	0.0789
OVERALL	resp	RES_STA	0.016075	7.7274	0.0054
OVERALL	resp	DEPC	0.011413	3.8952	0.0484
OVERALL	resp	GENDER	0.008698	2.2624	0.1325
OVERALL	resp	MS	0.007168	1.5366	0.4638
OVERALL	resp	MOB	0.002769	0.2294	0.6320

Variable Selection and Transformation of Variables

3.1 Introduction

This chapter will familiarize you with the SAS Enterprise Miner **Variable Selection** and **Transform Variables** nodes, which are essential for building predictive models. The **Variable Selection** node is useful when you want to identify important variables (inputs) for building predictive models. The **Transform Variables** node provides a wide variety of transformations that can be applied to the inputs for improving the precision of the predictive models.

The following topics are covered:

- **Variable Selection** node
  - variable selection for categorical (class) targets
  - variable selection for continuous (interval-scaled) targets
  - binning continuous inputs for variable selection
  - grouping or collapsing the categories of categorical inputs for variable selection
  - SAS code generated by the **Variable Selection** node

- **Transform Variables** node
    - mathematical transformations of continuous inputs
    - binning continuous variables
    - creating dummy variables for categorical (class) inputs
    - how to make more than one type of transformation of an input
    - SAS code generated by the **Transform Variables** node

- using the two nodes in alternative sequences
    - In some cases, particularly when there are a large number of inputs, it may be best to make the variable selection first and then transform the selected variables.
    - In other cases it may be best to use the **Transform Variables** node first and then the **Variable Selection** node. Both these sequences are illustrated.

3.2 Variable Selection

In this section, I discuss how variables are selected by the **Variable Selection** node for different measurement scales of the inputs and the target, and I discuss the available selection criteria.

In predictive modeling and data mining we are often confronted with a large number of inputs (explanatory variables). The number of potential inputs to choose from may be as large as 2000 or higher. Some of these inputs may not have any relation to the target. An initial screening is therefore necessary to eliminate irrelevant variables to keep the number of inputs to a manageable size. A final selection can then be made from the remaining variables using either a stepwise regression or a decision tree.

The **Variable Selection** node performs the initial selection as well as a final variable selection. I will discuss the way that the **Variable Selection** node handles interval-, nominal-, and ordinal-scaled inputs, as well as the selection criteria for categorical and continuous targets. The discussion will focus on four common situations (Cases 1–4):

- The target is continuous (interval-scaled) and the inputs are numeric, consisting of interval-scaled variables (Case 1).
- The target is continuous and the inputs are categorical and nominal-scaled. These are referred to as class variables (Case 2).
- The target is binary and the inputs are numeric interval-scaled (Case 3).
- The target is binary and the inputs are categorical and nominal-scaled (Case 4).

To focus on how each case is handled, I have used separate data sets for illustration purposes. In addition, to keep the discussion general, I have indicated all numeric inputs as NVAR1, NVAR2, etc., and all nominal (character) inputs as CVAR1, CVAR2, etc.

Here are some examples of a continuous target:

- change in savings deposits of a hypothetical bank after a rate change
- customer's asset value obtained from a survey of a brokerage company
- loss frequency of an auto insurance company

All of the data that I use here is all simulated, so it does not refer to any real bank, etc.

The binary target used in the following examples will represent a response to a direct mail campaign by a hypothetical auto insurance company. The binary target, as its name suggests, takes only two values: response (represented by 1), and no response (represented by 0).

In Enterprise Miner, both ordinal and nominal inputs are treated as class variables.

3.2.1 Continuous Target with Numeric Interval-scaled Inputs (Case 1)

In this example, a hypothetical bank wants to measure the effect of a promotion that involves an increase in the interest rate offered on savings deposits in the bank. Therefore, the target variable is the change in a customer's deposits in response to the increase.

When the target is continuous (interval-scaled), the **Variable Selection** node uses R-Square as the default criterion of selection. For each interval-scaled input the **Variable Selection** node calculates two measures of correlation between each input and the target. One is the R-Square between the target and the original input. The other is the R-Square between the target and the *binned* version of the input variable. The binned variable is a categorical variable created by the **Variable Selection** node from each continuous (interval-scaled) input. The levels of this categorical variable are the bins. In Enterprise Miner, this binned variable is referred to as an **AOV16** variable. The number of levels or categories of the binned variable (**AOV16**) is at most 16, corresponding to 16 intervals that are equal in width. This does not mean that the number of records in each bin is equal.

To get a better picture of these binned variables, let us take an example. Suppose we are binning the AGE variable, and it takes on values from 18–81. The first bin includes all individuals with ages 18–21, the second with ages 22–25, and so on. If there are no cases in a given bin, that bin is automatically eliminated, and the number of bins will be fewer than 16. The binned variable from AGE is called AOV16_AGE. The binned variable for INCOME is called AOV16_INCOME, and so on.

In the data set used in this chapter, there are 230 interval-scaled inputs and 1 binary input. Some of these interval variables were originally ordinal, but I converted them into interval scale. See Section 3.5, "Appendix to Chapter 3," for information on changing the measurement scales of variables in a data source.

The 230 variables are named NVAR001, NVAR002, etc. The **Variable Selection** node created 230 binned variables (AOV16 variables) from the interval-scaled variables, and then included all 461 variables in the selection process. At the end of the Variable Selection process, I had two original variables, NVAR133 and NVAR239, and three binned variables, AOV16_NVAR133, AOV16_NVAR049, and AOV16_NVAR239.

The following is a description of the variables that were selected:

NVAR133 average balance in the customer's checking account for the 12-month period prior to the promotional increase in the interest rate at the hypothetical bank

NVAR049 number of transactions during the last 12 months by the customer

NVAR239 other deposit balance 3 months prior to the promotional rate change

Note that variables NVAR133 and NVAR239 are selected in both their original form and in their binned form. As discussed in Chapter 2, only **AOV16** variables are passed to the next node if you set the value of **Use AOV16 Variables** property to **Yes**. As I have set the **Use AOV16 Variables** property to **Yes** in this example, NVAR133 and NVAR239 are not passed to the next node as inputs.

As we discussed in Chapter 2, Section 2.6.6.1, the R-Square method uses a two-step selection procedure. In the first step, a preliminary selection is made, based on **Minimum R-Square**. For the original variables the R-Square is calculated from a regression of the target on each input; for the binned variables it is calculated from a one-way Analysis of Variance (ANOVA).

The default value of the **Minimum R-Square** property is 0.005, but the user can set the **Minimum R-Square** property to a different value if desired. The inputs that have R-Square values above the lower bound set by the **Minimum R-Square** property are included in the second step of the variable selection process.

In the second step, a sequential forward selection process is used. This process starts by selecting the input variable that has the highest correlation coefficient with the target. A regression equation (model) is estimated with the selected input. At each successive step of the sequence, an additional input variable that provides the largest incremental contribution to the Model R-Square is added to the regression. If the lower bound for the incremental contribution to the Model R-Square is reached, the selection process stops. The lower bound for the incremental contribution to the Model R-Square can be specified by setting the **Stop R-Square** property to the desired value.

In this example, the **Minimum R-Square** property is set to 0.05 and the **Stop R-Square** property is set to 0.005. In this example, 92 variables, some original and some binned, were selected in the first step and, of these, only five variables were selected in the second step.

The smaller the values to which **Minimum R-Square** and **Stop R-Square** properties are set, the larger the number of variables passed to the next node with the **Role** of **Input**. Letting more variables into the next node makes more sense if there is a **Decision Tree** or **Regression** node downstream, as both these nodes can themselves perform rigorous methods of selection. Variable selection is automatic in the **Decision Tree** node but needs to be explicitly specified in the case of the **Regression** node.

To use the **Variable Selection** node for preliminary variable selection only, set the minimum for the **Stop R-Square** property to a very low value such as 0.00001. This, in effect, bypasses Step 2 of the variable selection process, which is okay since Step 2 can always be performed in the **Decision Tree** node or the **Regression** node.

The R-Square for the original raw inputs captures the linear relationship between the target and the original input, while the R-Square between a binned (**AOV16**) variable and the target captures the non-linear relationship. In our example, NVAR049 was not selected in its original form, but the binned version of it, namely AOV16_NVAR049, was selected. By plotting the target mean for each bin against the binned variable, you can see if there exists a non-linear relationship between the input and the target.

In the following pages, I show how to set up the properties of the **Variable Selection** node in Enterprise Miner 5.2. Display 3.1 shows the process flow for the variable selection.

Display 3.1

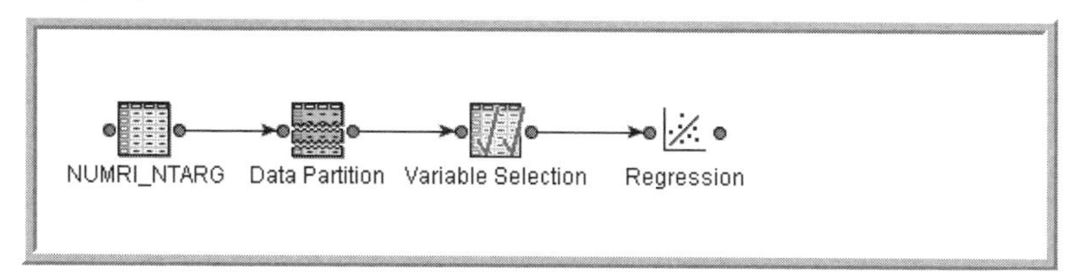

The first tool in the process flow (Display 3.1) is the **Input Data** node, which reads the SAS modeling data set. The data set is then partitioned into *training*, *validation*, and *test* data sets. The **Variable Selection** node selects variables using only the **training** data set. The selected variables are then passed to a **Regression** node, with the **Role** of **Input**.

The selection criterion is specified by setting the **Target Model** property. Display 3.2 shows the properties panel of the **Variable Selection** node. In the case of a continuous target variable, only the R-Square method is available. When the **Variable Selection** node detects that the target is continuous, it selects the R-Square method.

Display 3.2

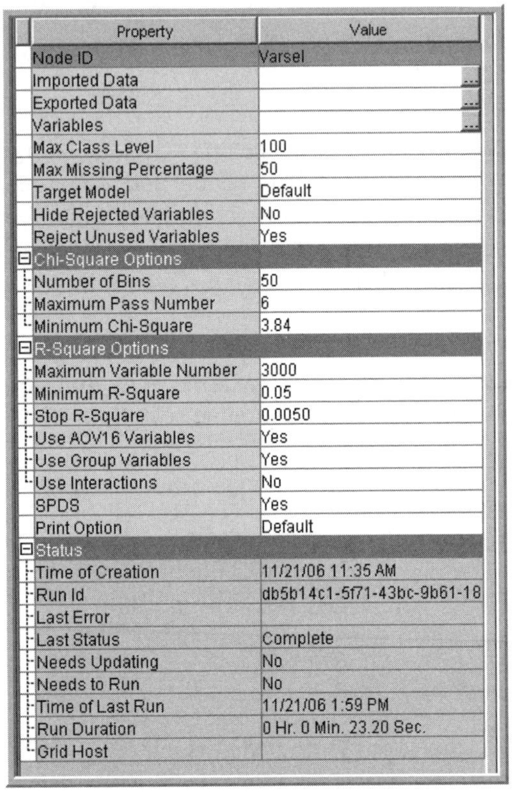

Property	Value
Node ID	Varsel
Imported Data	...
Exported Data	...
Variables	...
Max Class Level	100
Max Missing Percentage	50
Target Model	Default
Hide Rejected Variables	No
Reject Unused Variables	Yes
⊟Chi-Square Options	
Number of Bins	50
Maximum Pass Number	6
Minimum Chi-Square	3.84
⊟R-Square Options	
Maximum Variable Number	3000
Minimum R-Square	0.05
Stop R-Square	0.0050
Use AOV16 Variables	Yes
Use Group Variables	Yes
Use Interactions	No
SPDS	Yes
Print Option	Default
⊟Status	
Time of Creation	11/21/06 11:35 AM
Run Id	db5b14c1-5f71-43bc-9b61-18
Last Error	
Last Status	Complete
Needs Updating	No
Needs to Run	No
Time of Last Run	11/21/06 1:59 PM
Run Duration	0 Hr. 0 Min. 23.20 Sec.
Grid Host	

In the properties panel, I have set the **Use AOV16 Variables** property to **Yes** and the **Use Group Variables** property to **Yes**. As a result, the AOV16 variables are created and passed to the next node, and all class variables are grouped (their levels collapsed). The **Variable Selection** node considers both the original and AOV16 variables in the two-step selection process, but only the selected AOV16 variables are assigned the role of **Input** and passed to the next node. Output 3.1 shows a section of the output window within the **Results** window of the **Variable Selection** node. The output shows the variables selected in Step 1.

Output 3.1

R-Squares for Target Variable: DEPV

Effect	DF	R-Square
Var: NVAR133	1	0.214804
Class: NVAR125	1	0.170772
AOV16: NVAR239	15	0.168009
AOV16: NVAR133	9	0.152164
AOV16: NVAR009	10	0.141193
AOV16: NVAR010	9	0.140946
AOV16: NVAR008	9	0.139130
AOV16: NVAR007	10	0.137466
AOV16: NVAR135	12	0.137062
AOV16: NVAR232	12	0.137062
AOV16: NVAR011	9	0.135750
AOV16: NVAR134	11	0.135647
AOV16: NVAR231	11	0.135647
AOV16: NVAR233	10	0.131550
AOV16: NVAR136	10	0.131550
AOV16: NVAR235	11	0.131436
AOV16: NVAR138	11	0.131436
AOV16: NVAR139	11	0.129268
AOV16: NVAR236	11	0.129268
AOV16: NVAR241	11	0.128856
AOV16: NVAR137	11	0.127621
AOV16: NVAR234	11	0.127621
Var: NVAR011	1	0.125407
Var: NVAR009	1	0.125106

Output 3.1 (*continued*)

Effect	DF	R-Square	
Var: NVAR198	1	0.082589	
Var: NVAR199	1	0.082346	
Var: NVAR196	1	0.082067	
Var: NVAR200	1	0.080211	
Var: NVAR084	1	0.076044	
AOV16: NVAR033	12	0.073751	
AOV16: NVAR012	11	0.067226	
AOV16: NVAR014	10	0.066869	
AOV16: NVAR013	10	0.064891	
AOV16: NVAR049	13	0.064249	
AOV16: NVAR166	8	0.061181	
AOV16: NVAR169	12	0.057662	
Var: NVAR033	1	0.057612	
AOV16: NVAR178	12	0.057351	
AOV16: NVAR179	12	0.056720	
AOV16: NVAR177	8	0.051881	
AOV16: NVAR180	11	0.051642	
AOV16: NVAR167	12	0.050377	
AOV16: NVAR173	12	0.049710	R2 < MINR2
AOV16: NVAR086	12	0.048313	R2 < MINR2
AOV16: NVAR207	12	0.048313	R2 < MINR2
AOV16: NVAR168	9	0.048211	R2 < MINR2
AOV16: NVAR061	15	0.047419	R2 < MINR2
AOV16: NVAR048	12	0.047174	R2 < MINR2
AOV16: NVAR103	15	0.046800	R2 < MINR2
AOV16: NVAR063	11	0.046161	R2 < MINR2
AOV16: NVAR090	8	0.045998	R2 < MINR2
Var: NVAR086	1	0.045849	R2 < MINR2
Var: NVAR207	1	0.045849	R2 < MINR2

A number of variables have met the **Minimum R-Square** criterion in the preliminary variable selection (Step 1). Those that have an R-Square below the minimum threshold are not included in Step 2 of the variable selection process.

In Step 2, all the variables that are selected in Step 1 are subjected to a sequential forward regression selection process, as discussed before. In the second step, only a handful of variables met the **Stop R-Square** criterion, as can be seen from Output 3.2. This output contains a table taken from the output window within the **Results** window of the **Variable Selection** node.

Output 3.2

The DMINE Procedure

Effects Chosen for Target: DEPV

Effect	DF	R-Square	F Value	p-Value	Sum of Squares	Error Mean Square
Var: NVAR133	1	0.214804	398.860949	<.0001	854.145803	2.141463
AOV16: NVAR133	9	0.382606	153.008124	<.0001	1521.393734	1.104802
AOV16: NVAR239	15	0.024565	6.212188	<.0001	97.678241	1.048243
AOV16: NVAR049	13	0.012379	3.700517	<.0001	49.222481	1.023194
Var: NVAR239	1	0.008199	32.572863	<.0001	32.603923	1.000954

The variables NVAR133 and NVAR239 are selected both in their original form and in their binned forms. All other continuous inputs were selected in only their binned forms. Although a great number of variables were selected in the preliminary variables assessment (Step 1), most of them did not meet the **Stop R-Square** criterion used in Step 2.

The next node in the process flow is the **Regression** node. I right-clicked on it and selected **Update** so that the path is updated. This means that the variables selected by the **Variable Selection** node are now passed to the **Regression** node with the role of **Input**. To verify this, I clicked on [...], which is located to the right of the **Variables** property of the **Regression** node. The **Variables** window opened as shown in Display 3.3.

Display 3.3

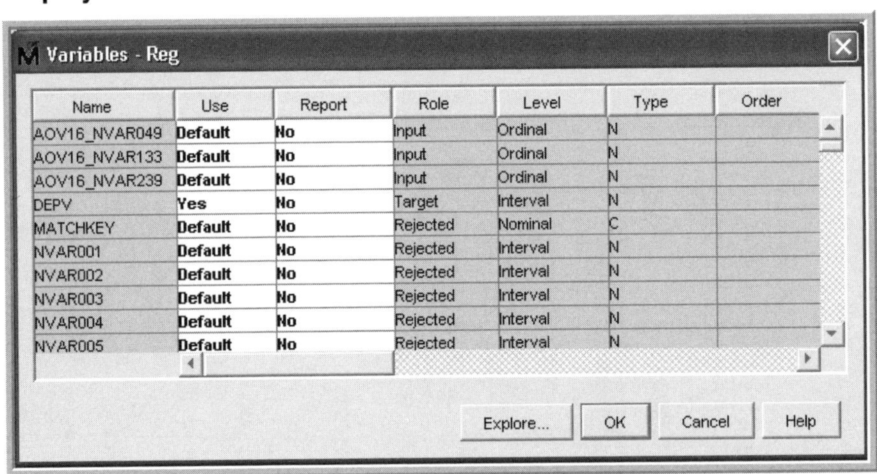

Name	Use	Report	Role	Level	Type	Order
AOV16_NVAR049	Default	No	Input	Ordinal	N	
AOV16_NVAR133	Default	No	Input	Ordinal	N	
AOV16_NVAR239	Default	No	Input	Ordinal	N	
DEPV	Yes	No	Target	Interval	N	
MATCHKEY	Default	No	Rejected	Nominal	C	
NVAR001	Default	No	Rejected	Interval	N	
NVAR002	Default	No	Rejected	Interval	N	
NVAR003	Default	No	Rejected	Interval	N	
NVAR004	Default	No	Rejected	Interval	N	
NVAR005	Default	No	Rejected	Interval	N	

At the top of this window, the selected AOV16 variables are listed. Under the column **Role**, notice that the variables are given the role of **Input**. To see the **Role** assigned to the variables NVAR133 and NVAR239, scroll down the **Variables** window. As you scroll, NVAR133 becomes visible as shown in Display 3.4. Upon scrolling down further, NVAR239 becomes visible, as shown in Display 3.5.

Display 3.4

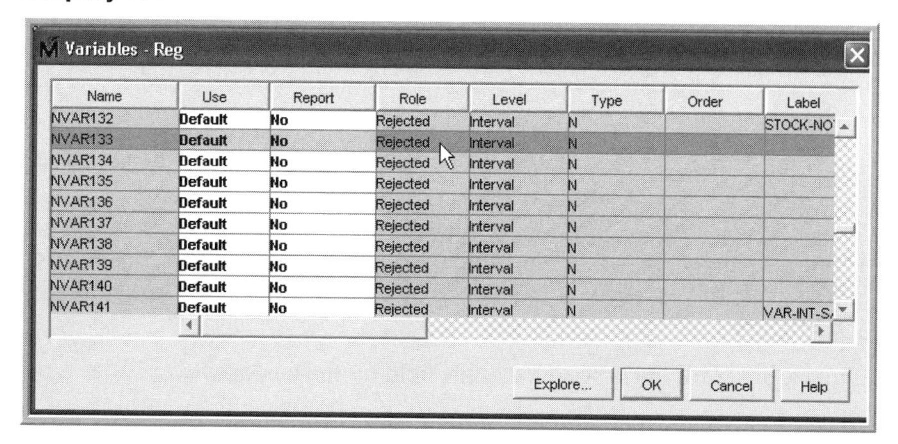

Display 3.5

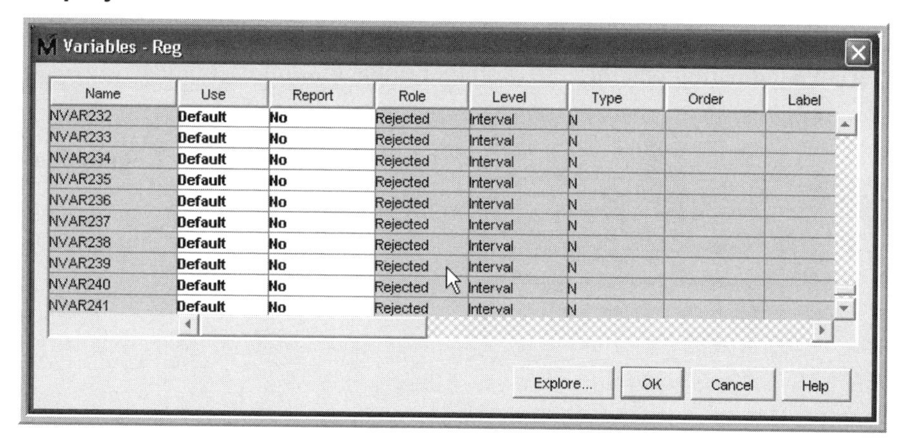

From Displays 3.4 and 3.5, it is clear that the variables NVAR133 and NVAR239 are given the **Role** of **Rejected**, despite the fact that they were selected by the **Variable Selection** node. This is because I have set the **Use AOV16 Variables** property to **Yes** in the **Variable Selection** node. Note also that, as Output 3.1 shows, AOV16_NVAR133 and AOV16_NVAR239 have met both the **Minimum R-Square** and **Stop R-Square** criteria and hence are given the role of **Input** in the **Regression** node.

3.2.2 Continuous Target with Nominal Categorical[1] Inputs (Case 2)

The target variable here is the same as in Case 1, and all the inputs are nominal-scaled categorical.

In the case of nominal-scaled categorical inputs with a continuous target, R-Square is calculated using one-way ANOVA. Here you have the option of using either the original or the grouped variables. I decided to use only grouped class variables. If a variable cannot be grouped, it will still be included in the selection process. Hence, I have set the **Use Group Variables** property to **Yes** in the properties panel (Display 3.6). I made this choice because some of the class variables have too many categories. The term *grouped variables* refers to those variables whose categories (levels) are collapsed or combined. For example, suppose there is a categorical (nominal) variable

[1] The SAS manuals often use the term “Class” instead of “Categorical.” I use these two terms synonymously.

called LIFESTYLE, which indicates the lifestyle of the customer. It may take on values such as "Urban Dweller," "Foreign Traveler," etc. If the variable LIFESTYLE has 100 levels or categories, it can be collapsed to fewer levels or categories by setting the **Group Variables** property to **Yes**.

The descriptions of some of the nominal-scaled class inputs in my example data set are given below:

CVR14	a customer segment, based on the value of the customer, where value is based on factors such as interest revenue generated by the customer. This takes only two values: High or Low.
CVAR03	a segment, based on the type of accounts held by the customer
CVR13	a variable indicating whether the customer uses the Internet
CVAR07	a variable indicating whether the customer contacts by phone
CVAR06	a variable indicating whether the customer visits the branch frequently
CVAR11	a variable indicating whether the customer holds multiple accounts

Display 3.6

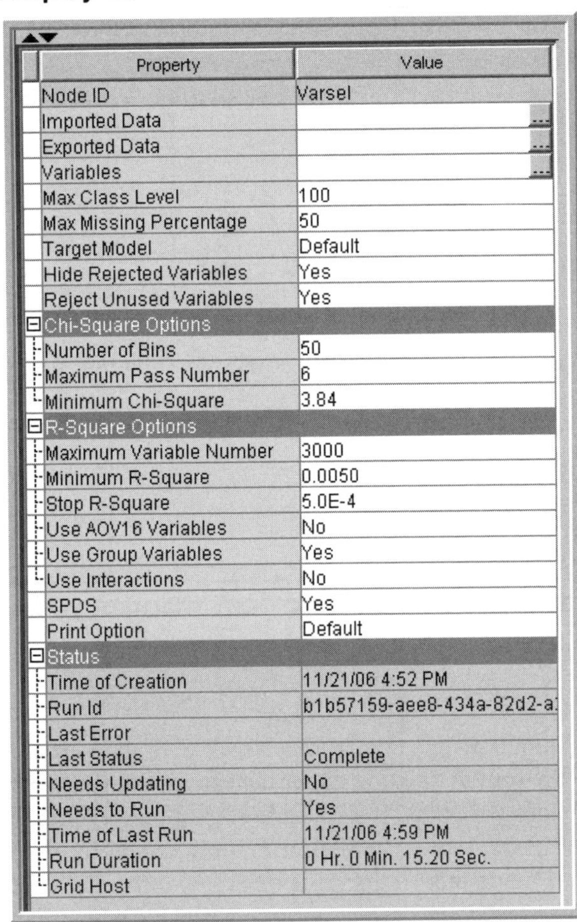

Property	Value
Node ID	Varsel
Imported Data	...
Exported Data	...
Variables	...
Max Class Level	100
Max Missing Percentage	50
Target Model	Default
Hide Rejected Variables	Yes
Reject Unused Variables	Yes
⊟ Chi-Square Options	
Number of Bins	50
Maximum Pass Number	6
Minimum Chi-Square	3.84
⊟ R-Square Options	
Maximum Variable Number	3000
Minimum R-Square	0.0050
Stop R-Square	5.0E-4
Use AOV16 Variables	No
Use Group Variables	Yes
Use Interactions	No
SPDS	Yes
Print Option	Default
⊟ Status	
Time of Creation	11/21/06 4:52 PM
Run Id	b1b57159-aee8-434a-82d2-a
Last Error	
Last Status	Complete
Needs Updating	No
Needs to Run	Yes
Time of Last Run	11/21/06 4:59 PM
Run Duration	0 Hr. 0 Min. 15.20 Sec.
Grid Host	

Output 3.3 shows the variables selected in Step 1 of the **Variable Selection** node.

Output 3.3

The DMINE Procedure

R-Squares for Target Variable: DEPV

Effect	DF	R-Square	
Class: CVR14	1	0.173336	
Class: CVAR03	10	0.048790	
Group: CVAR03	4	0.048027	
Class: CVR13	2	0.020582	
Group: CVR13	1	0.020577	
Class: CVAR07	2	0.014605	
Group: CVAR07	1	0.014560	
Class: CVAR06	2	0.014177	
Group: CVAR06	1	0.013946	
Class: CVAR11	2	0.007594	
Class: CVAR16	4	0.004338	R2 < MINR2
Class: CVAR05	5	0.004122	R2 < MINR2
Group: CVAR05	4	0.004120	R2 < MINR2
Class: CVR08	4	0.003097	R2 < MINR2
Group: CVR08	3	0.003067	R2 < MINR2
Class: CVAR10	2	0.000041720	R2 < MINR2

To see how Enterprise Miner groups or collapses categorical variables, consider the nominal-scaled variable CVAR03. By using PROC FREQ in the **SAS Code** node, we can see that the nominal-scaled variable CVAR03 has 11 categories, including the missing or unlabeled category, as shown in Output 3.4.

Output 3.4

The FREQ Procedure

CVAR03	Frequency	Percent	Cumulative Frequency	Cumulative Percent
	101	8.65	101	8.65
01	52	4.45	153	13.10
02	60	5.14	213	18.24
03	126	10.79	339	29.02
04	228	19.52	567	48.54
05	156	13.36	723	61.90
06	148	12.67	871	74.57
07	130	11.13	1001	85.70
08	104	8.90	1105	94.61
09	54	4.62	1159	99.23
10	9	0.77	1168	100.00

The **Variable Selection** node created a new variable called G_CVAR03 by collapsing the 11 categories of CVAR03 into 5 categories. Using PROC FREQ again, we can see the frequency distribution of the categories of G_CVAR03, which is shown in Output 3.5. For the SAS code that generated Outputs 3.4 and 3.5, see Display 3.29 in the appendix to this chapter.

Output 3.5

The FREQ Procedure

G_CVAR03	Frequency	Percent	Cumulative Frequency	Cumulative Percent
0	436	37.33	436	37.33
1	257	22.00	693	59.33
2	237	20.29	930	79.62
3	186	15.92	1116	95.55
4	52	4.45	1168	100.00

Display 3.7 shows which categories of CVAR03 are grouped (collapsed) to create the variable G_CVAR03.

Display 3.7

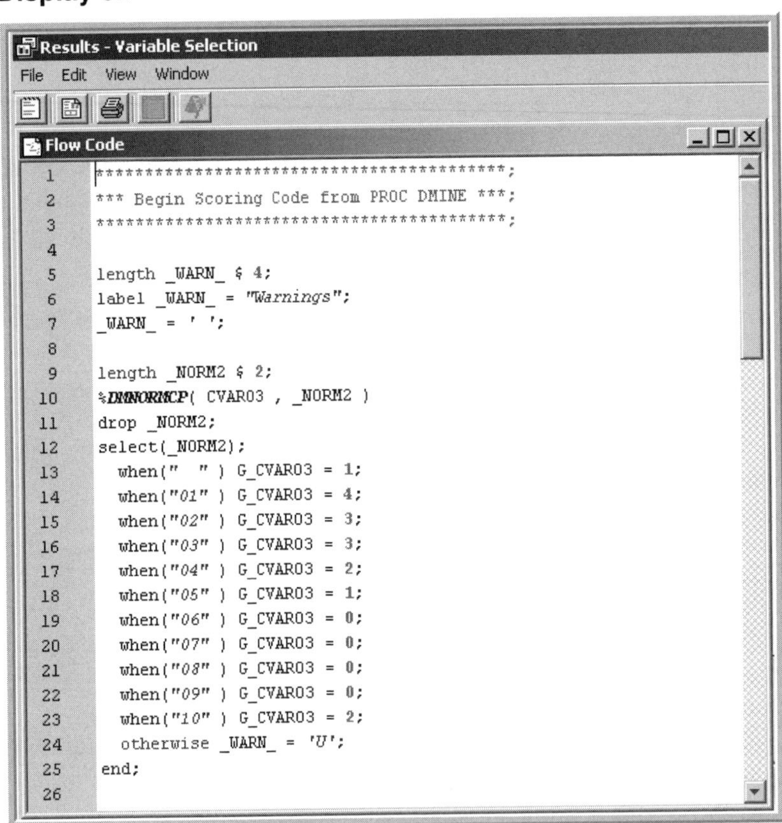

The grouping procedure Enterprise Miner uses in collapsing the categories of a class variable can be illustrated by a simple example. Suppose a class variable has five categories: A, B, C, D, and E. For each level, the mean of the target for the records in that level is calculated. In the case of a binary target, the target mean for a given level is the proportion of responders within that level, if I represent the response by 1 and non-response by 0. The categories are then sorted in descending order by their target means. That is, the category that has the highest mean for the target becomes the first category; the one with the next highest target mean comes next, and so on. Suppose this ordering resulted in the sequence: C, E, D, A, and B. Then the procedure combines the categories starting from the top. First it calculates the R-Square between the input and the target (prior to grouping). Suppose the R-Square is 0.20. Next, the procedure combines the categories C and E and recalculates the R-Square. Now the input has four categories, and the R-Square between the input and the target will now be less than 0.20. If the reduction in R-Square is less than 5%, then C and E remain combined. A reduction of less than 5% in an R-Square of 0.20 means that the new grouped input must have an R-square of at least 0.19 with the target. If not, C and E are not combined. Then an attempt is made to combine E and D using this same method, and so on. Output 3.3 shows the R-Square between the target and the original inputs before grouping (labeled "Class") as well as the R-Square between the target and the grouped version of the inputs (labeled "Group").

The nominal-scaled variable CVAR03 has an R-Square of 0.048790 with the target before its categories are combined or grouped, and an R-Square of 0.048027 afterward. Both G_CVAR03 and CVAR03 met the R-Square criterion and were selected in Step 1. However, only the grouped variable is kept in the final Step 2 selection, as can be seen in Output 3.6. The R-Square shown in Output 3.6 is an 'incremental' or 'sequential' R-Square as distinct from the bivariate R-Square (coefficient of determination) shown in Output 3.3.

Output 3.6

The DMINE Procedure

Effects Chosen for Target: DEPV

Effect	DF	R-Square	F Value	p-Value	Sum of Squares	Error Mean Square
Class: CVR14	1	0.173336	244.487591	<.0001	544.655917	2.227745
Group: CVAR03	4	0.009016	3.203197	0.0125	28.329505	2.211033
Group: CVR13	1	0.008175	11.725295	0.0006	25.687918	2.190812
Group: CVAR07	1	0.006723	9.715610	0.0019	21.126479	2.174488
Group: CVAR06	1	0.003592	5.210083	0.0226	11.288295	2.166625

The first variable, CVAR14, is a two-level class variable, and hence its levels are not collapsed (see Output 3.3 and note that no Group version of CVR14 appears). As before, the final selection is based on the stepwise forward selection procedure. The procedure starts with the best variable, i.e., the variable with the highest correlation, followed by the other variables according to their contribution to the improvement of the stepwise **R-Square.**

Next the **Regression** node is connected to the **Variable Selection** node, and Display 3.8 shows the variables that are passed on to the **Regression** node with the **Role** of **Input**.

Display 3.8

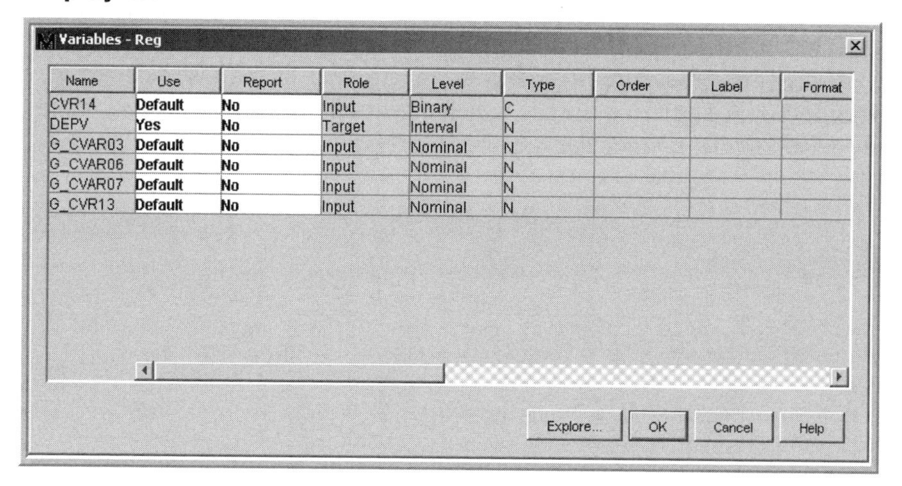

Variables - Reg

Name	Use	Report	Role	Level	Type	Order	Label	Format
CVR14	Default	No	Input	Binary	C			
DEPV	Yes	No	Target	Interval	N			
G_CVAR03	Default	No	Input	Nominal	N			
G_CVAR06	Default	No	Input	Nominal	N			
G_CVAR07	Default	No	Input	Nominal	N			
G_CVR13	Default	No	Input	Nominal	N			

Explore... OK Cancel Help

3.2.3 Binary Target with Numeric Interval-scaled Inputs (Case 3)

In this example, a hypothetical bank wants to cross-sell a mortgage refinance product. The bank sends mail to its existing customers. The target variable is response to the mail, taking the value of 1 for response and 0 for non-response.

The following is a description of some of the inputs in the data set used in this example:

NVAR004	credit card balance
NVAR014	auto loan balance
NVAR016	auto lease balance
NVAR030	mortgage loan balance
NVAR044	other loan balances
NVAR075	value of monthly transactions in checking and savings accounts
NVAR137	number of direct payments by the bank per month
NVAR148	available credit
NVAR187	number of investment-related transactions per month
NVAR193	percentage of households with children under age 10 in the block group
NVAR195	percentage of households living in rented homes in the block group
NVAR210	length of residence
NVAR286	age of the customer
NVAR315	number of products owned by the customer
NVAR318	number of teller-transactions per month

In the case of binary targets, inputs can be selected by using either the R-Square criterion, as in Case 1, or the Chi-Square criterion.

3.2.3.1 R-Square Criterion

When the R-Square criterion is used, the variable selection is similar to the case in which the target is continuous (Case 1, above). The target variable is treated like a continuous variable, and the R-Square with the target is computed for each original input and for each binned input (**AOV16** variable.) Then the usual two-step procedure is followed: in the first step, an initial variable set is selected based on the **Minimum R-Square** criterion; in the second step, a sequential forward selection procedure is used to select variables from those selected in Step 1 on the basis of a **Stop R-Square** criterion. Inputs which meet both these criteria receive the **Role** of **Input** in the subsequent modeling tool, such as the **Regression** node.

3.2.3.2 Chi-Square Criterion

This criterion can be used only when the target is binary. When this criterion is selected, the selection process does not have two distinct steps, as in the case of the **R-Square** criterion. Instead, a tree is constructed. The inputs selected in the construction of the tree are passed to the next node with the assigned **Role** of **Input**. In this example, there are 302 inputs, and the **Variable Selection** node selected 18 inputs. This selection is done by developing a tree based on the Chi-Square criterion, as explained below. A tree constructed using the Chi-Square criterion is sometimes called CHAID (Chi-Squared Automatic Interaction Detection).

The tree development proceeds as follows. First, the records of the data set are divided into two groups. Each group is further divided into two more groups, and so on. This process of dividing is called *recursive partitioning*. Each group created during the recursive partitioning is called a *node*. The nodes that are not divided further are called *terminal nodes,* or *leaf nodes,* or *leaves* of the tree. The nodes that are divided further are called *intermediate nodes*. The node with all the records of the data set is called the *root node*. All these nodes form the tree. Display 3.8A shows such a tree.

In order to split the records in a given node into two new nodes, the algorithm evaluates all inputs (302 in this example) one at a time, and finds the best split for each input. To find the best split for an interval input, the algorithm first divides the input into intervals, or bins, of equal length. The number of bins into which the interval input is divided is determined by the value to which the **Number of Bins** property is set. The default value is 50. However, suppose the user sets it to 10 and that the interval input to be binned is **income** with values that range from $10,000 to $110,000 in the data set. In that case the 10 equal intervals will be $10,000– $20,000, $20,000–$30,000, $30,000–$40,000... and $100,000–$110,000. Any node split based on this input will be partitioned at one of the split values of $20,000, $30,000, $40,000... etc. A 2x2 contingency table is created for each of these nine split values and a Chi-Square statistic is computed for each split. In this example, there are nine Chi-Square values. The split value with the highest Chi-Square value is the best split based on the input **income**. Then the algorithm does a similar calculation to find the best split of the records on the basis of each of the other available inputs. After finding the best split for each input, the algorithm compares the Chi-Square values of all these best splits and picks the one that gives the highest Chi-Square value. It then splits the records into two nodes on the basis of that input that gave the *best* split. This process is repeated at each node. All those inputs that are selected in this process, i.e. those that are chosen to be the basis of the split at any node of the tree, are passed to the next node with the assigned **Role** of **Input**.

The tree generated by the **Variable Selection** node for our example data is shown in Displays 3.8A and 3.8B. Display 3.8A shows the root node and the left-hand half of the tree while Display 3.8B shows the root node and the right-hand half of the tree, because the whole tree is too wide to fit in one display.

Display 3.8A

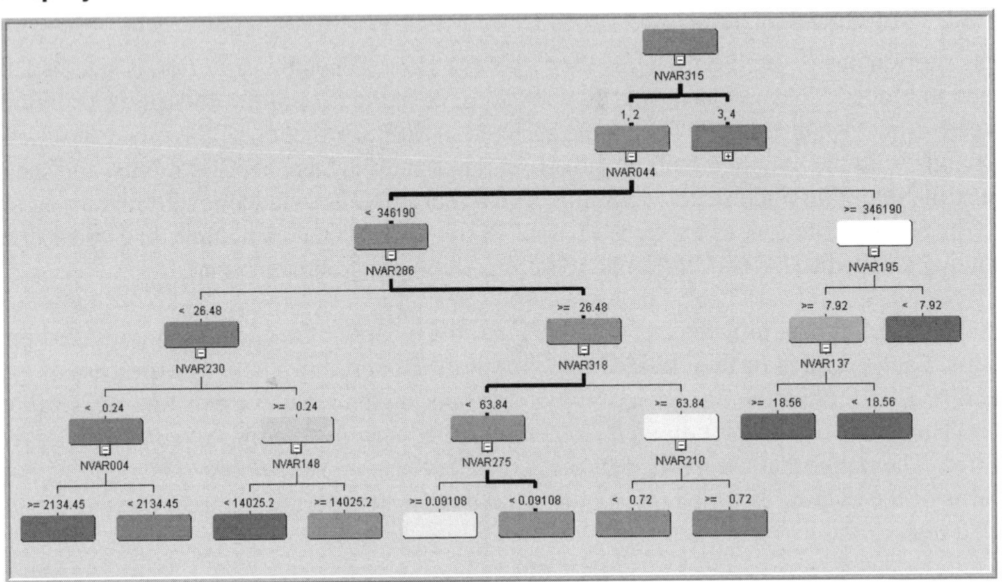

Display 3.8B

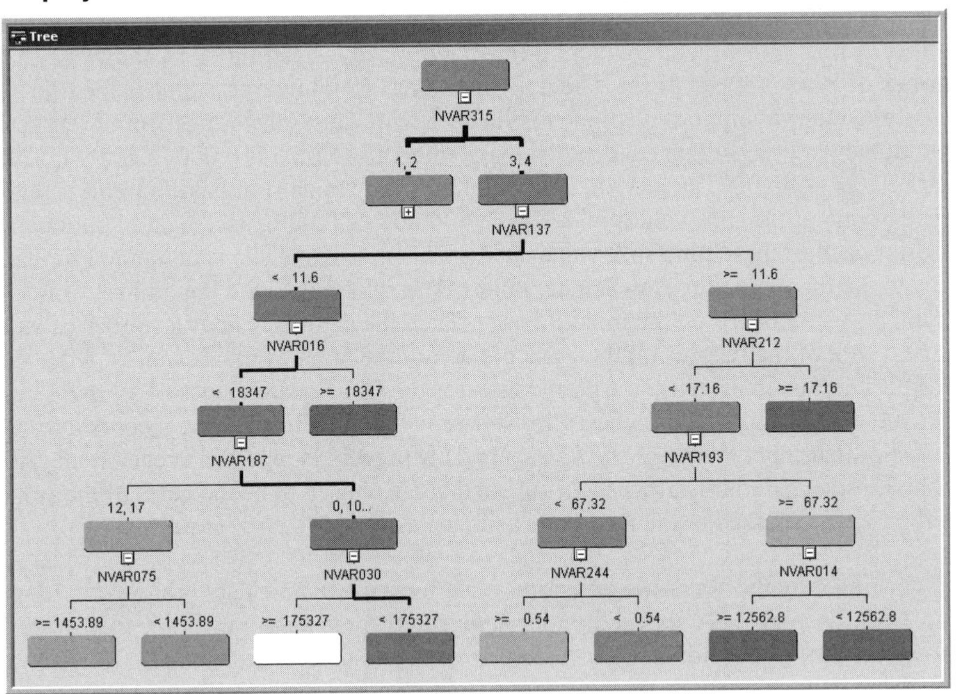

Output 3.7 shows the relative importance of the selected variables.

Output 3.7

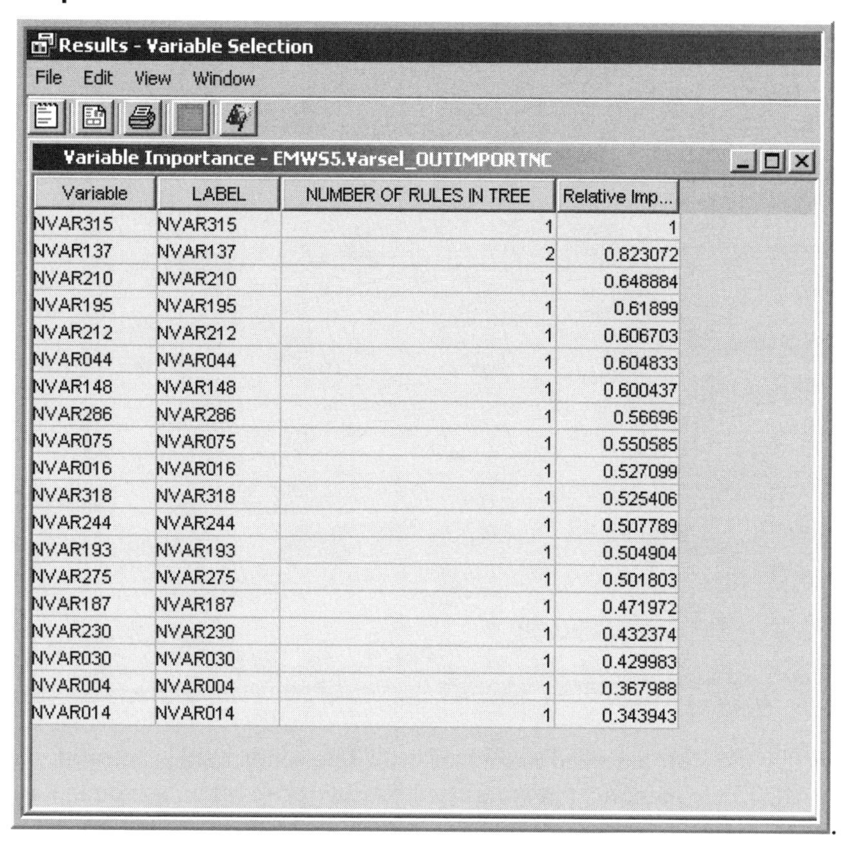
Results - Variable Selection

File Edit View Window

Variable Importance - EMWS5.Varsel_OUTIMPORTNC

Variable	LABEL	NUMBER OF RULES IN TREE	Relative Imp...
NVAR315	NVAR315	1	1
NVAR137	NVAR137	2	0.823072
NVAR210	NVAR210	1	0.648884
NVAR195	NVAR195	1	0.61899
NVAR212	NVAR212	1	0.606703
NVAR044	NVAR044	1	0.604833
NVAR148	NVAR148	1	0.600437
NVAR286	NVAR286	1	0.56696
NVAR075	NVAR075	1	0.550585
NVAR016	NVAR016	1	0.527099
NVAR318	NVAR318	1	0.525406
NVAR244	NVAR244	1	0.507789
NVAR193	NVAR193	1	0.504904
NVAR275	NVAR275	1	0.501803
NVAR187	NVAR187	1	0.471972
NVAR230	NVAR230	1	0.432374
NVAR030	NVAR030	1	0.429983
NVAR004	NVAR004	1	0.367988
NVAR014	NVAR014	1	0.343943

The variables that appear in Displays 3.8A and 3.8B are passed to the next node with the **Role** of **Input**. You can verify this by first connecting a **Regression** node to the **Variable Selection** node, as shown in Display 3.9.

Display 3.9

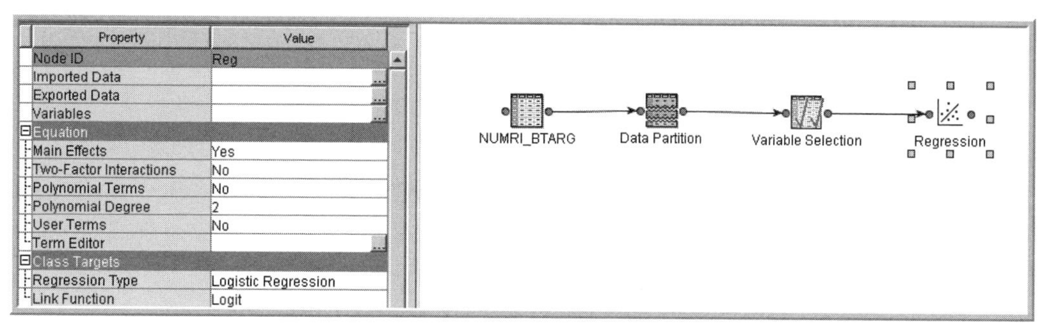
Property	Value
Node ID	Reg
Imported Data	...
Exported Data	...
Variables	...
Equation	
Main Effects	Yes
Two-Factor Interactions	No
Polynomial Terms	No
Polynomial Degree	2
User Terms	No
Term Editor	...
Class Targets	
Regression Type	Logistic Regression
Link Function	Logit

Then if you select the **Regression** node, update the path and click on [...] to the right of the **Variables** property in the properties panel, a window will appear displaying the inputs passed to the **Regression** node. This is shown in Display 3.10.

Display 3.10

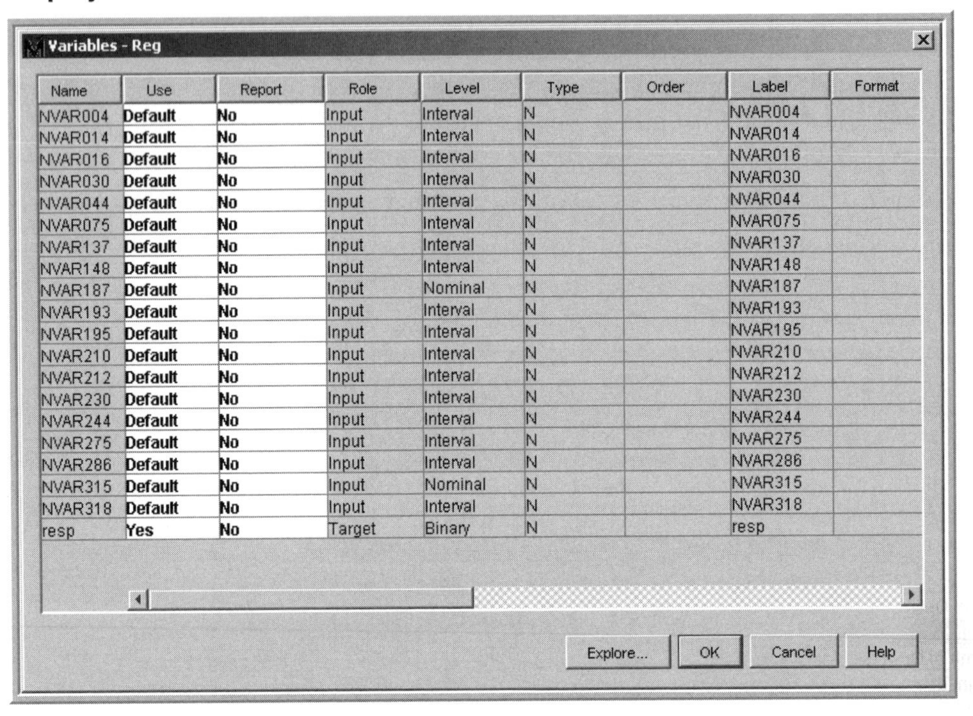

Variables - Reg

Name	Use	Report	Role	Level	Type	Order	Label	Format
NVAR004	Default	No	Input	Interval	N		NVAR004	
NVAR014	Default	No	Input	Interval	N		NVAR014	
NVAR016	Default	No	Input	Interval	N		NVAR016	
NVAR030	Default	No	Input	Interval	N		NVAR030	
NVAR044	Default	No	Input	Interval	N		NVAR044	
NVAR075	Default	No	Input	Interval	N		NVAR075	
NVAR137	Default	No	Input	Interval	N		NVAR137	
NVAR148	Default	No	Input	Interval	N		NVAR148	
NVAR187	Default	No	Input	Nominal	N		NVAR187	
NVAR193	Default	No	Input	Interval	N		NVAR193	
NVAR195	Default	No	Input	Interval	N		NVAR195	
NVAR210	Default	No	Input	Interval	N		NVAR210	
NVAR212	Default	No	Input	Interval	N		NVAR212	
NVAR230	Default	No	Input	Interval	N		NVAR230	
NVAR244	Default	No	Input	Interval	N		NVAR244	
NVAR275	Default	No	Input	Interval	N		NVAR275	
NVAR286	Default	No	Input	Interval	N		NVAR286	
NVAR315	Default	No	Input	Nominal	N		NVAR315	
NVAR318	Default	No	Input	Interval	N		NVAR318	
resp	Yes	No	Target	Binary	N		resp	

Explore... OK Cancel Help

Display 3.10 shows that the **Role** assigned to each of these selected variables is **Input**. Note that NVAR187 and NVAR315 are numeric, but are treated by Enterprise Miner as nominal since each of them has fewer than 20 distinct values. NVAR187 has 14 distinct values, and NVAR315 has 7 distinct values.

Unlike the **Decision Tree** node, the **Variable Selection** node does not create dummy variables from the terminal nodes under the Chi-Square criterion. The Chi-Square criterion is discussed in more detail in Chapter 4 in the context of decision trees.

Display 3.11 shows the properties panel where the properties for the **Chi-Square** criterion are set.

Display 3.11

Property	Value
Node ID	Varsel
Imported Data	...
Exported Data	...
Variables	...
Max Class Level	100
Max Missing Percentage	50
Target Model	Chi-Square
Hide Rejected Variables	Yes
Reject Unused Variables	Yes
⊟Chi-Square Options	
Number of Bins	50
Maximum Pass Number	6
Minimum Chi-Square	3.84
⊟R-Square Options	
Maximum Variable Number	3000
Minimum R-Square	0.0050
Stop R-Square	5.0E-4
Use AOV16 Variables	No
Use Group Variables	Yes
Use Interactions	No
SPDS	Yes
Print Option	Default
⊟Status	
Time of Creation	11/22/06 4:28 PM
Run Id	00424662-6800-460b-910b-9
Last Error	
Last Status	Complete
Needs Updating	No
Needs to Run	No
Time of Last Run	11/22/06 4:38 PM
Run Duration	0 Hr. 0 Min. 38.78 Sec.
Grid Host	

3.2.4 Binary Target with Nominal-scaled Categorical Inputs (Case 4)

In this case the target variable is the same as in Case 3, and all the inputs are nominal-scaled categorical. Here are descriptions of some of the nominal-scaled class inputs in the data set:

CVAR08	dwelling size code
CVAR11	income code
CVAR20	indicator of home ownership
CV30	demographic cluster
CVAR34	indicator of Internet banking
CVAR35	customer type
CVAR40	life cycle code
CVAR47	indicator of Internet usage

Here also you can use either the R-Square or the Chi-Square criterion to select the variables.

3.2.4.1 R-Square Criterion

With this criterion the selection procedure is similar to the procedure in Case 2 described in Chapter 2 (Section 2.11.1) in which two steps are needed. I set the **Target Model** property to **R-Square** and ran the **Variable Selection** node. The variables selected in Step 1 are shown in Output 3.8.

Output 3.8

The DMINE Procedure

R-Squares for Target Variable: resp

Effect	DF	R-Square	
Class: CVAR47	1	0.014025	
Class: CVAR40	14	0.002226	
Group: CVAR40	6	0.002188	
Class: CVAR35	6	0.002118	
Group: CVAR35	3	0.002084	
Class: CV30	10	0.001715	R2 < MINR2
Group: CV30	6	0.001694	R2 < MINR2

Output 3.9 shows the results of the forward selection process (Step 2).

Output 3.9

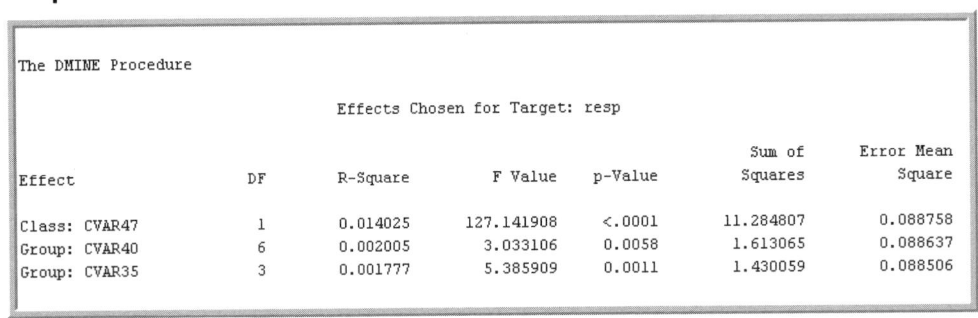

The DMINE Procedure

Effects Chosen for Target: resp

Effect	DF	R-Square	F Value	p-Value	Sum of Squares	Error Mean Square
Class: CVAR47	1	0.014025	127.141908	<.0001	11.284807	0.088758
Group: CVAR40	6	0.002005	3.033106	0.0058	1.613065	0.088637
Group: CVAR35	3	0.001777	5.385909	0.0011	1.430059	0.088506

When you compare Outputs 3.8 and 3.9, you can see that the class variables CVAR40 and CVAR35 met the **Minimum R-Square** criterion in the first step in their original and grouped form. But in the second step, only the grouped variables survived. Display 3.12 shows the **Variables** window of the next node, which is the **Regression** node.

Display 3.12

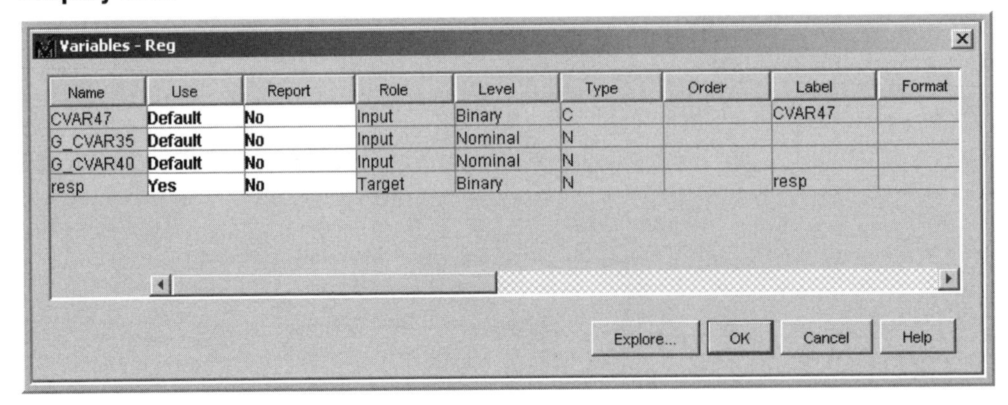

Variables - Reg

Name	Use	Report	Role	Level	Type	Order	Label	Format
CVAR47	Default	No	Input	Binary	C		CVAR47	
G_CVAR35	Default	No	Input	Nominal	N			
G_CVAR40	Default	No	Input	Nominal	N			
resp	Yes	No	Target	Binary	N		resp	

3.2.4.2 Chi-Square Criterion

Using the same data set, I selected the Chi-Square criterion, shown in Display 3.13, by setting the value of the **Target Model** property to **Chi-Square**.

Display 3.13

Property	Value
Node ID	Varsel
Imported Data	...
Exported Data	...
Variables	...
Max Class Level	100
Max Missing Percentage	50
Target Model	Chi-Square
Hide Rejected Variables	Yes
Reject Unused Variables	Yes
⊟ Chi-Square Options	
Number of Bins	50
Maximum Pass Number	6
Minimum Chi-Square	3.84
⊟ R-Square Options	
Maximum Variable Number	3000
Minimum R-Square	0.0020
Stop R-Square	5.0E-4
Use AOV16 Variables	No
Use Group Variables	Yes
Use Interactions	No
SPDS	Yes
Print Option	Default
⊟ Status	
Time of Creation	11/23/06 6:19 AM
Run Id	a2c5353b-3d22-47b2-bf60-7d
Last Error	
Last Status	Complete
Needs Updating	No
Needs to Run	No
Time of Last Run	11/23/06 6:53 AM
Run Duration	0 Hr. 0 Min. 16.30 Sec.
Grid Host	

Display 3.14 shows the tree constructed by this method.

Display 3.14

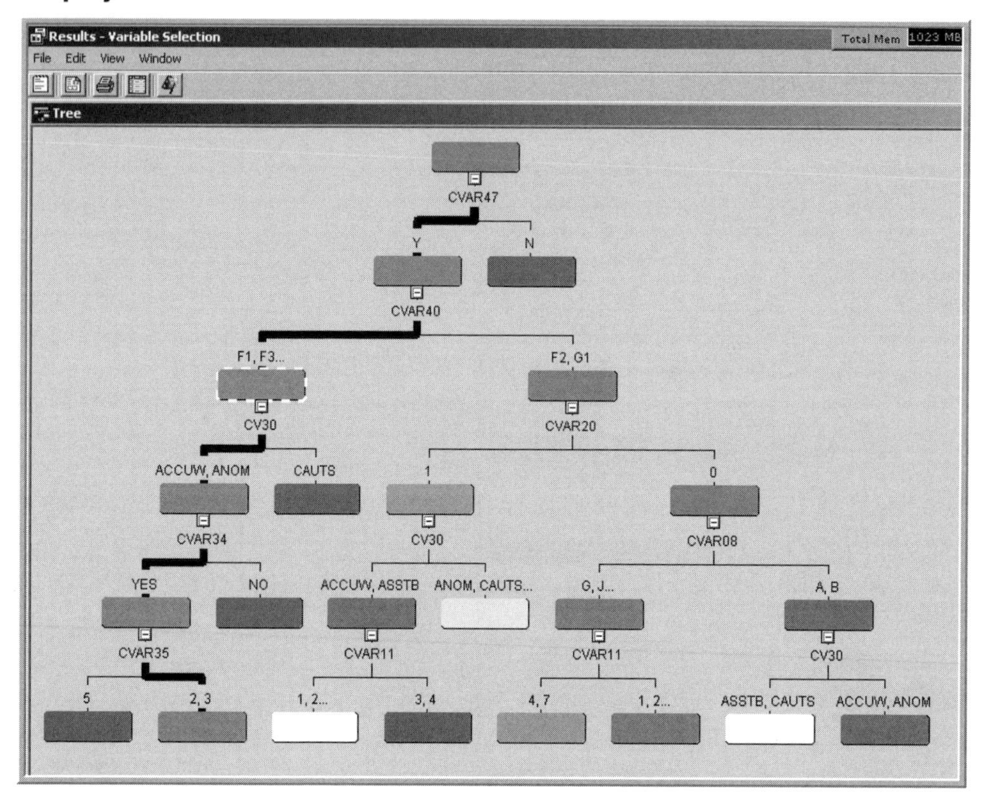

The names of the variables selected appear below the intermediate (non-terminal) nodes of the tree in Display 3.14. Display 3.14A shows the variable importance.

Display 3.14A

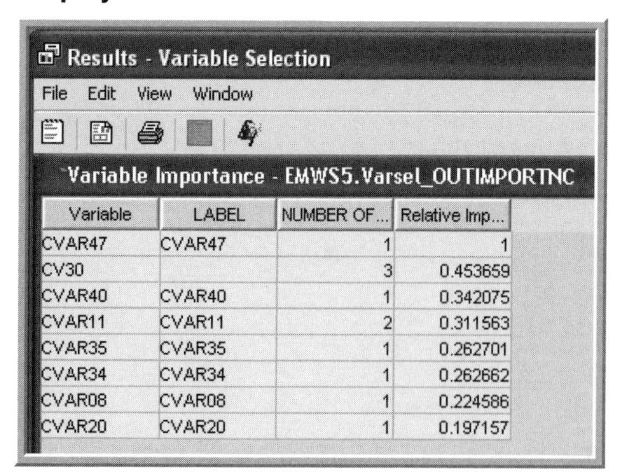
Results - Variable Selection

File Edit View Window

Variable Importance - EMWS5.Varsel_OUTIMPORTNC

Variable	LABEL	NUMBER OF...	Relative Imp...
CVAR47	CVAR47	1	1
CV30		3	0.453659
CVAR40	CVAR40	1	0.342075
CVAR11	CVAR11	2	0.311563
CVAR35	CVAR35	1	0.262701
CVAR34	CVAR34	1	0.262662
CVAR08	CVAR08	1	0.224586
CVAR20	CVAR20	1	0.197157

When you compare the variables given in Display 3.14A with those in Outputs 3.8 and 3.9, you can see that CVAR47 emerges as the most important variable by both the R-Square and Chi-Square criteria. As you might recall, this variable is an indicator of Internet usage.

These variables in Display 3.14A are exported to the next node. In this example, the next node is the **Regression** node. By opening its **Variables** window, you will see the same variables with the **Role** of **Input.** The **Variables** window of the **Regression** node is shown in Display 3.15.

Display 3.15

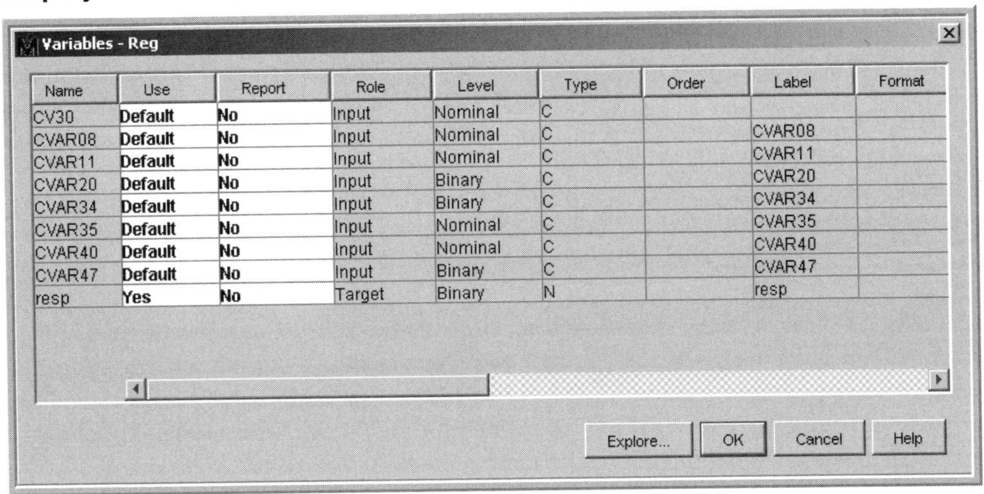

Variables - Reg

Name	Use	Report	Role	Level	Type	Order	Label	Format
CV30	Default	No	Input	Nominal	C			
CVAR08	Default	No	Input	Nominal	C		CVAR08	
CVAR11	Default	No	Input	Nominal	C		CVAR11	
CVAR20	Default	No	Input	Binary	C		CVAR20	
CVAR34	Default	No	Input	Binary	C		CVAR34	
CVAR35	Default	No	Input	Nominal	C		CVAR35	
CVAR40	Default	No	Input	Nominal	C		CVAR40	
CVAR47	Default	No	Input	Binary	C		CVAR47	
resp	Yes	No	Target	Binary	N		resp	

Explore... OK Cancel Help

Eight variables were selected, which were those that produced the best splits in the development of the tree. The selected variables are given the **Role** of **Input** in the **Regression** node, used next.

3.3 Transformation of Variables

3.3.1 Transform Variables Node

The **Transform Variables** node can make a variety of transformations of interval-scaled variables. These were described in Section 2.6.7.1. In addition, there are two types of transformations for the categorical (class) variables:

- Group Rare Levels
- Dummy Indicators Transformation for the categories (see Section 2.6.7.2)

If you want to transform variables prior to variable selection, you can set up the path as **Input Data** node → **Data Partition** node → **Transform Variables** node → **Variable Selection** node, as shown in the upper path in Display 3.16. If you do not want to reject inputs on the basis of linear relationship alone, you can use this process to capture non-linear relationships between the inputs and the target. However, with large data sets this process may become quite tedious.

When there are a large number of inputs (e.g., 2000 or more), it may be more practical to eliminate irrelevant variables first, and then define transformations on important variables only. The process flow for this is **Input Data** node → **Data Partition** node → **Variable Selection** node→ **Transform Variables** node. This sequence is represented by the lower path of Display 3.16. I will examine the results of both of these sequences with alternative transformations using a small data set. Display 3.16 does not include the **Impute** node, but in any given project it might be necessary to utilize this node before applying transformations or selecting variables.

Display 3.16

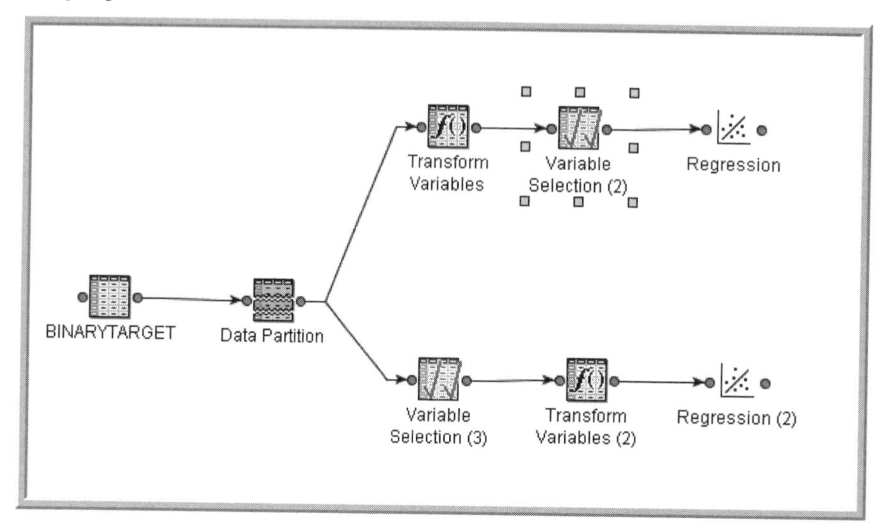

To demonstrate the transformations, I have selected a small number of inputs and the binary target RESP, which represents the response to a solicitation by a hypothetical bank. Here are these inputs, which are shown in Display 3.17:

VAR1N	number of on-line inquiries by the customer per month
VAR2O	number of ATM deposits during the past three months
VAR3N	number of ATM withdrawals during the past three months
VAR4C	lifestyle indicator
VAR5O	number of asset types
VAR6C	an indicator of wealth
VAR7C	life stage segment
VAR8O	total number of bank products owned by the customer
CREDSCORE	credit score
RESP	target variable; indicates response to an offer

Display 3.17

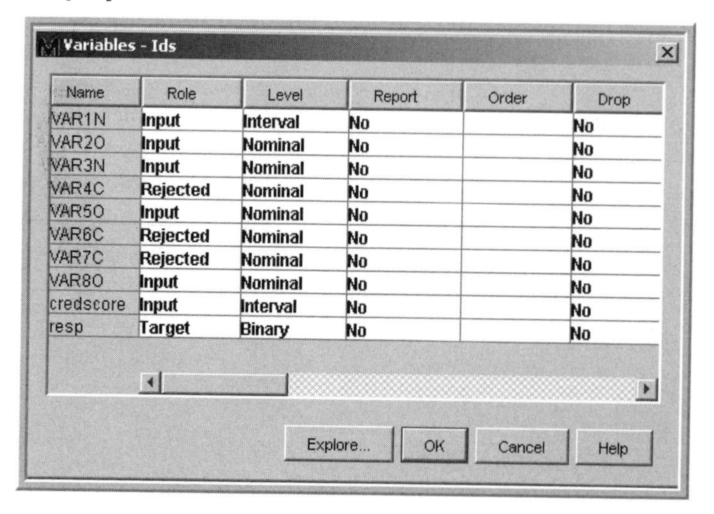

Name	Role	Level	Report	Order	Drop
VAR1N	Input	Interval	No		No
VAR2O	Input	Nominal	No		No
VAR3N	Input	Nominal	No		No
VAR4C	Rejected	Nominal	No		No
VAR5O	Input	Nominal	No		No
VAR6C	Rejected	Nominal	No		No
VAR7C	Rejected	Nominal	No		No
VAR8O	Input	Nominal	No		No
credscore	Input	Interval	No		No
resp	Target	Binary	No		No

3.3.2 Transformation before Variable Selection

In the upper path of Display 3.18 the variables are transformed using the **Transform Variables** node and then passed to the **Variable Selection** node.

In the **Transform Variables** node, I selected the **Maximum Normal** transformation as the default method for all continuous variables. To make this selection, I set the **Interval Inputs** property to **Maximum Normal**. When I select this transformation, the node generates different power transformations of the variable, and picks the transformation that transforms the input into a new variable whose distribution is closest to normal bell-shaped distribution. (Section 2.6.7.1 gives the details of how it compares the distributions of all transformations with the normal distribution and selects the one closest to it.) For class inputs, I selected the **Dummy Indicators** transformation by setting the **Class Inputs** property to **Dummy Indicators**. These settings are shown in Display 3.18.

Display 3.18

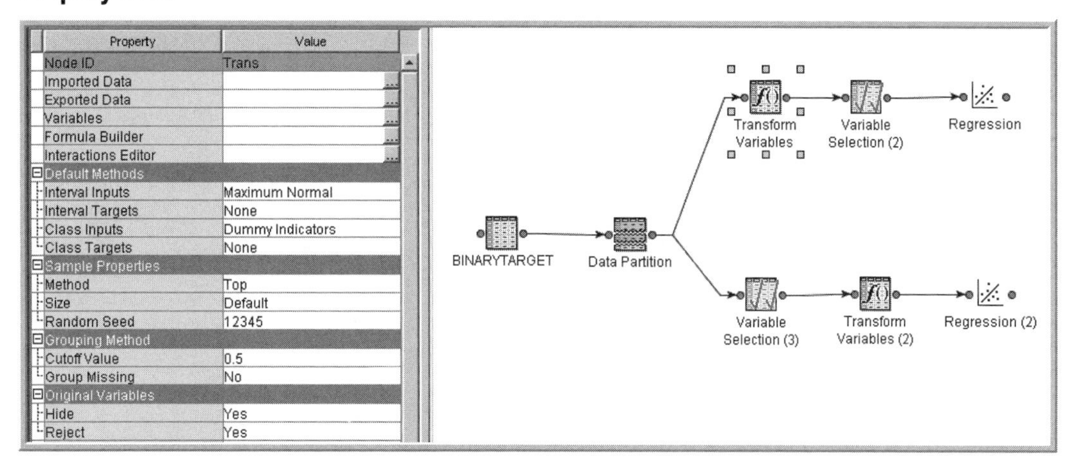

In this data set there are only two interval-scaled variables. They are VAR1N and CREDSCORE. If a numeric variable has only a few values, then Enterprise Miner sets its measurement level to **Nominal** if the **Metadata Advisor Option** is set to **Advanced** when creating the **data source** (see Display 2.14A). VAR3N, VAR2O, VAR5O, and VAR8O are **Ordinal**, and are set to **Nominal** by Enterprise Miner. Then, for the variables set to interval-scaled, the node searches and finds the **Maximum Normal**, as I specified earlier, and for the rest of the variables, which are now nominal-scaled, it creates a dummy variable for each level. The transformed variables are then passed to the next node in the process flow. In this example, the **Variable Selection** node follows the **Transform Variables** node. When you open the **Variables** window of the **Variable Selection** node, you can see the variables passed to it by the **Transform Variables** node. These variables are shown in Displays 3.19A and 3.19B.

Display 3.19A

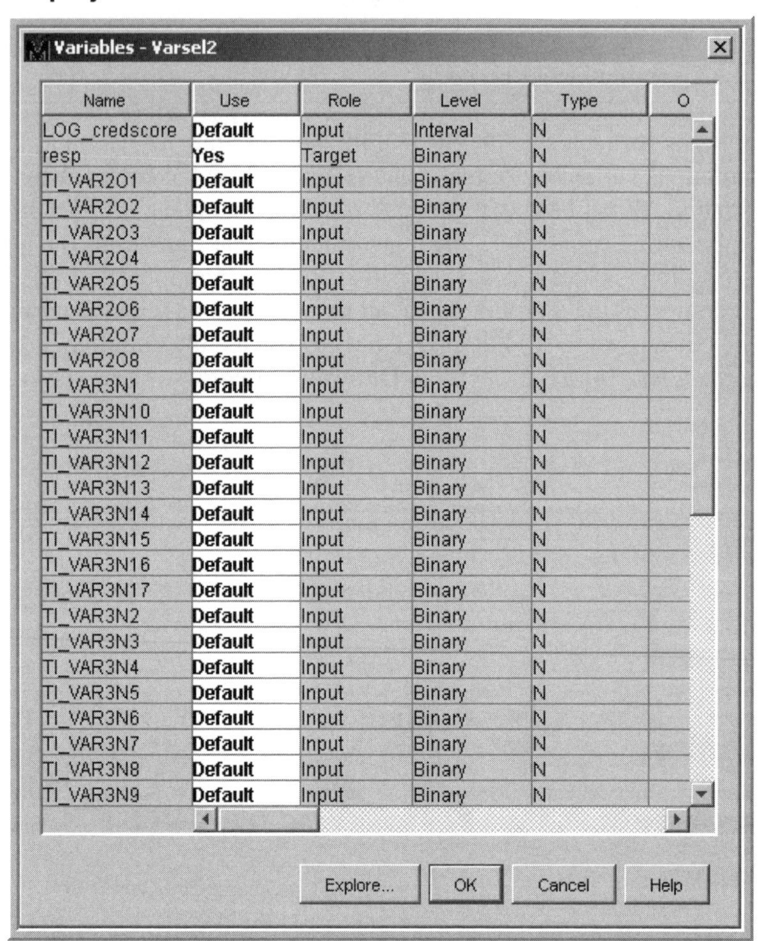

Variables - Varsel2

Name	Use	Role	Level	Type	O
LOG_credscore	Default	Input	Interval	N	
resp	Yes	Target	Binary	N	
TI_VAR2O1	Default	Input	Binary	N	
TI_VAR2O2	Default	Input	Binary	N	
TI_VAR2O3	Default	Input	Binary	N	
TI_VAR2O4	Default	Input	Binary	N	
TI_VAR2O5	Default	Input	Binary	N	
TI_VAR2O6	Default	Input	Binary	N	
TI_VAR2O7	Default	Input	Binary	N	
TI_VAR2O8	Default	Input	Binary	N	
TI_VAR3N1	Default	Input	Binary	N	
TI_VAR3N10	Default	Input	Binary	N	
TI_VAR3N11	Default	Input	Binary	N	
TI_VAR3N12	Default	Input	Binary	N	
TI_VAR3N13	Default	Input	Binary	N	
TI_VAR3N14	Default	Input	Binary	N	
TI_VAR3N15	Default	Input	Binary	N	
TI_VAR3N16	Default	Input	Binary	N	
TI_VAR3N17	Default	Input	Binary	N	
TI_VAR3N2	Default	Input	Binary	N	
TI_VAR3N3	Default	Input	Binary	N	
TI_VAR3N4	Default	Input	Binary	N	
TI_VAR3N5	Default	Input	Binary	N	
TI_VAR3N6	Default	Input	Binary	N	
TI_VAR3N7	Default	Input	Binary	N	
TI_VAR3N8	Default	Input	Binary	N	
TI_VAR3N9	Default	Input	Binary	N	

Explore... OK Cancel Help

Display 3.19B

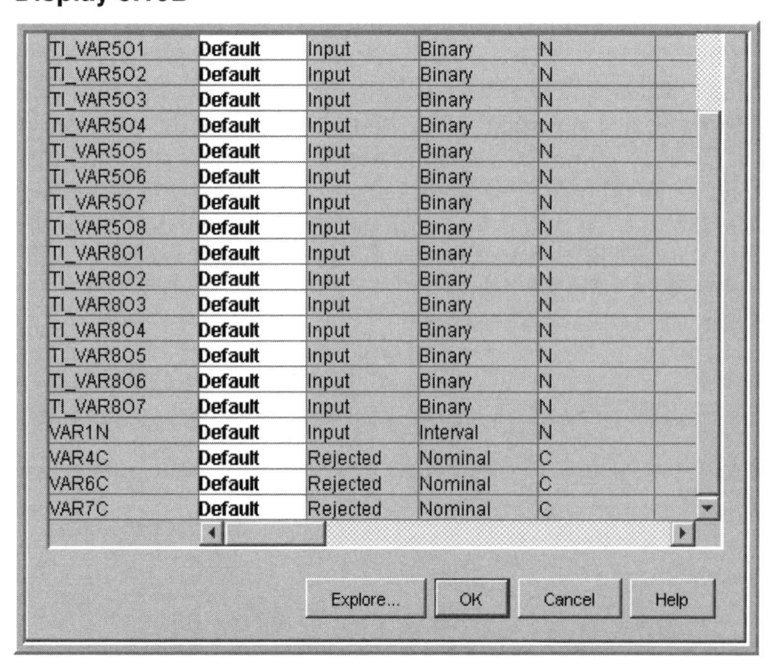

TI_VAR5O1	Default	Input	Binary	N	
TI_VAR5O2	Default	Input	Binary	N	
TI_VAR5O3	Default	Input	Binary	N	
TI_VAR5O4	Default	Input	Binary	N	
TI_VAR5O5	Default	Input	Binary	N	
TI_VAR5O6	Default	Input	Binary	N	
TI_VAR5O7	Default	Input	Binary	N	
TI_VAR5O8	Default	Input	Binary	N	
TI_VAR8O1	Default	Input	Binary	N	
TI_VAR8O2	Default	Input	Binary	N	
TI_VAR8O3	Default	Input	Binary	N	
TI_VAR8O4	Default	Input	Binary	N	
TI_VAR8O5	Default	Input	Binary	N	
TI_VAR8O6	Default	Input	Binary	N	
TI_VAR8O7	Default	Input	Binary	N	
VAR1N	Default	Input	Interval	N	
VAR4C	Default	Rejected	Nominal	C	
VAR6C	Default	Rejected	Nominal	C	
VAR7C	Default	Rejected	Nominal	C	

Explore... OK Cancel Help

From Display 3.19A, it can be seen that a log transformation made the CREDSCORE variable closest to the normal distribution. No transformation could be found for VAR1N; hence, it is passed to the next node as is. If a transformation is made for a variable, then by default only the transformed variable is passed to the next node, not the original variable. If you set the **Hide** property to **No** and the **Reject** property to **No,** then both the original and transformed variables will be passed to the next node with the **Role** set to **Input**.

The dummy variables created for the class variable levels can also be seen in Displays 3.19A and 3.19B. For example, VAR2O has eight categories. They are 0,1,2,3,4,5,6, and 8. Hence, eight dummy variables (binary variables) are created. They are TI_VAR2O1 through TI_VAR2O8. The variable VAR2O has some missing values. However, no separate category is created for it. Therefore, no dummy variable is created for the missing values. The dummy variable TI_VAR2O1 takes the value of 1 if VAR2O is 0, and 0 otherwise. TI_VAR2O2 takes the value of 1 if VAR2O is 1, and 0 otherwise. TI_VAR2O3 takes the value of 1 if VAR2O is 2, and 0 otherwise, etc.

The variables selected by the **Variable Selection** node are given in Output 3.10. Note that only two variables were selected, including one of the new dummy variables created by the transformation. The **Variable Selection** node created the variable AOV16_VAR1N because I set the **Use AOV16 Variables** property of the **Variable Selection** node to **Yes**.

Output 3.10

The DMINE Procedure

Effects Chosen for Target: resp

Effect	DF	R-Square	F Value	p-Value	Sum of Squares	Error Mean Square
Class: TI_VAR2O1	2	0.007478	33.667288	<.0001	6.016813	0.089357
AOV16: VAR1N	14	0.004542	2.930341	0.0002	3.654798	0.089088

You may want to pass more than one type of transformation for each interval input to the **Variable Selection** node or the **Regression** node in order to select the better of the two. For example, you may want to pass **Maximum Normal** and **Optimal** transformations. This can be done by applying two instances of the **Transform Variables** node. In the first instance (represented by **Trans** in Display 3.20) of the **Transform Variables** node, I transform all the interval inputs using the **Maximum Normal** transformation by setting the **Interval Inputs** property to **Maximum Normal**. In the second instance (represented by **Trans2** in Display 3.20), I transform all interval inputs using the **Optimal** transformation by setting the **Interval Inputs** property to **Optimal**. Then, using the **SAS Code** node, I merge the data sets created by the **Transform Variables** nodes so that I can pass both types of transformation to the **Regression** node for final selection of the variables considering both types of transformation.

Display 3.20 shows the process flow for combining the two types of transformations.

Display 3.20

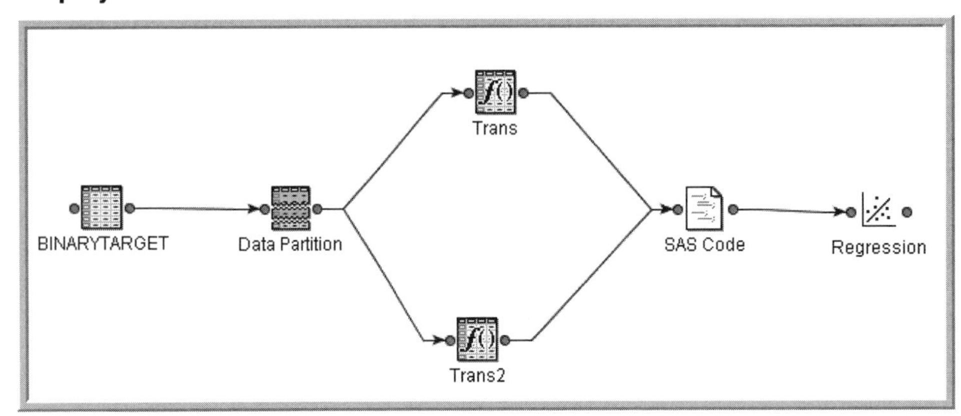

The **Maximum Normal** transformation is done by the **Transform Variables** node **Trans**, and the **Optimal** transformation is done by **Transform Variables** node **Trans2**. The **SAS Code** node is used to combine the data sets generated by these two nodes, and the combined data set is passed to the **Regression** node.

Display 3.21 shows the SAS code used to combine the data sets created by the **Transform Variables** nodes **Trans** and **Trans2**.

Display 3.21

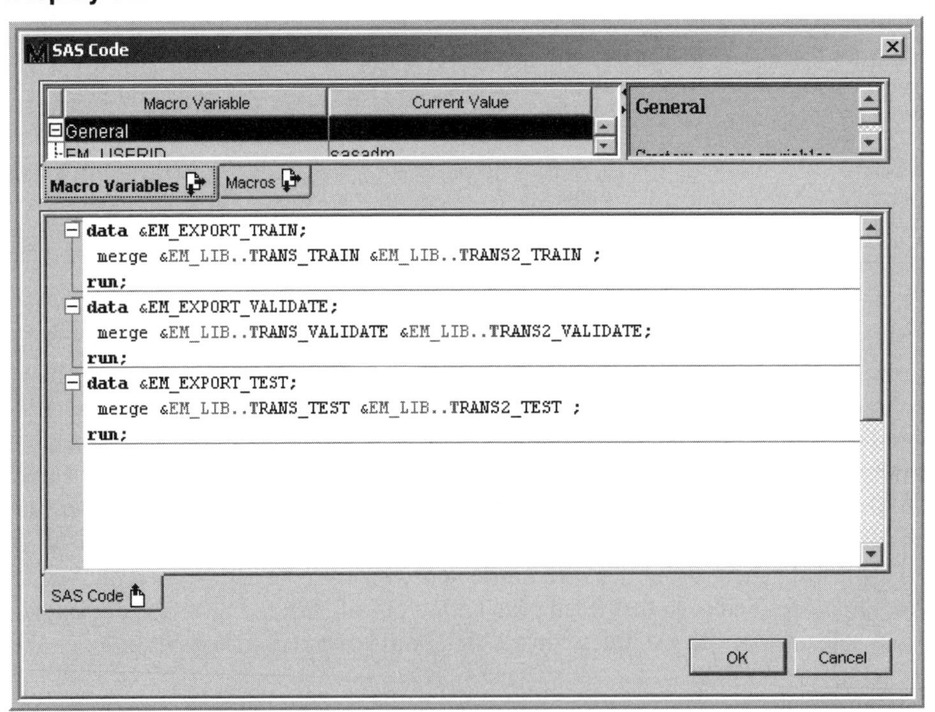

Note that the data sets are merged without a **BY** statement, because there is a one-to-one correspondence between the records of the data sets merged. The macro variables containing the names of the export data sets are created by the **SAS Code** node. Note that the **Node ID** property of the first instance of the **Transform Variables** node is **Trans**, and that of the second instance is **Trans2**.

The **Variables** window of the **Regression** node is shown in Display 3.22.

Display 3.22

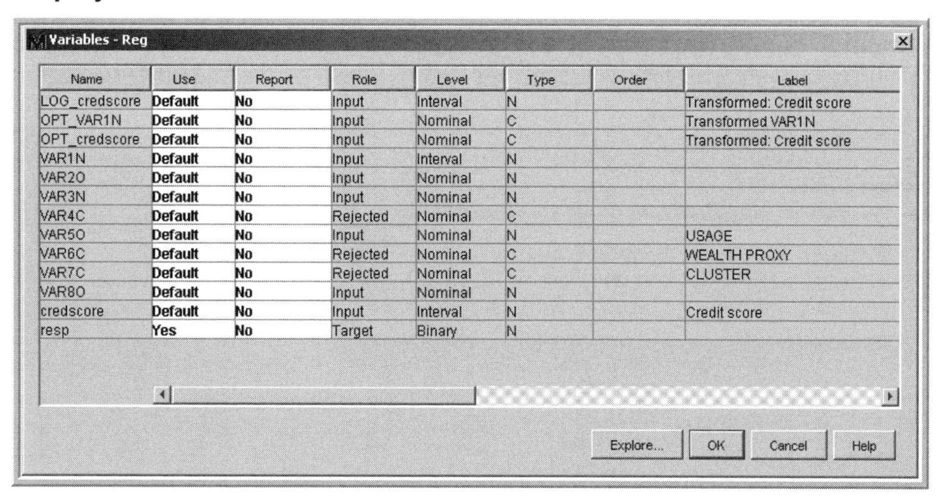

Variables - Reg

Name	Use	Report	Role	Level	Type	Order	Label
LOG_credscore	Default	No	Input	Interval	N		Transformed: Credit score
OPT_VAR1N	Default	No	Input	Nominal	C		Transformed VAR1N
OPT_credscore	Default	No	Input	Nominal	C		Transformed: Credit score
VAR1N	Default	No	Input	Interval	N		
VAR2O	Default	No	Input	Nominal	N		
VAR3N	Default	No	Input	Nominal	N		
VAR4C	Default	No	Rejected	Nominal	C		
VAR5O	Default	No	Input	Nominal	N		USAGE
VAR6C	Default	No	Rejected	Nominal	C		WEALTH PROXY
VAR7C	Default	No	Rejected	Nominal	C		CLUSTER
VAR8O	Default	No	Input	Nominal	N		
credscore	Default	No	Input	Interval	N		Credit score
resp	Yes	No	Target	Binary	N		

Explore... OK Cancel Help

The transformations done by both the **Transform Variables** nodes (shown in Display 3.20) are now available to the **Regression** node. From Display 3.22, it can be seen that the variable CREDSCORE has two types of transformation. The first one is LOG_CREDSCORE, which is created by transforming the variable credscore using the **Maximum Normal** method by the **Transform Variables** node **Trans**. The second transformation of the variable CREDSCORE is OPT_CREDSCORE, created by transforming the variable CREDSCORE using the **Optimal** transformation method that is performed with the **Transform Variables** node **Trans2**. In addition, the original interval variable CREDSCORE is also passed to the **Regression** node with its **Role** assigned to **Input** since I set the **Hide** and **Reject** properties to **No** in both the **Transform Variables** nodes **Trans** and **Trans2**.

We can also see here that the variable VAR1N has only one kind of transformation, and that it is an Optimal transformation. The **Maximum Normal** method did not produce any transformation of VAR1N, because the original variable is closer to normal than any power transformation of the original.

Output 3.11 shows the variables selected by the **Regression** node. Variable selection by the **Regression** node is covered in Chapter 6.

Output 3.11

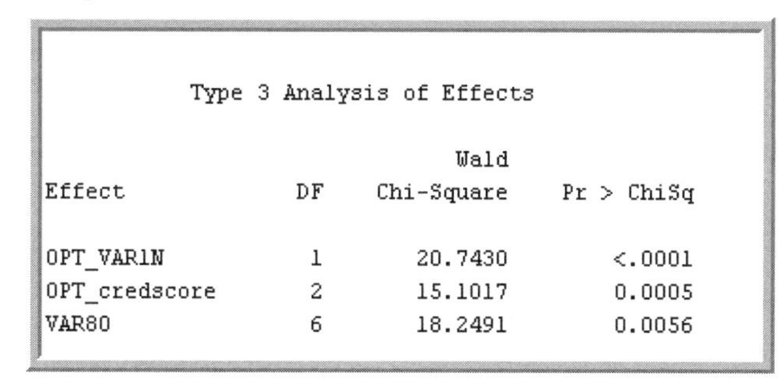

Type 3 Analysis of Effects

Effect	DF	Wald Chi-Square	Pr > ChiSq
OPT_VAR1N	1	20.7430	<.0001
OPT_credscore	2	15.1017	0.0005
VAR80	6	18.2491	0.0056

Note that VAR80 is a nominal variable. The **Class Inputs** property for this case was set to **None**. Hence, the variable was not transformed.

3.3.3 Transformation after Variable Selection

If you have a very large number of variables in your data set, you may want to first eliminate those that have extremely low linear correlation with the target. To achieve this, I set the values of the **Minimum R-Square** and **Stop R-Square** properties of the **Variable Selection** node to 0.005 and 0.0005, respectively. The variables selected in Step 1 of the **Variable Selection** are shown in Output 3.12.

Output 3.12

R-Squares for Target Variable: resp

Effect	DF	R-Square	
Class: VAR3N	17	0.010908	
Group: VAR3N	5	0.010724	
Class: VAR5O	8	0.009589	
Group: VAR5O	3	0.009517	
Class: VAR2O	8	0.009424	
Group: VAR2O	4	0.009410	
Class: VAR8O	6	0.008450	
Group: VAR8O	2	0.008392	
AOV16: VAR1N	14	0.006479	
Var: VAR1N	1	0.003508	R2 < MINR2
AOV16: credscore	3	0.000289	R2 < MINR2
Var: credscore	1	0.000162	R2 < MINR2

Output 3.13 shows the variables selected in Step 2.

Output 3.13

The DMINE Procedure

Effects Chosen for Target: resp

Effect	DF	R-Square	F Value	p-Value	Sum of Squares	Error Mean Square
Group: VAR3N	5	0.010724	19.369933	<.0001	8.628783	0.089095
AOV16: VAR1N	14	0.003912	2.529776	0.0013	3.147905	0.088882
Group: VAR5O	3	0.001891	5.715847	0.0007	1.521686	0.088741
Group: VAR2O	4	0.000673	1.525318	0.1918	0.541304	0.088720
Group: VAR8O	2	0.000567	2.574017	0.0763	0.456572	0.088689

Displays 3.23A and 3.23B show the variables passed to the **Regression** node.

Display 3.23A

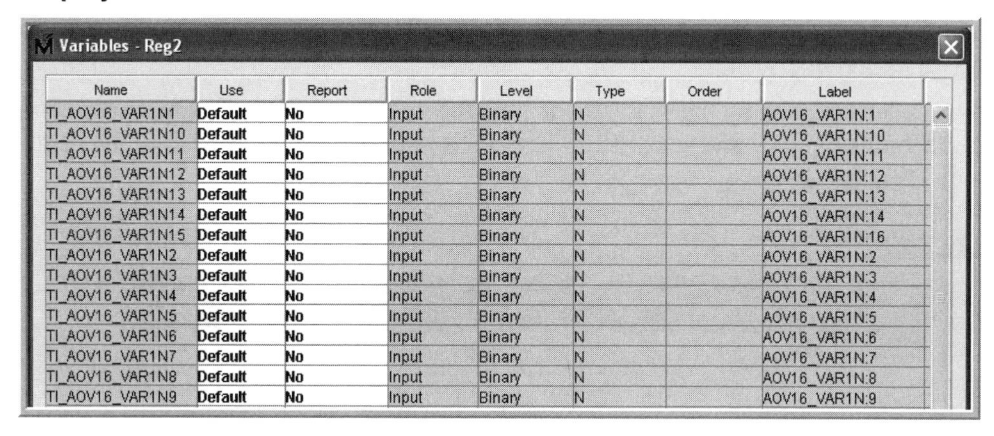

Name	Use	Report	Role	Level	Type	Order	Label
TI_AOV16_VAR1N1	Default	No	Input	Binary	N		AOV16_VAR1N:1
TI_AOV16_VAR1N10	Default	No	Input	Binary	N		AOV16_VAR1N:10
TI_AOV16_VAR1N11	Default	No	Input	Binary	N		AOV16_VAR1N:11
TI_AOV16_VAR1N12	Default	No	Input	Binary	N		AOV16_VAR1N:12
TI_AOV16_VAR1N13	Default	No	Input	Binary	N		AOV16_VAR1N:13
TI_AOV16_VAR1N14	Default	No	Input	Binary	N		AOV16_VAR1N:14
TI_AOV16_VAR1N15	Default	No	Input	Binary	N		AOV16_VAR1N:16
TI_AOV16_VAR1N2	Default	No	Input	Binary	N		AOV16_VAR1N:2
TI_AOV16_VAR1N3	Default	No	Input	Binary	N		AOV16_VAR1N:3
TI_AOV16_VAR1N4	Default	No	Input	Binary	N		AOV16_VAR1N:4
TI_AOV16_VAR1N5	Default	No	Input	Binary	N		AOV16_VAR1N:5
TI_AOV16_VAR1N6	Default	No	Input	Binary	N		AOV16_VAR1N:6
TI_AOV16_VAR1N7	Default	No	Input	Binary	N		AOV16_VAR1N:7
TI_AOV16_VAR1N8	Default	No	Input	Binary	N		AOV16_VAR1N:8
TI_AOV16_VAR1N9	Default	No	Input	Binary	N		AOV16_VAR1N:9

Display 3.23B

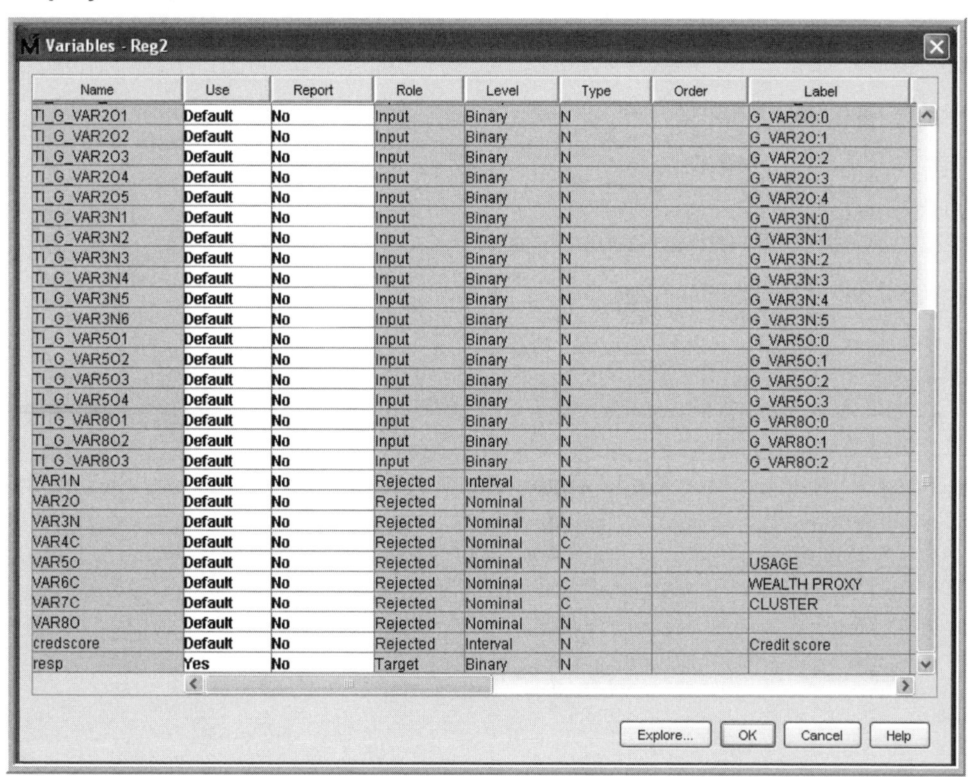

Name	Use	Report	Role	Level	Type	Order	Label
TI_G_VAR2O1	Default	No	Input	Binary	N		G_VAR2O:0
TI_G_VAR2O2	Default	No	Input	Binary	N		G_VAR2O:1
TI_G_VAR2O3	Default	No	Input	Binary	N		G_VAR2O:2
TI_G_VAR2O4	Default	No	Input	Binary	N		G_VAR2O:3
TI_G_VAR2O5	Default	No	Input	Binary	N		G_VAR2O:4
TI_G_VAR3N1	Default	No	Input	Binary	N		G_VAR3N:0
TI_G_VAR3N2	Default	No	Input	Binary	N		G_VAR3N:1
TI_G_VAR3N3	Default	No	Input	Binary	N		G_VAR3N:2
TI_G_VAR3N4	Default	No	Input	Binary	N		G_VAR3N:3
TI_G_VAR3N5	Default	No	Input	Binary	N		G_VAR3N:4
TI_G_VAR3N6	Default	No	Input	Binary	N		G_VAR3N:5
TI_G_VAR5O1	Default	No	Input	Binary	N		G_VAR5O:0
TI_G_VAR5O2	Default	No	Input	Binary	N		G_VAR5O:1
TI_G_VAR5O3	Default	No	Input	Binary	N		G_VAR5O:2
TI_G_VAR5O4	Default	No	Input	Binary	N		G_VAR5O:3
TI_G_VAR8O1	Default	No	Input	Binary	N		G_VAR8O:0
TI_G_VAR8O2	Default	No	Input	Binary	N		G_VAR8O:1
TI_G_VAR8O3	Default	No	Input	Binary	N		G_VAR8O:2
VAR1N	Default	No	Rejected	Interval	N		
VAR2O	Default	No	Rejected	Nominal	N		
VAR3N	Default	No	Rejected	Nominal	N		
VAR4C	Default	No	Rejected	Nominal	C		
VAR5O	Default	No	Rejected	Nominal	N		USAGE
VAR6C	Default	No	Rejected	Nominal	C		WEALTH PROXY
VAR7C	Default	No	Rejected	Nominal	C		CLUSTER
VAR8O	Default	No	Rejected	Nominal	N		
credscore	Default	No	Rejected	Interval	N		Credit score
resp	Yes	No	Target	Binary	N		

From Displays 3.23A and 3.23B you can see that the **Variable Selection** node created both the AOV16 variables and grouped class variables (indicated by the prefix G_), while the **Transform Variables** node created the dummy variables (indicated by the prefix TI) for each level of these AOV16 and grouped class variables.

3.3.4 Saving and Exporting the Code Generated by the Transform Variables Node

The SAS code generated by the **Transform Variables** node can be saved and exported. Simply open the **Results** window and select **View → SAS Results → Flow Code** from the menu bar as shown in Display 3.23C.

Display 3.23C

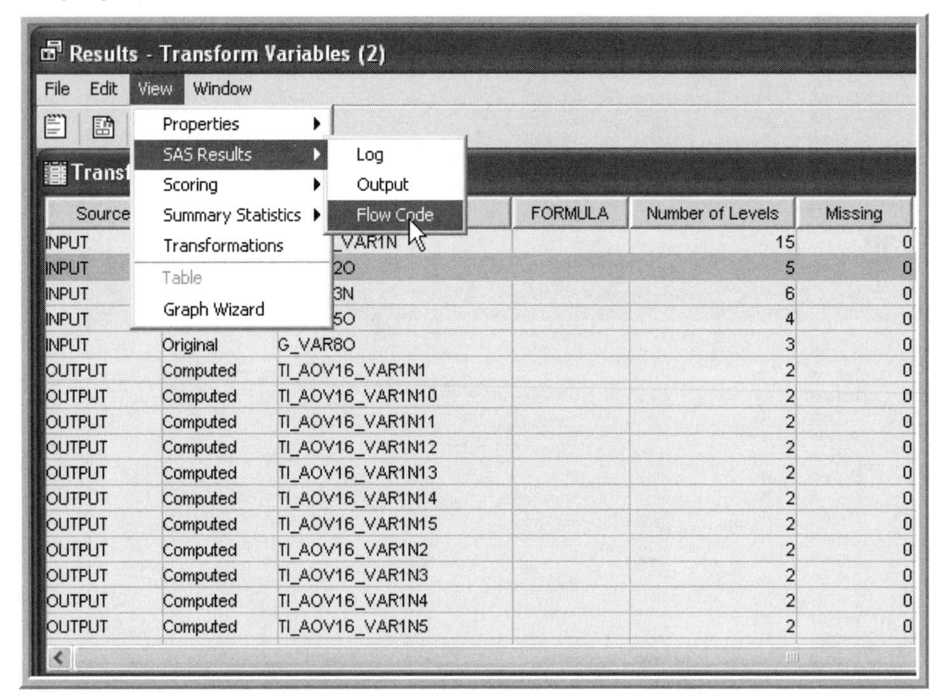

To save the file, open the **Flow Code** window, select **File** from the menu bar, click on **Save As**, and select a name for the saved file. The SAS Flow Code saved in my example is shown in Display 3.24. It is now available as a file for export.

Display 3.24

Flow Code

```
1   *------------------------------------------------------------*;
2   * Computed Code;
3   *------------------------------------------------------------*;
4   length _DC_Format $200;
5   _DN_Format=.;
6   drop _DC_Format _DN_Format;
7   *------------------------------------------------------------*;
8   * Dummy for AOV16_VAR1N;
9   *------------------------------------------------------------*;
10  * Dummy for level 1;
11  label TI_AOV16_VAR1N1='AOV16_VAR1N:1';
12  _DN_Format = AOV16_VAR1N;
13  if AOV16_VAR1N eq . then TI_AOV16_VAR1N1 = .;
14  else
15  if _DN_Format = 1 then TI_AOV16_VAR1N1 = 1;
16  else TI_AOV16_VAR1N1 = 0;
17  * Dummy for level 2;
18  label TI_AOV16_VAR1N2='AOV16_VAR1N:2';
19  _DN_Format = AOV16_VAR1N;
20  if AOV16_VAR1N eq . then TI_AOV16_VAR1N2 = .;
21  else
22  if _DN_Format = 2 then TI_AOV16_VAR1N2 = 1;
23  else TI_AOV16_VAR1N2 = 0;
24  * Dummy for level 3;
25  label TI_AOV16_VAR1N3='AOV16_VAR1N:3';
26  _DN_Format = AOV16_VAR1N;
27  if AOV16_VAR1N eq . then TI_AOV16_VAR1N3 = .;
28  else
29  if _DN_Format = 3 then TI_AOV16_VAR1N3 = 1;
30  else TI_AOV16_VAR1N3 = 0;
31  * Dummy for level 4;
```

3.4 Summary

- This chapter showed how the **Variable Selection** node selects categorical and continuous variables to be used as inputs in the modeling process when the target is binary or continuous.
- Two criteria are available for variable selection when the target is binary: R-Square and Chi-Square.
- The R-Square criterion can be used with binary as well as continuous targets, but the Chi-Square criterion can be used only with binary targets.
- The **Variable Selection** node transforms inputs into binned variables, called AOV16 variables, which are useful in capturing non-linear relationships between inputs and targets.
- The **Variable Selection** node also simplifies categorical variables by collapsing the categories.
- The **Transform Variables** node offers a wide of range choices for transformations, including multiple transformations of an input, and the option of creating dummy variables for the categories of categorical variables.

- If there are a large number of inputs, you can use the **Variable Selection** node to first make a preliminary variable selection, and then do transformations of the selected variables. If the number of inputs is relatively small, then you can transform all of the inputs prior to variable selection, if so desired.
- The code generated by the **Variable Selection** node and the **Transform Variables** node can be saved and exported.

3.5 Appendix to Chapter 3

3.5.1 Changing the Measurement Scale of a Variable in a Data Source

A data source is created with the SAS table BINARYTARGET_B. In creating the data source, the **Advanced Metadata Advisor Option** was selected (see Displays 2.14A and 2.14B in Section 2.5 of Chapter 2). When the **Advanced Metadata Advisor Option** is selected, Enterprise Miner 5.2 marks the interval variables with less than 20 distinct levels as Nominal, unless the **Class Levels Count Threshold** property is set to some value other than 20 (see Display 2.14B in Chapter 2).

In the example data set BINARYTARGET_B, the variable VAR3N has 19 distinct values. Hence, it is marked as Nominal. It is a numeric variable, as can be seen in the **Type** column in Displays 3.26 through 3.28. To change the measurement scale of this variable to interval, follow the steps given below.

Click on the table name under **Data Sources** in the project panel, as shown in Display 3.25.

Display 3.25

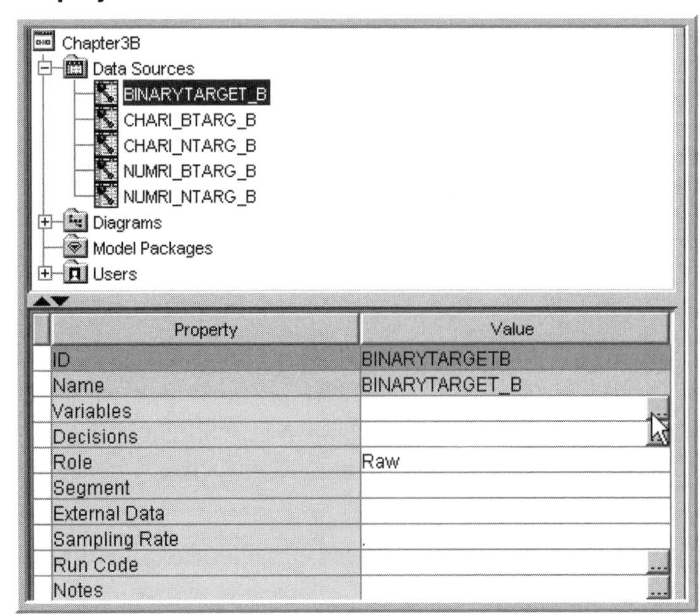

Then click on the **Value** column of the **Variables Property** as shown by the mouse pointer in Display 3.25. This opens the **Variables** window, as shown in Display 3.26.

Display 3.26

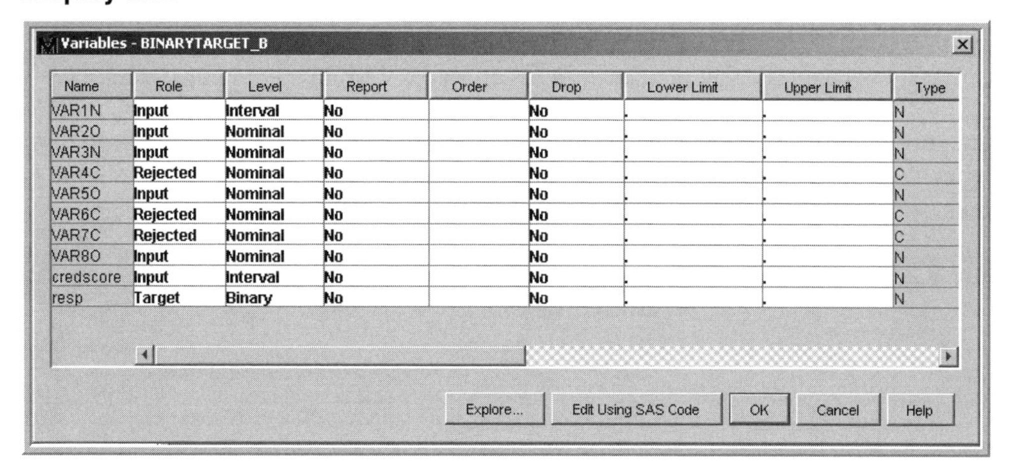

When you click on the arrow in the Level column corresponding to the variable name VAR3N, select **Interval** (Display 3.27) and click **OK**, the measurement scale of the input VAR3N is changed from **Nominal** to **Interval** (Display 3.28).

Display 3.27

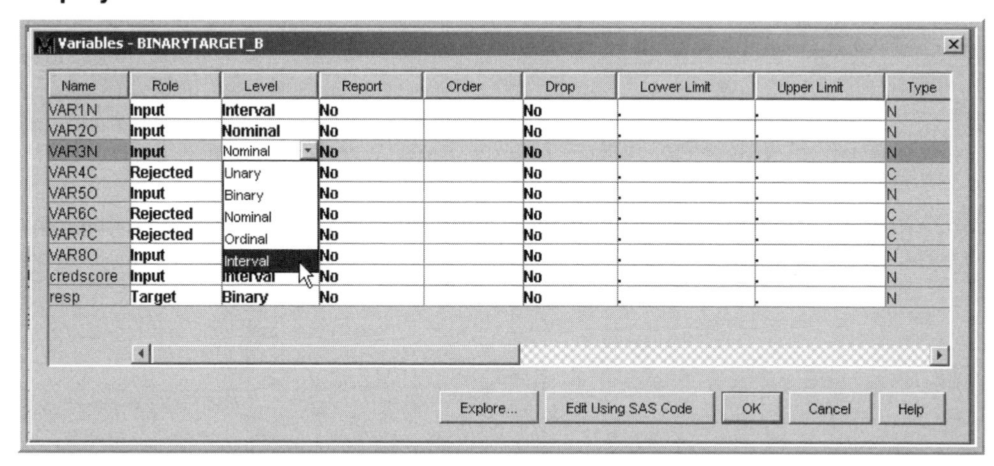

Display 3.28

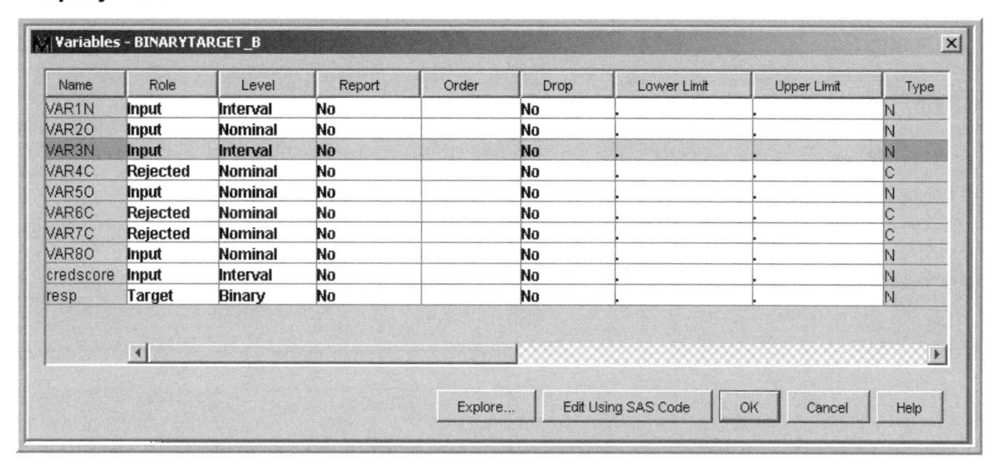

For a full discussion of the **Metadata Advisor Options** in creating a data source, see Section 2.5 of Chapter 2.

3.5.2 SAS Code for Comparing Grouped Categorical Variables with the Ungrouped Variables

In Display 3.7 you can see that the categorical variable CVAR03 has 11 categories.

The categories of CVAR03 are grouped to create the variable G_CVAR03 with only four categories. You can compare the frequency distribution of the original variable with the grouped variables using the SAS code shown in Display 3.29.

Display 3.29

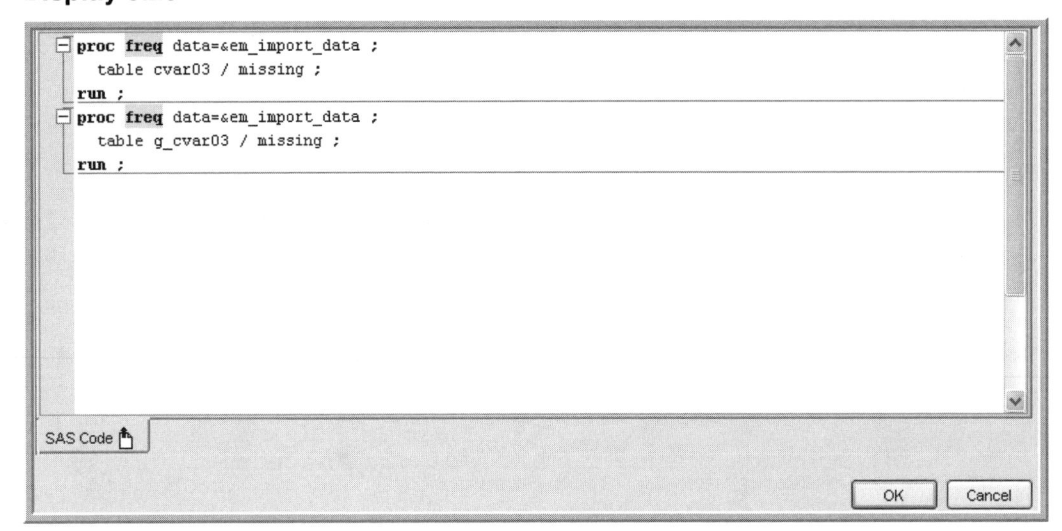

Building Decision Tree Models to Predict Response and Risk

4.1 Introduction

This chapter shows you how to build *decision tree models* to predict a categorical target and how to build *regression tree models* to predict a continuous target. Two examples are presented. The first example shows how to build a *decision tree model* to predict *response* to direct mail. In this example, the target variable is binary, taking on the values *response* and *no response*. The second example shows how to build a regression tree model to forecast a continuous (but interval-scaled) target often used in the auto insurance industry, namely *loss frequency* (described in Chapter 1). *Loss frequency* can also be modeled as a categorical target variable if it takes on only a few values but in this example it is treated as a continuous target.

The following topics are covered:

- What is a decision tree?
  - What are a root node, an intermediate node, and a terminal node/leaf?
- What is a decision tree model? What are its components?
  - node definitions
  - posterior probabilities
  - assignment of target levels to nodes
- How does Enterprise Miner arrive at node definitions?
  - which part of the data is used for building node definitions?
  - methods for splitting the records at each node
  - selecting splits
  - measuring the worth of a split
- What are posterior probabilities?
  - Which part of the data is used for estimating these probabilities?
- How to assign each node of a tree to a target class
  - Which part of the data is used for this?
- How to prune or select the right-sized tree
  - What part of the data is used for pruning?
  - What criteria are used for pruning?
- How to develop a tree
- Using a decision tree model to predict response to direct marketing
- Using a regression tree model to assess risk in auto insurance
- Using a regression tree model to develop profiles of risky customers and not-so-risky customers
- Adjusting the predicted probabilities for over-sampling (in the chapter appendix)

4.2 An Overview of the Tree Methodology in SAS Enterprise Miner

4.2.1 Decision Trees

A decision tree represents a hierarchical segmentation of the data. The *original segment* is the entire data set, and it is called the *root node* of the tree. The original segment is first partitioned into two or more *segments* by applying a series of simple rules. Each rule assigns an observation to a segment based on the value of an *input* (*explanatory variable*) for that observation. In a similar fashion, each resulting segment is further partitioned into *sub-segments* (segments within a segment); each sub-segment is further partitioned into more sub-segments, and so on. This process continues until no more partitioning is possible. This process of segmenting is called *recursive partitioning,* and it results in a hierarchy of segments within segments. The hierarchy is called a *tree*, and each segment or sub-segment is called a *node*.

Sometimes the term *parent node* is used to refer to a segment that is partitioned into two or more sub-segments. The sub-segments are called *child nodes*. When a child node itself is partitioned further, it becomes a parent node.

Any segment or sub-segment that is further partitioned into more sub-segments can also be called an *intermediate node*. A node with all its successors forms a *branch* of the tree. The final segments that are not partitioned further are called *terminal nodes* or *leaf nodes* or *leaves* of the tree. Each leaf node is defined by a unique combination of ranges of the values of the input variables. The leaf nodes are disjoint subsets of the original data. There is no overlap between them, and each record in the data set belongs to one and only one leaf node. In this book, I sometimes refer to a leaf node as a *group*.

4.2.2 Decision Tree Models

A decision tree includes intermediate as well as terminal nodes (leaf nodes). In contrast, the *decision tree model*, which you deploy for prediction or classification, includes only the leaf nodes.

A decision tree model is composed of several parts, such as

- node definitions, or rules, to assign each record of a data set to a leaf node
- *posterior probabilities* of each leaf node
- the assignment of a *target level* to each leaf node

Node definitions are developed using the *training data set* and are stated in terms of input ranges. *Posterior probabilities* are calculated for each node using the *training data set*. The assignment of the target level to each node is also done during the training phase using the *training data set*.

Posterior probabilities are *observed proportions* of target levels within each node in the training data set. Take the example of a binary target. A binary target has two levels which can be represented by *response* and *no response,* or 1 and 0. The *posterior probability* of response in a node is the proportion of records with the target level equal to response or 1 within that node. Similarly, the *posterior probability* of no response of a node is the proportion of records with the target level equal to no response, or 0, within that node. These *posterior probabilities* are determined during the training of the tree and they become part of the decision tree model. As mentioned above, they are calculated using the training data.

The assignment of a target level to an individual record, or to a node as a whole, is called a *decision*. The decisions are made in order to maximize profit, minimize cost, or minimize misclassification error. To illustrate a decision based on profit maximization, consider the example of a binary target. In order to maximize profits, you need a profit matrix. For illustrative purposes, I devised the profit matrix given in Table 4.1.

Table 4.1

Actual target level/class	Decision1	Decision2
1	$10	0
0	−$1	0

In Table 4.1, Decision1 means assigning a record to target level 1 (response), and Decision2 means assigning a record to target level 0 (non-response). This profit matrix indicates that if a true responder is correctly classified as a responder, then the profit earned is $10. If a true non-responder is classified as a responder, a loss of $1 is incurred. If a true responder is classified as a non-responder, then the profit is zero[1]. If a true non-responder is classified as a non-responder, then the profit is also zero.

Suppose E_1 represents the expected profit under Decision1, and E_2 represents the expected profit under Decision2. If $E_1 > E_2$, then Decision 1 is taken, and the record is assigned to the target level 1. If $E_2 > E_1$, then Decision 2 is taken, and the record is assigned to the target level 0. Given a profit matrix, E_1 and E_2 are based on the *posterior probabilities* of the node to which the record belongs. In general, each record is assigned the posterior probabilities of the node in which it resides. The expected profits (E_1 and E_2) are calculated as $E_1 = 10 * \hat{P}_1 - 1 * \hat{P}_0$ and $E_2 = 0 * \hat{P}_1 + 0 * \hat{P}_0$ for each record, where $\hat{P}_1$ and $\hat{P}_0$ represent the posterior probabilities of response and no response for the record.

It follows from the profit equations above that all the records in a node have the same posterior probabilities, and are therefore assigned to the same target level.

An example of a tree is shown in Figure 4.1. The tree shown in Figure 4.1 is handcrafted[2] in order to highlight the process of assigning target levels to the nodes.

---

[1] There might be a loss when a true responder is classified as a non-responder. However, in this example I am assuming it is zero for simplicity. This assumption can be relaxed, and all the procedures shown here would apply.

[2] The decision tree shown in Figures 4.1, 4.3, and 4.4 is developed using the **Decision Tree** node. However, the tree diagrams shown are handcrafted for illustrating various steps of the tree development process.

Figure 4.1

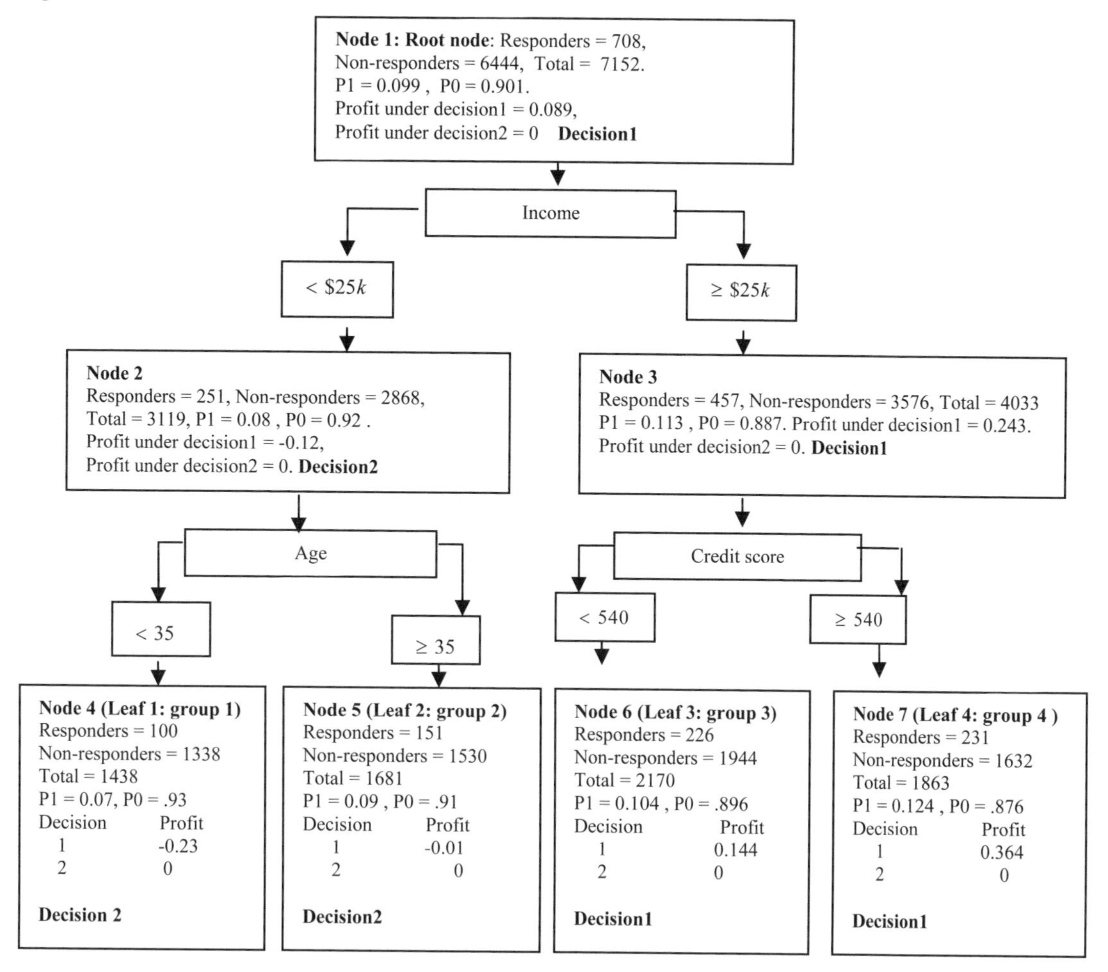

Figure 4.1 is an example of a tree for the target variable response. This variable has two levels: 1 for response and 0 for no response. Each node gives the information on the number of responders and non-responders, the proportion of responders (P1) and non-responders (P0), the decision regarding the allocation of a target level to the node, and the profit under each decision for that node. Section 4.3.4 gives a step-by-step illustration of how Enterprise Miner derives the information given in each node of the tree.

4.2.3 Decision Tree Models vs. Logistic Regression Models

A decision tree model is composed of a set of rules which can be applied to partition the data into disjoint groups. Unlike the logistic regression model, the tree model does not have any equations or coefficients. Instead, it contains, for each disjoint group or *leaf node*, posterior probabilities which are themselves used as predicted probabilities. These *posterior probabilities* are developed during the *training* of the tree.

4.2.4 Applying the Decision Tree Model to Prospect Data

The SAS code generated by the **Tree** node can be applied to the prospect data set for purposes of prediction and classification. The code places each record in one of the predefined *leaf nodes*. The predicted probability of a target level (such as response) for each record is the *posterior probability* associated with the *leaf node* into which the record is placed. Since a target level is assigned to each leaf node, all records that fall into a leaf node are assigned the same target level. For example, the target level might be responder or non-responder.

4.2.5 Calculation of the Worth of a Tree

There are situations in which it is necessary to calculate the *worth* of a tree. The worth of a tree can be calculated using the validation data set, test data set, or any other data set where the target levels are known and the inputs necessary for defining the *leaf nodes* are available. Only leaf nodes are used for calculating the worth.

In Enterprise Miner, the worth of a tree is calculated by using the validation data set to compare trees of different sizes in order to pick the tree with the optimum number of leaves. The worth of a tree can also be calculated using the test data in order to compare the performance of the decision tree model with other models. In both the cases, the method of calculating the worth of a tree is the same.

Here I illustrate the computation of the worth of a tree using the validation data set. I use an example of a response model with a binary target. The binary target has two levels or classes. They are response (indicated by 1) and no response (indicated by 0). I use profit as a measure of worth, and show how the profits of the leaf nodes of a tree are calculated using the profit matrix introduced in Table 4.1. The calculation is a two-step procedure.

In step 1, each record from the validation data set is assigned to a *leaf node* based on rules or node definitions that are developed during the training phase. As noted above, all records that are placed in a particular leaf node are assigned the same posterior probabilities that were computed for that leaf node during the training phase using the training data set. Similarly, all records that fall into a particular leaf node are assigned the same target class/level (responder or non-responder) that was determined for that leaf node during training.

In step 2, the profit is calculated for each *leaf node* of the tree based on the *actual* value of the target in each record of the validation data set.

If a *leaf node* is classified as a responder node (assigned the target level of 1), having n_1 records with the *actual* target level of 1 and n_0 records with the *actual* target level of 0, then the profit of that leaf node is $n_1 * \$10 + n_0 * (-\$1)$. If, on the other hand, the leaf node is classified as a non-responder node, then the profit of that node is $n_1 * \$0 + n_0 * \$0$. Following this procedure, the profit of each leaf node of the tree is calculated and then summed to calculate the total profit of the tree. The *average profit* of the tree is the total profit divided by the total number of records in the tree.

As mentioned before, the total profit and average profit of a decision tree model can also be calculated using the test data set. This is normally done when you want to compare the performance of a decision tree model to other models.

In Enterprise Miner the profit matrix can be entered when you create a data source, as shown in Chapter 2, Displays 2.15, 2.16, 2.17, and 2.19. It can also be entered by selecting the **Input Data** node and clicking on [...] located to the right of the **Decisions** property. By selecting the **Decision Weights** tab, you can enter the profit matrix. Display 4.1 shows the window in which you enter the profit matrix.

Display 4.1

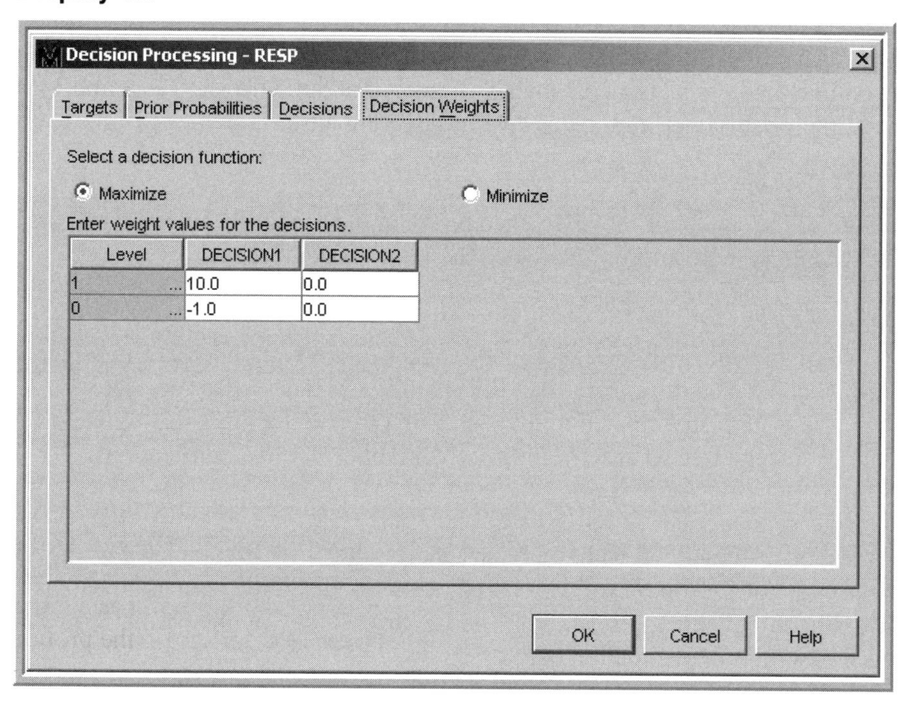

Display 4.1 also shows that I have selected the decision function **Maximize**. The next step is to tell the **Decision Tree** node to use the profit matrix and the decision function I have selected. To do this, I select the **Decision Tree** node, and then under **Subtree** in the properties panel of the **Decision Tree** node (shown in Display 4.2), I set the **Method** property to **Assessment** and the **Assessment Measure** property to **Decision**.

In Enterprise Miner, profit is not the only criterion that can be used for calculating the worth of a tree. There are a number of alternate criteria. The available criteria or measures can be seen by clicking in the value column of the **Assessment Measure** property. This brings up a menu of alternative assessment measures, as shown in Display 4.3.

Display 4.2

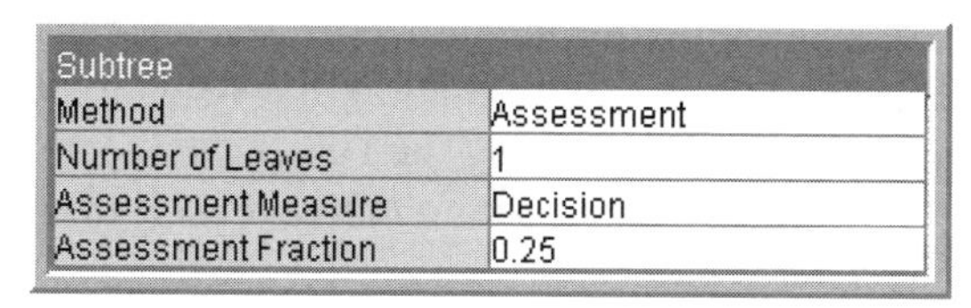

Display 4.3

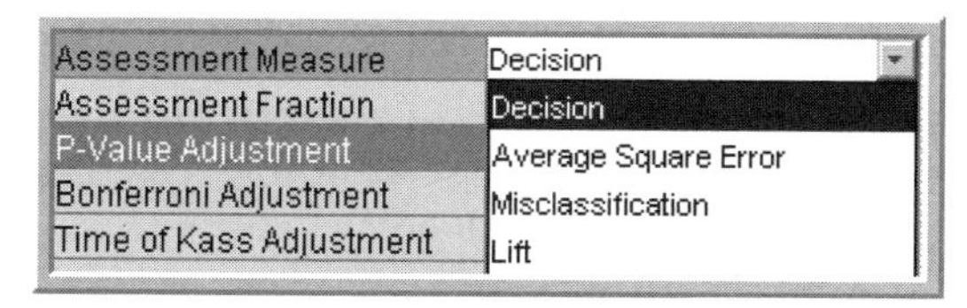

A brief description of each measure of model performance is given below. For further details, look in the Enterprise Miner Help facility.

Decision

If a profit matrix is entered and the decision function **Maximize** is selected (as in Display 4.1), then Enterprise Miner calculates average profit as a measure of the worth of a tree and selects a tree so as to maximize that value. If a cost matrix is entered and the decision function **Minimize** is selected, then Enterprise Miner minimizes cost in selecting a tree.

Average Squared Error

This is the average of the square of the difference between the predicted outcome and the actual outcome. This measure is more appropriate when the target is continuous.

Misclassification

Using the example of a response model, if the model classifies a responder as a non-responder or classifies a non-responder as a responder, then the result is a misclassification. The misclassification rate is inversely related to the worth of a tree. Selecting this option will instruct Enterprise Miner to select a tree so as to minimize the number of misclassifications.

Lift

In the case of a response model, which has a binary target, the "lift" is calculated as the ratio of the actual response rate of the top *n*% of the ranked observations in the validation data set to the overall response rate of the validation data set. The ranking is done by the predicted probability of response for each record in the validation data set.

The percentage of ranked data (*n*%) to be used as the basis for the left calculation is specified by setting the **Assessment Fraction** property of the **Decision Tree** node. The default value for this is 0.25 (25%).

4.2.6 Roles of the Training and Validation Data in the Development of a Decision Tree

To develop a tree model, you need two data sets. The first is for training the model, and the second is for pruning or fine-tuning the model. A third data set is optionally needed for an independent assessment of the model. These three data sets are referred to in Enterprise Miner as training, validation, and test, respectively. You can decide what percentage of the model data set is to be allocated to each of these three purposes. While there are no uniform rules for allocation, you can follow certain rules of thumb. If the data available for modeling is large, you can allocate it equally between the three data sets. If that is not the case, use something like 40%, 30%, and 30%, or 50%, 25%, and 25%. This partitioning is done by specifying the values for the **Training**, **Validation**, and **Test** properties of the **Data Partition** node. Reserving more data for the training data set generally results in more stable parameter estimates.

4.2.6.1 Training Data

The *training* data set is used to perform three main tasks:

- developing rules to assign records to nodes (*node definitions*)
- calculating the posterior probabilities (the proportion of cases or records in each level of the target) for each node
- assigning each node to a target level (decision)

4.2.6.2 Validation Data

The *validation* data set is used to *prune* the tree, i.e., select the *right-sized tree* or *optimal tree*. The initial tree is usually large. This is called the *maximal tree*. Some call it the *whole tree*. In this book, I use these two terms interchangeably.

By removing certain branches of the *maximal* tree, I can create smaller trees. By removing a different branch, I get a different tree. The smallest of these trees has only one leaf, called the *root node*, and the largest is the initial *maximal* tree, which can have many *leaves*. Thus removing different branches produces a number of possible sub-trees. It is necessary to select one of them. Validation data is used for the selection of a sub-tree.

As explained in Section 4.2.5, the worth of each tree of a different size is calculated using the validation data set, and one of the measures of worth is profit. Since profit is calculated using records of the validation data set, profit is referred to as the *validation profit* in Enterprise Miner.

Validation profit is used to select the *optimal* (right-sized) tree. The *size* of a tree is defined as the number of leaves in the tree. An *optimal tree* is defined as that tree that yields higher profit than any smaller tree (a tree with fewer leaves) and equal or higher profit than any tree that is larger.

Since it is easy to travel from node to node and construct trees that end up with, say, ten leaves, there might be more than one tree of a given size within the maximal (or whole) tree. Enterprise Miner searches through all of the possible trees to find the one with the highest validation profit. Details of this search process are discussed in Section 4.3.

4.2.6.3 Test Data

The *test* data set is used for an independent assessment of the final model, particularly when you want to compare the performance of a decision tree model to the performance of other models.

4.2.7 Regression Tree

When the target is continuous, the tree is called a *regression tree*. The regression tree model consists of the rules (node definitions) for partitioning the data set into leaf nodes and also the mean of the target for each leaf node.

4.3 Development of the Tree in SAS Enterprise Miner

Construction of a decision tree or regression tree involves two major steps. The first step is growing the tree. As mentioned before, many smaller trees (sub-trees) of different sizes can then be created by removing one or more branches of the large tree. The second step is selecting the optimal sub-tree. Enterprise Miner offers a number of choices for performing these steps. The following sections show how to use these methods by setting various properties.

4.3.1 Growing an Initial Tree

As described in Section 4.2.1, growing a tree involves successively partitioning the data set into segments using the method of *recursive partitioning*. Here I will discuss how the inputs are selected at each step of the sequence of recursive partitioning. I will focus on two-way, or binary, splits of the inputs, although three-way and multi-way splits can be performed by Enterprise Miner as well. In addition, the discussion assumes that the target is binary.

4.3.1.1 How a Node Is Split Using a Binary Split Search

In a binary split, two branches emerge from each node. When an input such as household income is used to partition the records into two groups, a *splitting value* such as the household income value of $10,000 may be chosen. The records with income less than the *splitting value* are sent to the left branch, and the records with income greater than or equal to the splitting value are sent to the right branch. In a multi-way split, more than two branches may emerge from any node. For example, the income variable could be divided into ranges such as $0–$10,000, $10,001–$20,000, $20,001– $30,000, etc. As mentioned earlier, for the purposes of simplicity and clarity, I will confine the discussion to two-way splits across an input.

In order to split any segment or sub-segment of the data set at a node, Enterprise Miner calculates the worth of all candidate splitting values for each of the inputs, and selects the split with the highest worth, which involves findings the best splitting value for each of the available inputs and then comparing all of the inputs to see which one's split is best. The method of calculating the worth of the split can be selected by the user. These methods are discussed in the following sections.

The process of selecting the best split consists of two steps. In the first step, the best splitting value for each input is determined. In the second step, the best input among all inputs is selected by comparing the worth of the best split of each variable with the best splits of the others, and selecting the input whose best split yields the highest worth. The node is split at the best *splitting value* on the best input. This process can be illustrated by the following example.

Suppose there are 100 inputs, represented by $X_1, X_2, ..., X_{100}$. The tree algorithm starts with input X_1 and examines all candidate splits of the form $X_1 < C$, where C is a splitting value somewhere between the minimum and maximum value of X_1. All records that have $X_1 < C$ go to the left child node, and all records that have $X_1 \geq C$ go to the right child node. The algorithm goes through all candidate splitting values on the same input and selects the best splitting value. Let the best splitting value on input X_1 be C_1. The algorithm repeats the same process on the next input X_2. This process is repeated on inputs $X_2, X_3, X_4, ..., X_{100}$ and the corresponding best splitting values $C_2, C_3, C_4, ..., C_{100}$ are found. Having found the best splitting value for each input, the algorithm then compares these splits to find the input whose best splitting value gives a better split of the data than the best splitting value of any other input. Suppose C_{10} is the best *splitting value* for input X_{10} and suppose X_{10} is chosen as the best input upon which to base a split of the node. Consequently, the node is partitioned using the input X_{10}. All records with $X_{10} < C_{10}$ are sent to the left child node, and all records with $X_{10} \geq C_{10}$ are sent to the right child node. This process is repeated for each node. A different input may be chosen at a different node. How are the splits compared to determine the best splits? To answer this, it is necessary to examine how we measure the worth of each split.

4.3.1.2 Measuring the Worth of a Split (Splitting Criterion Property of the Decision Tree Node in Enterprise Miner)

The splitting method used for partitioning the data is determined by the value of the **Criterion** property in the **Splitting Rule** section of the properties panel. Clicking in the column to the right of the **Criterion** property brings up a menu of the various methods of splitting, as shown in Display 4.4.

Display 4.4

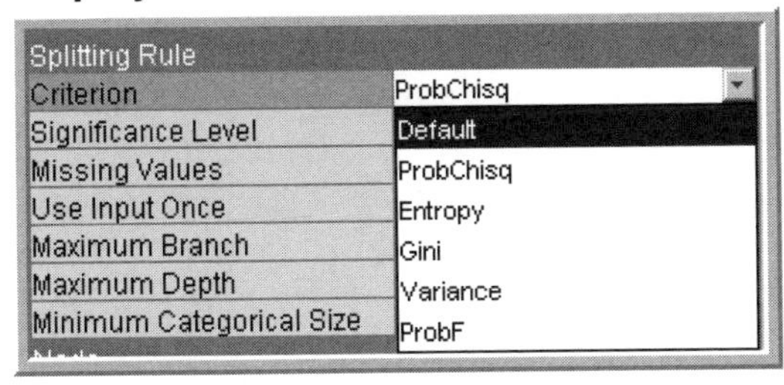

When the target variable is binary or categorical with more than two levels, there are two approaches for measuring the worth of a split:

- the degree of separation achieved by the split
- the impurity reduction achieved by the split. In Enterprise Miner, the degree of separation is measured by the *p*-value of the Pearson's Chi-Square test, and the impurity reduction is measured either by Entropy reduction or by Gini reduction.

When the target is continuous, Enterprise Miner measures the worth of a split by an F-test, which tests for the degree of separation of the child nodes. This chapter will look at these measures case by case.

4.3.1.2.A Measuring the Worth of a Split When the Target Is Binary

4.3.1.2.A.1 Degree of Separation

Any two-way split partitions a parent node into two child nodes. *Logworth* is a measure of how different these child nodes are from each other, and from the parent node, in terms of the composition of the classes. The greater the difference between the two child nodes, the greater the degree of separation achieved by the split, and the better the split is considered to be.

To see how logworth is calculated at each splitting value, consider this example. Let the target be *response*, and let *household income* be an input (explanatory variable). Let each row of the data set represent a record (or observation) from the data set. Table 4.2 shows a partial view of the data set in which I show only selected records from the data set and only two columns. I have also set up this table so that the records are in increasing order of income to help discuss the notion of splitting the records on the basis of income.

Table 4.2

Observation (record)	Response	Household income
1	1	$8,000
2	0	$10,000
...	...	...
100	1	$12,000
...	...	...
1200	0	$14,000
...	...	...
...	...	...
10,000	1	$150,000

The data shown in Table 4.2 can be split at different values of household income. At each splitting value, a 2x2 contingency table can be constructed, as shown in Table 4.3. Table 4.3 shows an example of one split. The columns represent the two child nodes that result from the split, and the rows represent the levels of the target.

Table 4.3

	Income $< \$10,000$	**Income** $\geq \$10,000$	**Total**
Responders	n_{11}	n_{12}	$n_{1\bullet}$
Non-responders	n_{01}	n_{02}	$n_{0\bullet}$
Total	$n_{\bullet 1}$	$n_{\bullet 2}$	n

To assess the degree of separation achieved by a split, it is necessary to calculate the value of the Chi-Square (χ^2) statistic and test the null hypothesis that the proportion of responders among those with income less than $10,000 is not different from those with income greater than or equal to $10,000. This can be written as

$$H_0 : P_1 = P_2 \text{ , where } \hat{P}_1 = \frac{n_{11}}{n_{\bullet 1}} \text{ , } \hat{P}_2 = \frac{n_{12}}{n_{\bullet 2}} \text{, and } \hat{P} = \frac{n_{1\bullet}}{n}$$

Under the null hypothesis, the expected values will be

	Income $< \$10,000$	**Income** $\geq \$10,000$
Responders	$E_{11} = \hat{P} \times n_{\bullet 1}$	$E_{12} = \hat{P} \times n_{\bullet 2}$
Non-responders	$E_{01} = (1 - \hat{P}) \times n_{\bullet 1}$	$E_{02} = (1 - \hat{P}) \times n_{\bullet 2}$

The Chi-Square statistic is calculated as follows:

$$\chi^2 = \sum_{i=0}^{1}\sum_{j=1}^{2}\frac{(n_{ij} - E_{ij})^2}{E_{ij}}$$

The p-value of χ^2 is found by solving the equation:

$P(\chi^2 > calculated\,\chi^2 \mid \text{null hypothesis is true}) = p\text{-value}$. And the *logworth* is simply calculated as $Logworth = -log_{10}(p\text{-value})$.

The larger the *logworth* (and therefore the smaller the p-value), the better the split.

Let the *logworth* calculated from the first split be LW_1. Another split is made at the next level of income (for example, $12,000), a contingency table is made, and the logworth is calculated in the same way. I call this LW_2. If there are 100 distinct values for income in the data set, 99 contingency tables can be created, and the logworth computed for each. The calculated values of the *logworth* are $LW_1, LW_2, ..., LW_{99}$. The split that generates the highest logworth is selected. Suppose the best splitting value for income is $15,000, with a logworth of 20.5. Now consider the next variable, AGE. If there are 67 distinct values of AGE in the data set, 66 splits are considered; the split with the maximum logworth is selected for AGE. Suppose the best splitting value for AGE is 35, with a logworth of 10.2. If age and income are the only inputs in the data set, then income is selected to split the node because it has the best logworth value. In other words, the best split on income ($15,000) has a higher logworth (20.5) than the best split on AGE (35) with a logworth of 10.2. Hence the data set would be split at $15,000 of income. This split can be called the best of the best.

If there were 200 inputs in the data set, the process of finding the best split would be performed 200 times (once for each input) and would be repeated again at each node that is further split. Each input would be examined, the best split found, and the one with the highest *logworth* chosen as the best of the best.

To use the Chi-Square method of choosing a split, set the **Splitting Rule Criterion** property of the **Decision Tree** node to **ProbChisq**. This is the default for binary targets.

4.3.1.2.A.2 Impurity Reduction as a Measure of Goodness of a Split

In Enterprise Miner, it is possible to choose maximizing *impurity reduction* as an alternative to maximizing the degree of separation for selecting the splits on a given input, and for selecting the best input. What is impurity? Impurity of a node is the degree of heterogeneity with respect to the composition of the levels of the target variable. If node v is split into child nodes a and b, and if ω_a and ω_b are the proportions of records in nodes a and b, then the decrease in impurity is $i(v) - \omega_a i(a) - \omega_b i(b)$, where $i(v)$ is the impurity index of node v, and $i(a)$ and $i(b)$ are the impurity indexes of the child nodes a and b, respectively. In Enterprise Miner there are two measures of impurity, one measure called the *Gini Impurity Index,* and the other called *Entropy*. I describe these measures below.

To split the node v into child nodes a and b based on the splitting value of an input X_1, the Tree algorithm examines all candidate splits of the form $X_1 < C_j$ and $X_1 \geq C_j$, where C_j is a real number between the minimum and maximum values of X_1. Those records that have $X_1 < C_j$ go to the left child node and $X_1 \geq C_j$ to the right. Suppose there are 100 candidate splitting values on the input X_1. The candidate splitting values are $C_j, j = 1,2,....100$. The algorithm compares impurity reduction on these 100 splits, and selects the one that achieves the greatest reduction as the best split.

4.3.1.2.A.2.1 Gini Impurity Index

If p_1 is the proportion of responders in a node, and p_0 is the proportion of non-responders, the *Gini Impurity Index* of that node is defined as $i(p) = 1 - p_1^2 - p_0^2$. If two records are chosen at random (with replacement) from a node, the probability that both are responders is p_1^2, while the probability that both are non-responders is p_0^2, and the probability that they are either both responders or both non-responders is $p_1^2 + p_0^2$. Hence $1 - p_1^2 - p_0^2$ can be interpreted as the probability that any two elements chosen at random (with replacement) are different. For binary targets, the Gini Index simplifies to $2p_1(1-p_1)$. A pure node has a *Gini Index* of zero. It has a maximum value of $1/2$ when both classes are equally represented. To use this criterion, set the **Splitting Rule Criterion** property to **Gini**.

4.3.1.2.A.2.2 Entropy

Entropy is another measure of the impurity of a node. It is defined as $i(p) = -\sum_{i=0}^{1} p_i \log_2(p_i)$ for binary targets. A node that has larger entropy than another node is more heterogeneous and therefore less pure.

The rarity of an event is measured as $-\log_2(p_i)$. If an event is rare, it means the probability of its occurrence is low. Consider the event *response*. Suppose the probability of response in a node is 0.005. Then the rarity of response is $-\log_2(0.005) = 7.644$. This is a rare event. The probability of non-response is conversely 0.995; hence the rarity of non-response is $-\log_2(0.995) = 0.0072$. A node that has a response rate of 0.005 is less impure than a node that has equal proportions of responders and non-responders. Thus $-\log_2(p_i)$ is high, when the rarity is high and low when the rarity of the event is low. The entropy of this node is as follows:

$$i(p) = -[0.005 * \log_2(0.005) + 0.995 * \log_2(0.995)] = 0.0454.$$

Consider another node in which the probability of response = probability of non-response = 0.5. The entropy of this node is $i(p) = -[0.5 * \log_2(0.5) + 0.5 * \log_2(0.5)] = 1$. The node that is predominantly non-responders (with a proportion of 0.995) has an entropy value of 0.0454. A node with an equal distribution of responders and non-responders has entropy equal to 1. A node that has all responders or all non-responders has an entropy equal to zero. Thus, entropy ranges between 0 to 1, with 0 indicating maximum purity and 1 indicating maximum impurity.

To use the Entropy measure, set the **Splitting Rule Criterion** property to **Entropy**.

4.3.1.3 Measuring the Worth of a Split When the Target Is Categorical with More Than Two Categories

If the target is categorical with more than two categories (levels), the procedures are the same. The Chi-Square statistics will be calculated from $r \times b$ contingency tables, where b is the number of child nodes being created on the basis of a certain input and r is the number of target levels (categories). The p-values are calculated from the Chi-Square distribution with degrees of freedom equal to $(r-1)(b-1)$.

The Gini Index and Entropy measures of impurity can also be applied in this case. They are simply extended for more than two levels (categories) of the target variable.

4.3.1.4 Measuring the Worth of a Split When the Target Is Continuous

If the target is continuous, an F-test is used to measure the degree of separation achieved by a split. Suppose the target is *loss frequency*, and the input is *income* having 100 distinct values. There are 99 possible two-way splits on income. At each split value, two groups are formed. One group has income level below the split value, and the other group has income greater than or equal to the split value. Here is an illustration. Consider a parent node with n records and a split value of \$10,000. All records with income less than \$10,000 are sent to the left node, and those with income greater than or equal to \$10,000 are sent to the right node, as shown in Figure 4.2. The mean of the target is calculated for each node.

Figure 4.2

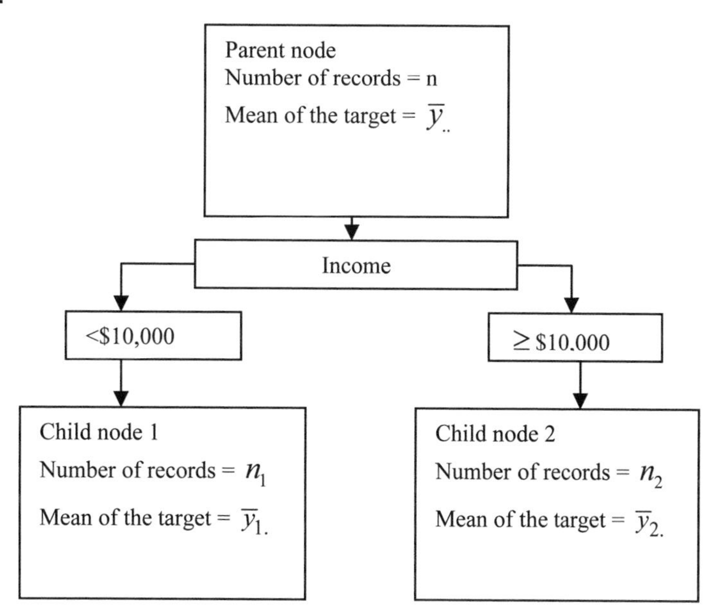

In order to calculate the worth of this split, a one-way ANOVA is first performed, and the *logworth* is then calculated from the F-test, as shown below.

The Sum of Squares between the groups (child nodes) is given by

$SS_{between} = \sum_{i=1}^{2} n_i(\bar{y}_{i.} - \bar{y}_{..})^2$, where $\bar{y}_{i.}$ is the mean of the target in the ith child node, and $\bar{y}_{..}$ is the mean of the target in the parent node.

The Sum of Squares within the groups (child nodes) is given by

$SS_{within} = \sum_{i=1}^{2}\sum_{j=1}^{n_i}(y_{ij} - \bar{y}_{i.})^2$, where y_{ij} is the value of the target variable for the jth record in the ith child node.

The Total Sum of Squares is given by

$SS_{total} = \sum_{i=1}^{2}\sum_{j=1}^{n_i}(y_{ij} - \bar{y}_{..})^2$, and the F-statistic is computed as

$F = \left(\frac{SS_{between}/(2-1)}{SS_{within}/(n-2)}\right)$. This statistic has an F-distribution with 1 and n – 2 degrees of freedom under the null hypohesis.

The null hypothesis says that there is no difference in the target mean between the child nodes.

The p-value is given by $P(F > calculated\ F \mid \text{null hypothesis is true}) = p\text{-value}$.

As before, the logworth is calculated from the p-value of the F-test as $-\log_{10}(p\text{-value})$. The larger the logworth (and therefore the smaller the p-value), the better the split.

One-way analysis is performed at each splitting value on a given input such as income, the logworth of each split is calculated, and the split with the highest logworth is selected for the input. (If income is split into more than two groups, then one-way analysis is done with more than two groups.) As with the Chi-Square method and all the other methods, the same analysis is applied to each input, and the best split for each input is found. From the best splits of each input, the best input (the one with the highest logworth) is found, and that will be the best of the best, that is, the best split on the best input.

To use the F-test shown earlier in this section, set the **Splitting Rule Criterion** property to **ProbF**, which is the default value for interval targets.

4.3.2 *P*-value Adjustment Options

Using the default settings, Enterprise Miner uses adjusted p-values in comparing splits on the same input and splits on different inputs. Two types of adjustments are made to p-values: Bonferroni adjustment and Depth adjustment.

4.3.2.1 Bonferroni Adjustment Property

When you are comparing the splits from different inputs, *p*-values are adjusted to take into account the fact that not all inputs have the same number of levels. In general, some inputs are binary, some are ordinal, some are nominal, and others are interval-scaled. For example, a variable such as *mail order buyer* takes on the value of 1 or 0, 1 being *mail order buyer* and 0 being *not a mail order buyer*. I will call this the *mail order buyer variable*. For this variable, only one split is evaluated, only one contingency table is considered, and only one test is performed. An input *like number of delinquencies in the past 12 months* can take any value from 0 to 10 or higher. I will call this the *delinquency variable*. This is an ordinal variable. Suppose it takes values from 0 to 10. Enterprise Miner compares the split values 0.5, 1.5, 2.5,..., 9.5. Ten contingency tables are constructed, and hence ten Chi-Square values are calculated. In other words, ten tests are performed on this input for selecting the best split.

Suppose the ith split on the delinquency variable has a p-value $= \alpha_i$, which means that $P(\chi^2 > calculated\,\chi_i^2 \mid \text{null hypothesis is true}) = \alpha_i$. In other words, the probability of finding a Chi-Square greater than the calculated χ_i^2 at random is α_i under the null hypothesis. The probability that, out of the ten Chi-Square tests on the delinquency variable, at least one of the tests yields a false positive decision (where we reject the null hypothesis when it is actually true) is $1-\prod_{i=1}^{10}(1-\alpha_i)$. This family-wise error rate is much larger than the individual error rate of α_i.

For example, if the individual error rate (α_i) on each test = 0.05, then the family-wise error rate is $1-0.95^{10} = 0.401$. This means that when you have multiple χ_i^2 comparisons (one for each possible split), the *p*-values underestimate the risk of rejecting the null hypothesis when it is true. Clearly the more possible splits a variable has, the less accurate the *p*-values will be. Therefore, when comparing the best split on the *delinquency variable* with the best split on the *mail order buyer variable*, the *logworths* need to be adjusted for the number of splits, or tests, on each variable. In the case of the *mail order buyer variable,* there is only one test, and hence no adjustment is necessary. But in the case of the *delinquency variable*, the best split is chosen from a set of ten splits. Therefore, $\log_{10}(10)$ is subtracted from the *logworth* of the delinquency variable's best split. In general, if an input has *m* splits, then $\log_{10}(m)$ is subtracted from the *logworth* of each split on that input. This adjustment is called the *Bonferroni Adjustment.*

To apply the *Bonferroni Adjustment,* set the **Bonferroni Adjustment** property to **Yes**, which is the default.

4.3.2.2 Time of Kass Adjustment Property

If you set the value of the **Time of Kass Adjustment** property to **After**, then the *p*-values of the splits are compared without *Bonferroni Adjustment.* If you set the property to **Before**, then the *p*-values of the splits are compared with *Bonferroni Adjustment.*

4.3.2.3 Depth Adjustment

In Enterprise Miner, if you set the **Split Adjustment** property of the **Decision Tree** node to **Yes**, then *p*-values are adjusted for the number of ancestor splits. You can call the adjustment for the number of ancestor splits the *depth adjustment*, because the adjustment depends on the depth of the tree at which the split is done. Depth is measured as the number of branches in the path from the current node, where the splitting is taking place, to the root node. The calculated *p*-value is multiplied by a depth multiplier, based on the depth in the tree of the current node, to arrive at the depth-adjusted *p*-value of the split. For example, let us assume that prior to the current node there were four splits (four splits were required to get from the root node to the current node) and that each split involved two branches (using binary splits). In this case, the depth multiplier is 2x2x2x2. In general the *depth multiplier* for binary splits is 2^d, where d is the depth of the current node. The calculated *p*-value is adjusted by multiplying by the depth multiplier. This means that at a depth of 4, if the calculated *p*-value is 0.04, the depth-adjusted *p*-value would be 0.04 x 16 = 0.64. Without the depth adjustment, the split would have been considered statistically significant. But after the adjustment, the split is not statistically significant.

The depth adjustment can also be interpreted as dividing the threshold *p*-value by the depth multiplier. If the threshold *p*-value specified by the **Significance Level** property is 0.05, then it is adjusted to be 0.05/16 = 0.003125. Any split with a *p*-value above 0.003125 will be rejected. In general, if α is the significance level specified by the **Significance Level** property, then any split which has a *p*-value above α / *depth multiplier* is rejected. The effect of the depth adjustment is to increase the threshold value of the *logworth* by $\log_{10}(2^d) = d\log_{10}(2)$. Hence, the deeper you go in a tree, the stricter the standard becomes for accepting a split as significant. This leads to the rejection of more splits than would have been rejected without the depth adjustment. Hence, the depth adjustment may also result in fewer partitions, thereby limiting the size of the tree.

4.3.3 Controlling Tree Growth: Stopping Rules

Stopping rules are applied during the training phase of tree development to decide when the recursive partitioning has gone far enough. This section discusses different ways of controlling the growth of the tree during the training phase of tree development in Enterprise Miner.

4.3.3.1 Controlling Tree Growth through the Threshold Significance Level Property

You can control the initial size of the tree by setting the threshold *p*-value. In Enterprise Miner the threshold *p*-value (*significance level*) is specified by setting the **Significance Level** property of the **Decision Tree** node to the desired *p*-value. From this *p*-value, Enterprise Miner calculates the threshold *logworth*. For example, if you set the **Significance Level** property to 0.05, then the threshold *logworth* is given by $-\log_{10}(0.05)$, or 1.30. If, at any node, none of the inputs has a split with *logworth* higher than or equal to the threshold, then the node is not partitioned further. By decreasing the threshold *p*-value, you increase the degree to which the two child nodes must differ in order to consider the given split to be significant. Thus, tree growth can be controlled by setting the threshold *p*-value.

4.3.3.2 Controlling Tree Growth through the Split Adjustment Property

As mentioned in the section on *p*-value adjustment options, if you set the **Split Adjustment** property of the **Decision Tree** node to **Yes**, then *p*-values are adjusted for the number of ancestor splits. In particular, if α is the significance level specified by the **Significance Level** property, then any split which has a *p*-value above α / *depth multiplier* is rejected. Hence, the deeper you go in a tree, the stricter the standard becomes for accepting a split as significant. This leads to the rejection of more splits than without the adjustment, resulting in fewer partitions. So, switching on the **Split Adjustment** property is another way of controlling tree growth.

4.3.3.3 Controlling Tree Growth through the Leaf Size Property

You can control the growth of the tree by setting the **Leaf Size** property. If you set the value of the property to a number such as 100, it means that if a split results in the creation of a leaf with fewer than 100 records, then that split should not be performed. Hence, the growth stops at the current node.

4.3.3.4 Controlling Tree Growth through the Split Size Property

If the value of the **Split Size** property is set to a number such as 300, it means that if a node has fewer than 300 records, then it should not be considered for splitting. The default value of this property is twice the leaf size, which is specified by the **Leaf Size** property.

4.3.3.5 Controlling Tree Growth through the Maximum Depth Property

The user can also choose a value for the **Maximum Depth** property. This determines the maximum number of generations of nodes. The root node is generation 0, and the children of the root node are the first generation, etc. The default setting of this property is 6. The **Maximum Depth** property can be set between 1 and 100.

4.3.4 Pruning: Selecting the Right-sized Tree Using Validation Data

Having grown the largest possible tree (*maximal tree*) under the constraints of stopping rules in the previous section, you now need to *prune* it to the *right size*. The pruning process proceeds as follows. Start with the *maximal tree*, and eliminate one split at each *step*. If the maximal tree has *M* leaves and if I remove one split at a given point, I will get a sub-tree of size *M*–1. If one split is removed at a different point I will get another sub-tree with of size *M*–1. Thus, there can be a number of sub-trees of size *M*–1. From these, Enterprise Miner selects the best sub-tree of size *M*–1. Similarly, by removing two splits from the maximal tree I will get several sub-trees of size *M*–2. From these, Enterprise Miner selects the best sub-tree of size *M*–2. This process continues until there is a tree with only one leaf. At the end of this process, there will be a sequence of trees of sizes *M*, *M*–1, *M*–2, *M*–3,...1. Display 4.7 shows a plot of the Average Profit associated with the sequence of the best of the sub-trees of the same size. Thus, the sequence is an optimum sequence. From the optimum sequence, the best tree is selected on the basis of average profit, or accuracy or, alternatively, the lift (I describe some of these methods later in the chapter). The process of selecting the optimum is illustrated in Section 4.3.4 with a concrete example.

Figure 4.3 shows how sub-trees are created by pruning a large tree at different points.

Figure 4.3

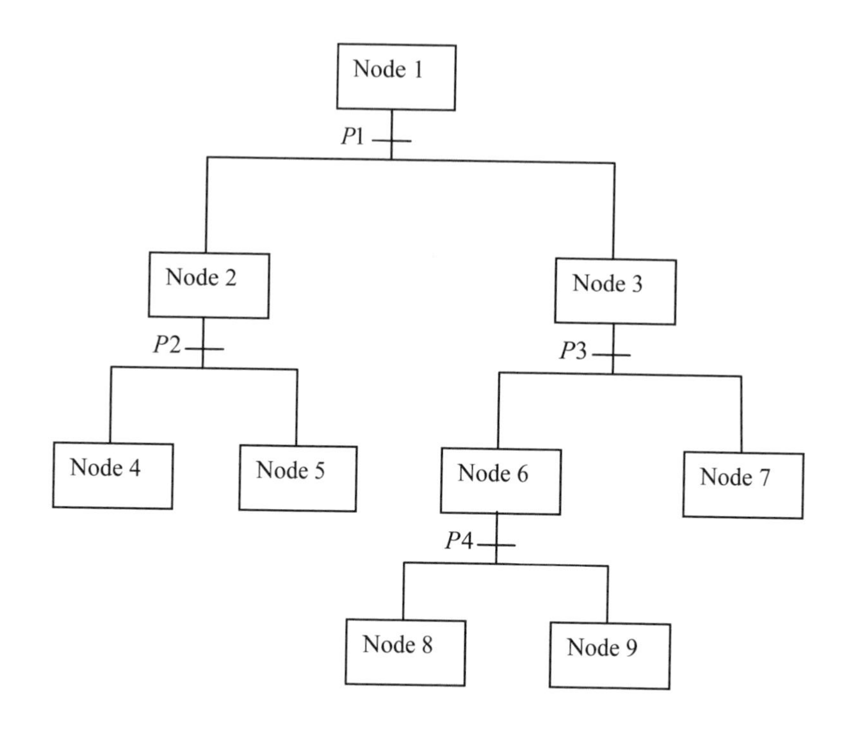

Figure 4.3 shows a hypothetical tree. It has nine nodes. In the maximal tree there are five leaf nodes (nodes 4, 5, 7, 8, and 9). By removing a split at the point *P*4, I get one tree with four leaves (nodes 4, 5, 6, and 7). By removing the split at *P*2, I get a tree with four leaves with nodes 2, 8, 9, and 7 as its leaves. Thus, there are two sub-trees of size 4.

By removing the splits at *P*3 and *P*4, I get a tree with three leaves (consisting of nodes, 4, 5, and 3 as its leaves). By removing the splits at *P*2 and *P*4, I will get another tree with three leaves consisting of nodes 2, 6, and 7 as its leaves. Thus, there are two sub-trees of size 3.

By removing the splits at *P*2, *P*3, and *P*4, I get one tree with two leaves consisting of nodes 2 and 3 as its leaves.

By removing the splits at *P*1, *P*2, *P*3, and *P*4, I get one tree with only one leaf, with node 1 as its leaf.

Thus the sequence consists of one tree with five leaves, two trees with four leaves, two trees with three leaves, one tree with two leaves, and one tree with a single leaf.

Since there two sub-trees of size 4, only one is selected. Similarly, one sub-tree of size 3 is selected. This selection is based on the user-supplied criterion discussed below.

The final sequence consists of one sub-tree of each size. This sequence, along with the profit or other assessment measure used to choose the best sub-tree of each size is displayed in a chart in the **Results** window of the Decision Tree. As mentioned previously, Display 4.7 shows an example of such a chart.

To choose the final model, Enterprise Miner selects the best tree from the sequence.

What is the criterion used for selecting the best tree from a number of trees of the same size at each step of the sequence described above? One possible criterion is to compare the profit of the sub-trees in each step of building the sequence. The profit associated with each tree is calculated using the *validation data set*.

Alternative criteria for selecting the best tree include cost minimization, minimization of miscalculation rate, minimization of average squared error, or maximization of lift. In the case of a continuous target, minimization of average squared error is used for selecting the best tree.

In Enterprise Miner, set the **Method** property under **Subtree** in the property panel of the **Decision Tree** node to **Assessment**, and the **Assessment Measure** property to **Decision**, **Average Square Error**, **Misclassification**, or **Lift**. These options are described Section 4.2.5.

If you set the value of the **Assessment Measure** property to **Decision**, **Decision Weights** must be entered at the time you create the **Data Source** or later by selecting the **Input Data** node and clicking on [...] located to the right of the **Decisions** property in the **Settings** group of the property panel. When you click on [...], the **Decision Processing** window opens as shown in Display 4.5.

Display 4.5

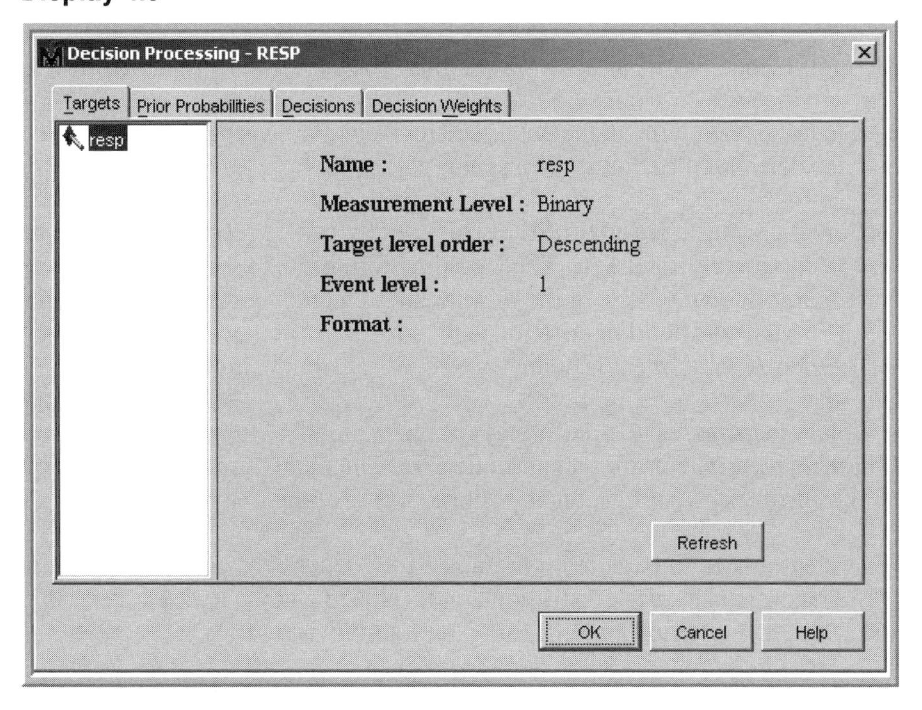

Then select the **Decision Weights** tab to enter the profit matrix. On the **Decision Weights** tab, enter the profit matrix and select **Maximize** under the label **Select a decision function**.

Display 4.6

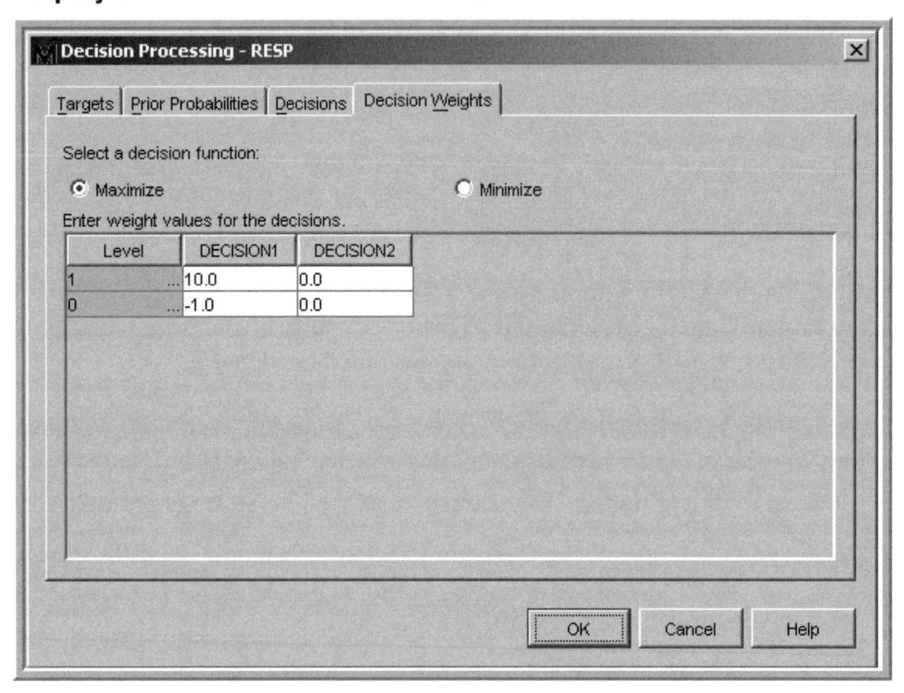

If you enter costs rather than profits as decision weights, you should select the **Minimize** option.

If a profit or loss matrix is not defined, then assessment is done by **Average Square Error** if the target is interval and **Misclassification** if the target is categorical.

As noted in Section 4.2.5, the **Assessment Measure** property can be set to **Decision Average Square Error**, **Misclassification**, or **Lift**. The **Decision** option calls for the assessment to be done by maximizing profit on the basis of the profit matrix entered by the user. This is illustrated in Section 4.3.5. The **Misclassification** criterion is illustrated in Section 4.3.7, and the **Average Square Error** criterion of assessment is demonstrated in an example in Section 4.3.8.

As the number of leaves increases, the profit may increase initially. However, beyond some point there is no additional profit. The profit may even decline. The point at which marginal, or incremental, profit becomes insignificant is the optimum size of the tree.

What follows is a step-by-step illustration of pruning a tree. The tree is grown using the training data set with 7152 records. The rules of partition are developed and the nodes are classified (so as to maximize profit or other criterion chosen) using the training data set.

The validation data set used for pruning consists of 5364 records. The rules of partition and the node classifications that were developed on the training data set are then applied to all 5364 records, in effect reproducing the tree with the validation data set. The node definitions and classification of the nodes are the same, but the records in each node are from the validation data set.

4.3.5 Step-by-Step Illustration of Growing and Pruning a Tree

The process begins by displaying the maximal tree constructed from the training data set.

Step 1A: A tree is developed using the training data set

Using the methods discussed in detail above, the initial tree is developed using the training data set. This tree is shown in Figure 4.4.

Figure 4.4

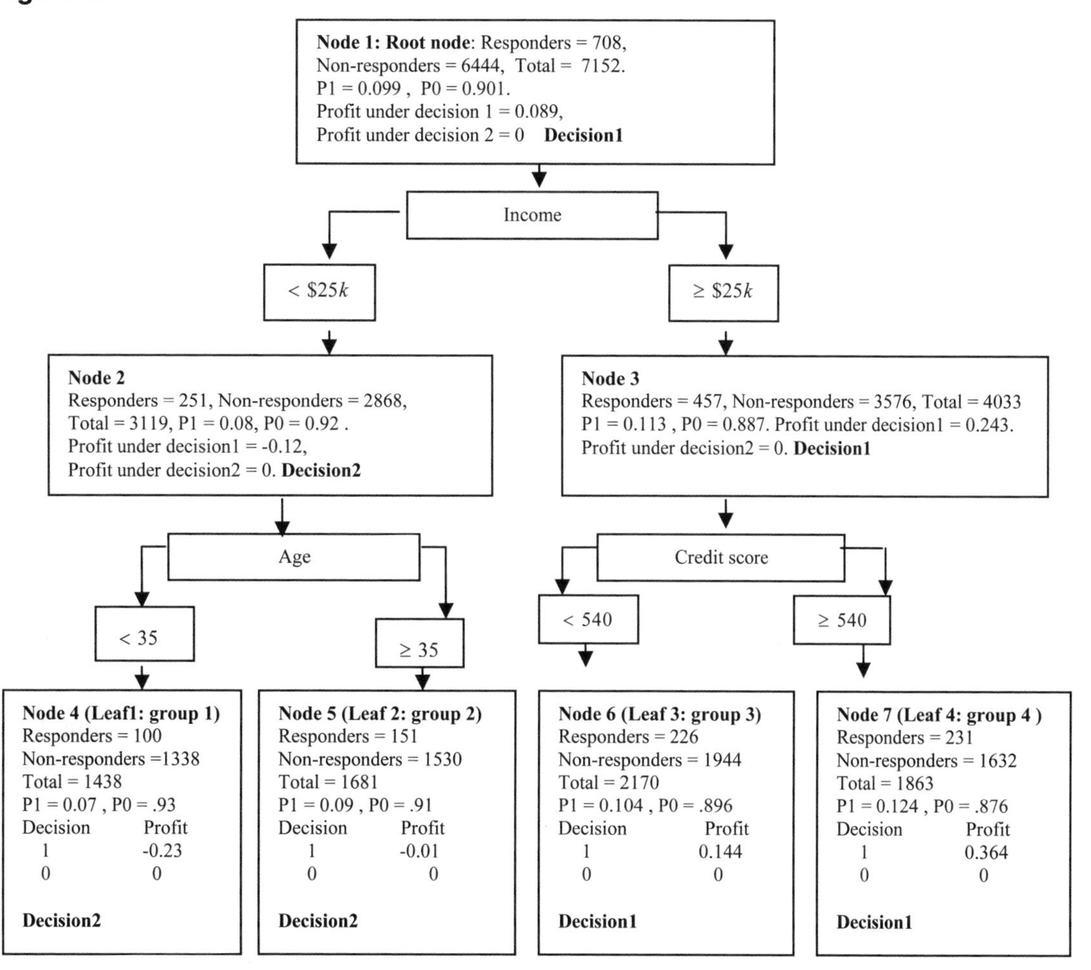

Figure 4.4 shows the *tree* developed from the *training data.* The above tree diagram gives information on: the node identification; the leaf identification, if the node is a terminal node; the number of responders in the node; the number of *non-responders*; the total number of records in each node; proportion of responders (*posterior probability* of response); proportion of non-responders (*posterior probability* of non-response); profit under Decision1; profit under Decision2; and the label of the decision under which the profits are maximized for the node.

The tree consists of the rules or definitions of the nodes. Starting from the *root node* and going down to a *terminal node*, you can read the definition of the *leaf node* of the *tree.* These definitions are stated in terms of the input ranges. The inputs selected by the tree algorithm in this example are age, income, and credit score. The *rules* or definitions of the *leaf nodes* are

Leaf 1: If income < $25,000 and age < 35 years, then the individual record goes to *group 1* (*leaf 1*).

Leaf 2: If the individual's income < $25,000 and age ≥ 35 years, then the record goes to *group 2* (*leaf 2*).

Leaf 3: If an individual's income ≥ $25,000 and credit score < 540, then the record goes to *group 3* (*leaf 3*).

Leaf 4: If the individual's income is ≥ $25,000 and credit score ≥ 540, then the record goes to *group 4* (*leaf 4*).

Step 1B: Posterior probabilities for the leaf nodes from the training data

When the target is response (which is a binary target), the *posterior probabilities* are the proportion of responders and the proportion of non-responders in each node. In applying the model to prospect data, these *posterior probabilities* are used as predictions of the probabilities. All customers (records) in a *leaf* are attributed the same predicted probability of response.

Step 1C: Classification of records and nodes by maximization of profits

Classification of records or nodes is the name given to the task of assigning a target level to a record. All records within a node are assigned the same target level. The decision to assign a class or target level to a record or *node* is based on both the *posterior probabilities* (Step 1B above) and the profit matrix. As described later, you can use a criterion other than profit maximization to classify a node or the records within a node. I will first demonstrate how classification is done using profit when the target is response. In Figure 4.4, each node contains a label: ***Decision1*** or ***Decision2***. A label of ***Decision1*** means the node is classified as a responder node. A label of ***Decision2*** means the node is classified as a non-responder node. In arriving at these decisions, the following *profit matrix* is used:

Table 4.4

Profit Matrix		
Target	***Decision***	
	1	0
1	$10	0
0	-$1	0

According to this *profit matrix*, if I correctly assign the target level 1 (***Decision1***) to a true responder (target = 1), then I make a $10 profit. If I incorrectly assign the target level 1 to a non-responder, I will lose $1 (a profit of –$1). For example, in node 4 the posterior probabilities are 0.07 for response and 0.93 for no response. Under ***Decision1*** the expected profit = 0.07*$10 + 0.93*(–$1) = –$0.23. Under ***Decision2*** (that is, if I decide to assign target level 0, or classify node 4 as a non-responder node) the expected profit = 0.07*0+0.93*0 = $0. Since $0 > –$0.23, I take ***Decision2;*** that is, I classify node 4 as a non-responder node. All records in a node are assigned the same target level since they all have the same posterior probabilities and the same profit matrix. Figure 4.4 shows the *posterior probabilities* for each node as P1 and P0. The following calculations show how these *posterior probabilities*, together with the *profit matrix*, are used to assign each node to a target level.

Node 1 (root node)

Under *Decision1* the expected profit is = $10*0.099 + (–$1)*0.901 = $0.089. Under *Decision2* the expected profit is = 0. Since the profit under *Decision1* is greater than the profit under *Decision2, Decision1* is taken. That is, this node is assigned to the target level 1. All members of this node are classified as responders.

Node 2
Under *Decision1* the expected profit is $10*0.08 + (–$1)*0.92 = – $0.12. Under *Decision2* the expected profit is 0*0.08 + 0*0.92 = $0. Since the profit under *Decision2* is greater than the profit under *Decision1*, this node is assigned to the target level 0. All members of this node are classified as non-responders.

Node 3
Under *Decision1* the expected profit is = $10*0.113 + (–$1)*0.887 = $0.243. Under *Decision2* the expected profit is 0*0.113 + 0*0.887 = $0. Since the profit under *Decision1* is greater than the profit under *Decision2*, this node is assigned the target level 1. All members of this node are classified as responders.

Node 4
Under *Decision1* the expected profit is = $10*0.07 + (–$1)*0.93 = –$0.23. Under *Decision2* the expected profit is 0*0.07 + 0*0.93 = $0. Since the profit under *Decision2* is greater than the profit under *Decision1*, this node is assigned the target level 0. All members of this node are classified as non-responders.

Node 5
Under *Decision1* the expected profit is = $10*0.09 + (–$1)*0.91 = –$0.01. Under *Decision2* the expected profit is 0*0.09 + 0*0.91 = $0. Since the profit under *Decision2* is greater than the profit under *Decision1*, this node is assigned the target level 0. All members of this node are classified as non-responders.

Node 6
Under *Decision1* the expected profit is = $10*0.104 + (–$1)*0.896 = $0.144. Under *Decision2* the expected profit is 0*0.104 + 0*0.896 = $0. Since the profit under *Decision1* is greater than the profit under *Decision2*, this node is assigned the target level 1. All members of this node are classified as responders.

Node 7
Under *Decision1* the expected profit is = $10*0.124 + (–$1)*0.876 = $0.364. Under *Decision2* the expected profit is 0*0.124 + 0*0.876 = $0. Since the profit under *Decision1* is greater than the profit under *Decision2*, this node is assigned the target level 1. All members of this node are classified as responders.

Step 2: Calculation of validation profits and selection of the right-sized tree (pruning)

First, the *node definitions* are used to partition the validation data into different nodes. Since each node is already assigned a target level based on the *posterior probabilities* and the profit matrix, I can calculate the total profit of each node of the tree using the validation data set. Since the profits are calculated from the validation data set, they are called *validation profits*. Figure 4.5 shows the application of the tree to the *validation* data set.

Figure 4.5

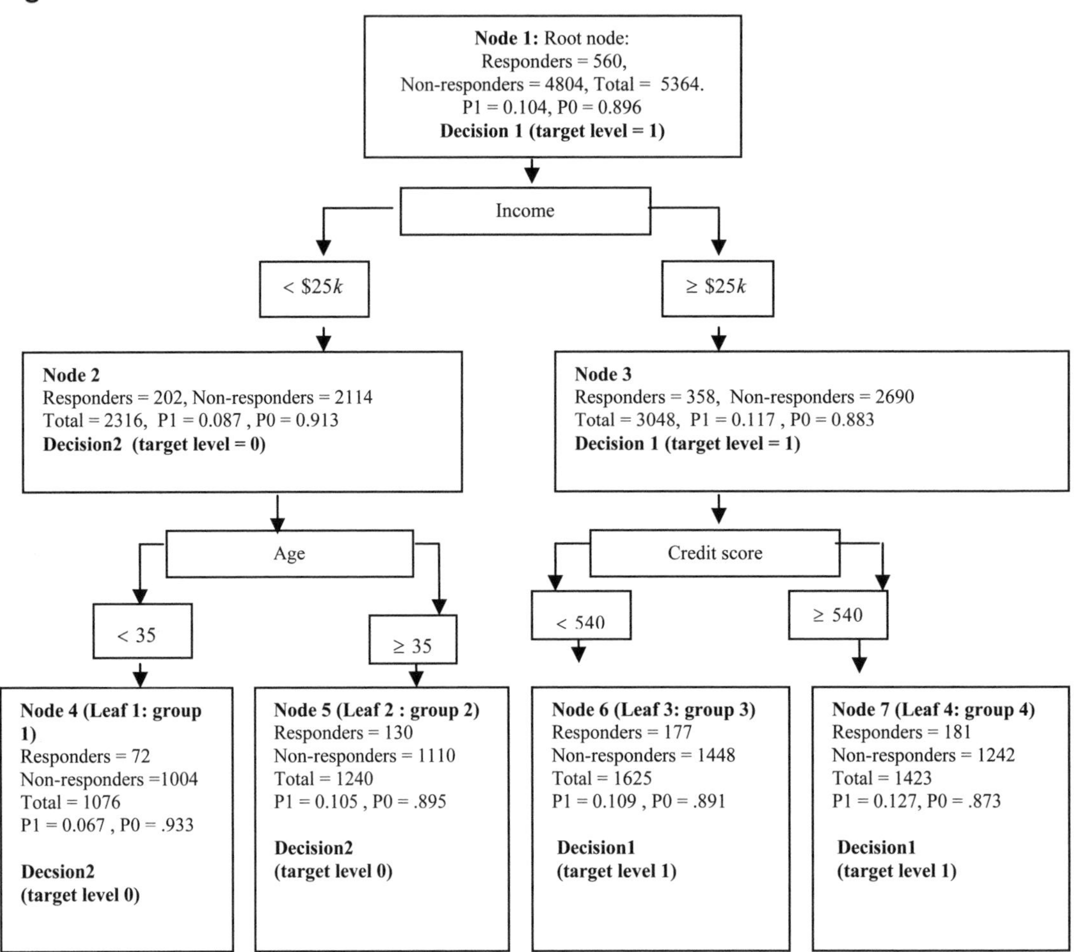

After applying the node definitions to the validation data set, I get the tree shown in Figure 4.5. Comparing the tree from the validation data (Figure 4.5) with the tree from the training data (Figure 4.4), notice that the decision labels (***Decision1*** or ***Decision2***) of each *node* are exactly the same in both the diagrams. This occurs because these decisions are based on the node definitions and the *posterior probabilities* generated during the training of the tree (using the training data set). They become part of the model and do not change when they are applied to a new data set.

Using these node definitions and this classification of the nodes, I will now demonstrate how an optimal tree is selected.

The tree in Figure 4.5 has four *leaf nodes*, and I refer to this tree as the *maximal tree*. However, within this tree there are several sub-trees of different sizes. There are two *trees* of size 3. That is, if I prune nodes 6 and 7, I will get a sub-tree with nodes 3, 4, and 5 as its *leaves*. I will call this sub-tree_3.1, indicating that it has three leaves and that it is the first one of the three-leaf trees. If I prune nodes 4 and 5, I get another sub-tree of size 3, this time with nodes 2, 6, and 7 as its *leaves*. I will call this sub-tree_3.2. Altogether, there are two sub-trees of size 3. From these two, I must select the one that yields the higher *profit*. Since the profit is calculated using the validation data set, it is called *validation profit* in Enterprise Miner. In the discussion below I use the terms *validation profit* and *profit* interchangeably.

The next step is to examine the profit of each *leaf* from sub-tree_3.1. Node 3 has 358 responders (from the validation data set), and 2690 non-responders (from the validation data set). Since this node was already classified as a *responder node* (during training), I must use the first column of the profit matrix given above to calculate the profit. Therefore, the validation profit of this leaf is $890 (= 358*$10 + 2690*($–1)). Node 4, on the other hand, is a non-responder node, so it has a validation profit of $0 (= 72 * $0 + 1004*$0). Node 5 is also a non-responder node, and so, it also yields a profit of $0. *Hence, the total validation profit of all the leaves of sub_tree_3.1 is $890.* Now I calculate the total *profit* of sub-tree_3.2, which has nodes 2, 6, and 7 as its *leaves*. Since node 2 is classified as a non-responder node, its profit is equal to $0 using the second column of the profit matrix. Node 6 is a responder node; it has177 responders and 1448 non-responders in *the validation data set*. So its total *profit* = 177*$10 + 1448*(–$1) = $322. Node 7 is also a responder node. It has 181 responders and 1242 non-responders. Total *profit* of this node = 181*($10) + 1242*(–$1) = $568. *Total profit of sub-tree_3.2 is $890 ($322 + $568).* So, between the two sub-trees of size 3 we have a tie in terms of their profits.

Now I must examine all sub-trees of size 2. In the example there is only one such sub-tree, and I will call it *sub-tree_2.1.* It has only two leaf nodes, namely node 2 and node 3. Node 2 is classified as a non-responder node; hence it will yield a profit of zero. Node 3 is classified as a responder node. It has 358 responders and 2690 non-responders from the validation data set. This yields a *profit = 358*($10) + 2690*(–$1) = $890.* The total profit of sub-tree_2.1 is $890.

Now, let us calculate the *profit* of the *root node*. The root node is classified as a responder, and it has 560 responders and 4804 non-responders in the validation data set, yielding a *profit* of $796.

I should also calculate the *validation profit* of the maximal tree, which is the tree with four leaf nodes. I call this sub-tree_4.1. This tree has nodes 4, 5, 6, and 7 as its leaves. Nodes 4 and 5 are classified as non-responder nodes, and hence yield zero profit. From the calculation above, node 6 is a responder node and yields a profit of $322, while node 7, also a responder node, yields a profit of $568, giving a total profit of $890 for the maximal tree.

Table 4.5 shows the summary of the *validation profits* of all sub-trees and the tree:

Table 4.5

Summary of Validation Profits		
Tree	Number of leaves	Validation Profit
Root node	1	$796
sub_tree_2.1 (nodes 2 and 3)	2	$890
sub_tree_3.1 (nodes 3,4 and 5)	3	$890
Sub_tree_3.2 (nodes 2,6 and 7)	3	$890
Sub_tree_4.1 (nodes 4,5,6, and 7)	4	$890

As can be seen from the above table, the optimal tree is sub-tree_2.1 with two leaves, because all of the trees with equally high profits are trees of larger size. As you go down the columns *Number of leaves* and *Validation profit,* notice that as the number of leaves increases, there is no increase in profit. So, another way to think of this comparison with larger trees is to say that beyond sub-tree_2.1, the marginal profit = 0. That is, the incremental profit per additional leaf is zero. Hence sub-tree_2.1 is the optimum, or right-sized, tree. Note that sub-tree_4.1 is the same as the maximal tree.

4.3.6 Average Profit vs. Total Profit for Comparing Trees of Different Sizes

In the above example I calculated the profit of a tree (or sub-tree) as the sum of the profits of its *leaf nodes*. The validation data has 5364 records, of which 560 are responders. Therefore *average profit* equals *total profit* divided by 5364. Since the denominator (number of records) is the same for all the trees compared, regardless of their number of *leaves*, I arrive at the same optimal tree using either *average* or *total profits* for comparison. Likewise, the sum of responders in all leaf nodes is equal to the total number of responders in the validation data set, which is 560 in this example. Hence calculating *average profit* as *total profit* / 560 for each (sub-)tree will produce the same optimal tree.

4.3.7 Accuracy/Misclassification Criterion in Selecting the Right-sized Tree (Classification of Records and Nodes by Maximizing Accuracy)

Thus far, I have described how to classify each node and how to *prune* a tree using profit maximization as the criterion. Another often-used criterion for both node classification and tree pruning is *validation accuracy*, which is described below. In order to use this criterion for sub-tree selection or pruning in Enterprise Miner, you must set the **Assessment Measure** property to **Misclassification.** Since misclassification rate = 1 – validation accuracy, I use the terms *validation accuracy* and *misclassification* interchangeably.

When the accuracy criterion is selected for assessment of the trees in pruning, the nodes are classified (assigned a target level) according to maximum *posterior probability*. This assignment is done during the training of the tree using the training data set.

4.3.7.1 Classification of the Nodes by Maximum Posterior Probability/Accuracy

This can be thought of as using a profit matrix of the following type:

Table 4.6

Accuracy		
Target	***Decision***	
	1	0
1	1	0
0	0	1

In this matrix, if a responder is correctly classified as a true responder, then one unit of accuracy is attained. If a non-responder is correctly classified as non-responder, then one unit of accuracy is gained. Otherwise there is no gain.

As before, the nodes are classified as responder or non-responder nodes based on the posterior probabilities calculated from the *training data set* and the above profit matrix. In the root node of the training data set the proportion of responders is 9.9%, and the proportion of non-responders is 90.1%. Hence, if the root node is classified as a responder node, the accuracy would be 9.9%; if it is classified as non-responder node, the accuracy would be 90.1%. Hence, the optimal decision is to classify the root node as a non-responder node.

Similarly, nodes 2, 3, 4, 5, 6, and 7 will all be classified as non-responder nodes, since they all have more non-responders than responders.

4.3.7.2 Calculation of the Misclassification Rate/Accuracy Rate and Selection of the Right-sized Tree

Since the misclassification rate is 1 – validation accuracy, maximizing the validation accuracy is the same as minimizing the misclassification rate.

Table 4.7 shows the accuracy of the different sub-trees when applied to the *validation data set*. Using rules and partitioning definitions derived from the *training data set*, all the nodes are classified as non-responder nodes. The tree has three sub-trees (not counting the root node as a tree): sub-tree_3.1 (with nodes 3, 4, and 5), sub-tree_3.2 (with nodes 2, 6, and 7), and sub-tree_2.1 (with nodes 2 and 3). The *validation accuracy* of sub-tree_3.1 = (2690 + 1004 + 1110)/5364 = 89.6%, and the misclassification rate is 10.4%. The *validation accuracy* of sub-tree_3.2 = (2114 + 1448 + 1242)/5364 = 89.6% and the corresponding misclassification rate is 10.4%. The *validation accuracy* of sub-tree_2.1 = (2114 + 2690)/5364 = 89.6%. The *validation accuracy* of the root node is (4804/5364) = 89.6%, and the misclassification rate is 10.4%. The *validation accuracy* of the maximal tree (with nodes 4, 5, 6, and 7) = (1004 + 1110 + 1448 + 1242)/5364 = 89.6% with a misclassification rate of 10.4%. There is no gain in accuracy by going beyond the root node. Hence, the optimal or right-sized tree is a tree with the root node as its only leaf. As the number of leaves increased, there was no gain in accuracy.

Table 4.7

Summary of Validation Accuracy		
Tree	Number of leaves	Validation Profit
Root node	1	89.6%
sub_tree_2.1 (nodes 2 and 3)	2	89.6%
sub_tree_3.1 (nodes 3,4 and 5)	3	89.6%
Sub_tree_3.2 (nodes 2,6 and 7)	3	89.6%
Sub_tree_4.1 (nodes 4,5,6, and 7)	4	89.6%

4.3.8 Assessment of a Tree or Sub-tree Using Average Square Error

When the target is continuous, the assessment of a tree is done using sums of squared errors. As in the case of a binary target, I use the *validation data set* to perform the assessment. Therefore, I begin by applying the node definitions or rules of partitioning developed using the training data set to the validation data set. The worth of a node is measured (inversely) by the Average Square Error for that node. For the tth node, the Average Square Error $= \frac{1}{n_t}\sum_{i=1}^{n_t}(y_{it} - \overline{y}_t)^2$, where y_{it} is the actual value of the target variable for the ith record in the tth node, and $\overline{y}_t$ is the expected value of the target for the tth node. The expected value $\overline{y}_t$ is calculated from the *training* data set, but the observations y_{it} are from the *validation* data set. The node definitions are constructed from the *training* data set.

The overall Average Square Error for a tree or sub-tree is the weighted average of the average squared errors of the individual leaves, and which can be as follows:

$$\sum_{t=1}^{T}\frac{n_t}{n}\left[\frac{1}{n_t}\sum_{i=1}^{n_t}(y_{it}-\bar{y}_t)^2\right].$$

Classification of nodes is not necessary for calculating the assessment value using Average Square Error (because the target in this case is continuous).

4.3.9 Selection of the Right-sized Tree

Display 4.7 gives another illustration of the selection of the *right-sized tree* from a sequence of optimal trees. In the graph shown in Display 4.7, the horizontal axis shows the number of leaves in each of the trees in the optimal sequence of trees. The vertical axis shows the average profit of each tree. From the display it is clear that the optimal tree has seven leaves. That is, a tree with seven leaf nodes has the same average validation profit as any tree with more nodes. Hence the tree with seven leaves is selected by the algorithm (see the vertical line corresponding to seven leaves on the horizontal axis).

Display 4.7 shows the initially constructed maximal tree of size 14, before pruning it to a tree of size 7 using the technique described earlier.

Display 4.7

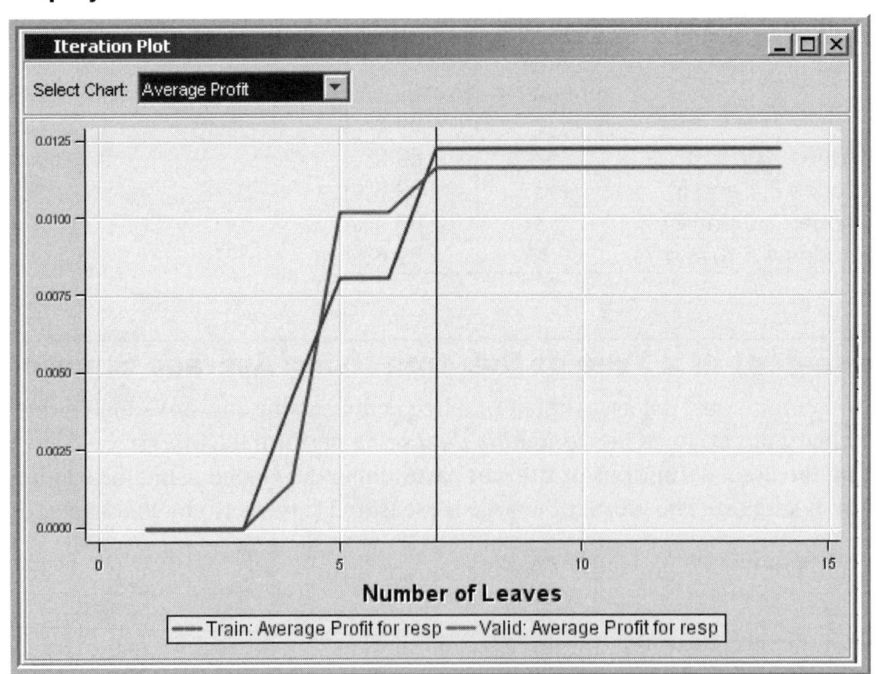

4.4 A Decision Tree Model to Predict Response to Direct Marketing

In this section, I develop a response model using the **Decision Tree** node. The model is based on simulated data representing the direct mail campaign of a hypothetical insurance company. The main purpose of this section is to show how to develop a response model using the **Decision Tree** node, and to make you familiar with the various components of the **Results** window.

Since this model predicts an individual's response to direct mailing, it is called a *response model.* In this model, the target variable (the variable that I want to predict) is binary: it takes on values of *response* and *no response.*

The process flow diagram for this model is shown in Display 4.8.

Display 4.8

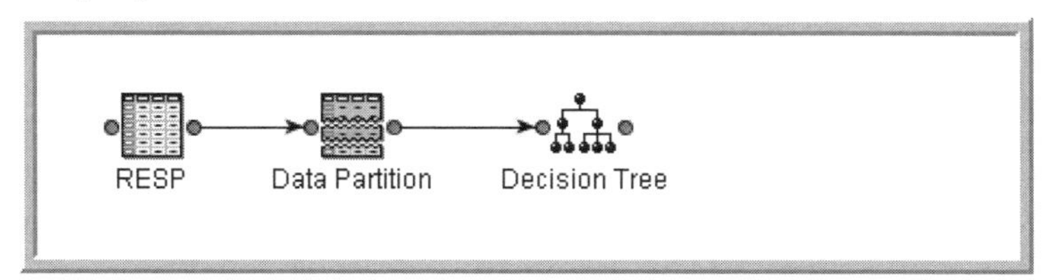

In the process flow diagram the first node is the **Input Data** node. The properties of this node are shown in Display 4.9.

Display 4.9

Property	Value
Node ID	Ids
Imported Data	
Exported Data	
Settings	
Variables	
Decisions	
Output Type	View
Role	Raw
Rerun	No
Data	
Data Selection	Data Source
New Table	
Metadata	
Data Source	RESP
Table	RESP
Library	THEBOOK
Description	
Number of Observations	29904
Number of Columns	26
External Data	No
Sampling Rate	.
Segment	
Scope	Local
Status	
Time of Creation	5/27/06 1:55 PM
Run Id	
Last Error	
Last Status	
Needs Updating	No
Needs to Run	Yes
Time of Last Run	
Run Duration	
Grid Host	

A profit matrix is entered for the **Decisions** property of the **Input Data** node. This matrix is shown in Display 4.10.

Display 4.10

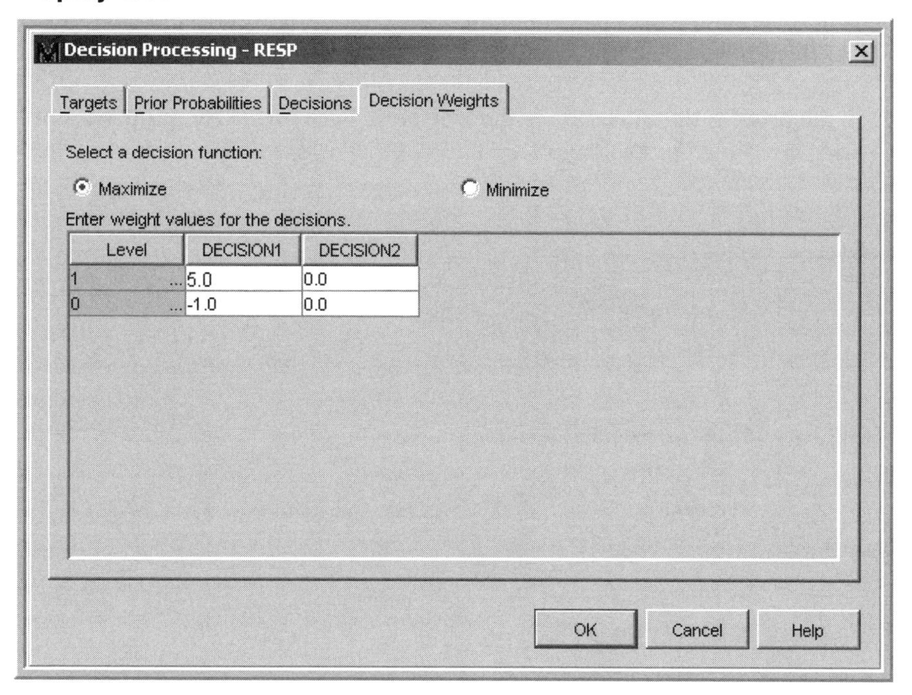

This matrix is used for making decisions about node classification and for making assessments of sub-trees.

The next node is the **Data Partition** node. Display 4.11 shows the settings of the properties of this node.

Display 4.11

Property	Value
Node ID	Part
Imported Data	...
Exported Data	...
Variables	...
Partitioning Method	Default
Random Seed	12345
Data Set Percentages	
Training	60.0
Validation	30.0
Test	10.0
Status	
Time of Creation	5/27/06 1:55 PM
Run Id	
Last Error	
Last Status	
Needs Updating	Yes
Needs to Run	Yes
Time of Last Run	
Run Duration	
Grid Host	

In this example, I allocated 60% of the data for training, 30% for validation, and 10% for testing. This allocation is quite arbitrary. It is possible to use different proportions and then to see how those proportions affect the results.

The third node in the process flow diagram is the **Decision Tree** node. The properties of the **Decision Tree** node are shown in Display 4.12.

Display 4.12

Property	Value
Node ID	Tree
Imported Data	...
Exported Data	...
Variables	...
Interactive	...
Splitting Rule	
Criterion	ProbChisq
Significance Level	0.2
Missing Values	Use in search
Use Input Once	No
Maximum Branch	2
Maximum Depth	6
Minimum Categorical Size	5
Node	
Leaf Size	5
Number of Rules	5
Number of Surrogate Rules	0
Split Size	
Split Search	
Exhaustive	5000
Node Sample	5000
Subtree	
Method	Assessment
Number of Leaves	1
Assessment Measure	Decision
Assessment Fraction	0.25
P-Value Adjustment	
Bonferroni Adjustment	Yes
Time of Kass Adjustment	Before
Inputs	No
Number of Inputs	1
Split Adjustment	Yes
Output Variables	
Variable Selection	Yes
Leaf Variable	Yes
Leaf Role	Segment
Performance	Disk
Status	
Time of Creation	5/27/06 1:56 PM

I kept the default values for all the properties. As a result, the nodes are split using the Chi-Square criterion. The threshold *p*-value for a split to be considered significant is set at 0.2. The sub-tree selection is done by comparing average profit since the **Sub Tree Method** property is set to **Assessment**, the **Assessment Measure** property is set to **Decision**, and a **profit matrix** is entered.

After running the **Decision Tree** node, you can open the **Results** window shown in Display 4.13.

Display 4.13

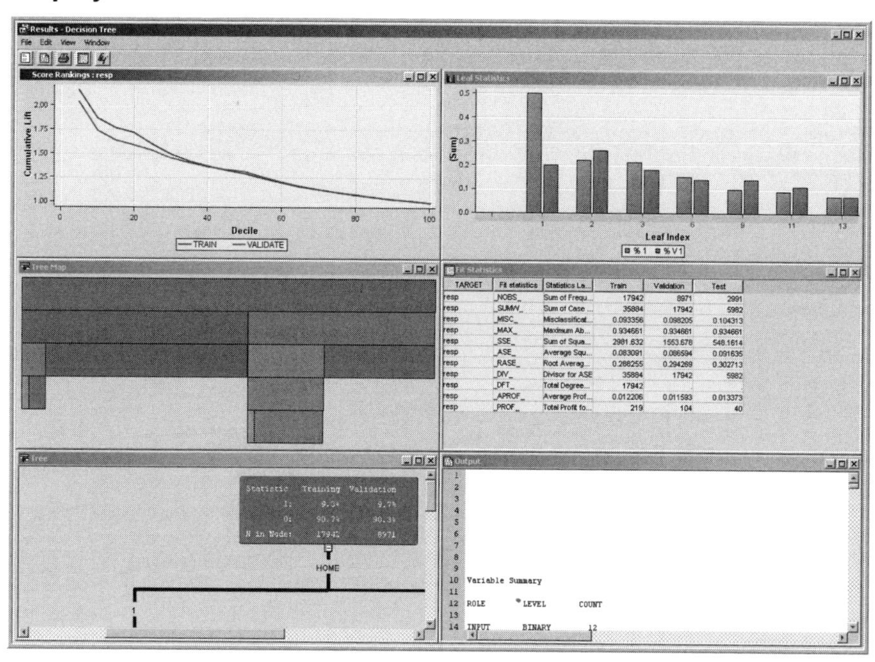

If you maximize the Tree pane in the **Results** window, you can see the tree as shown in Display 4.14.

Display 4.14

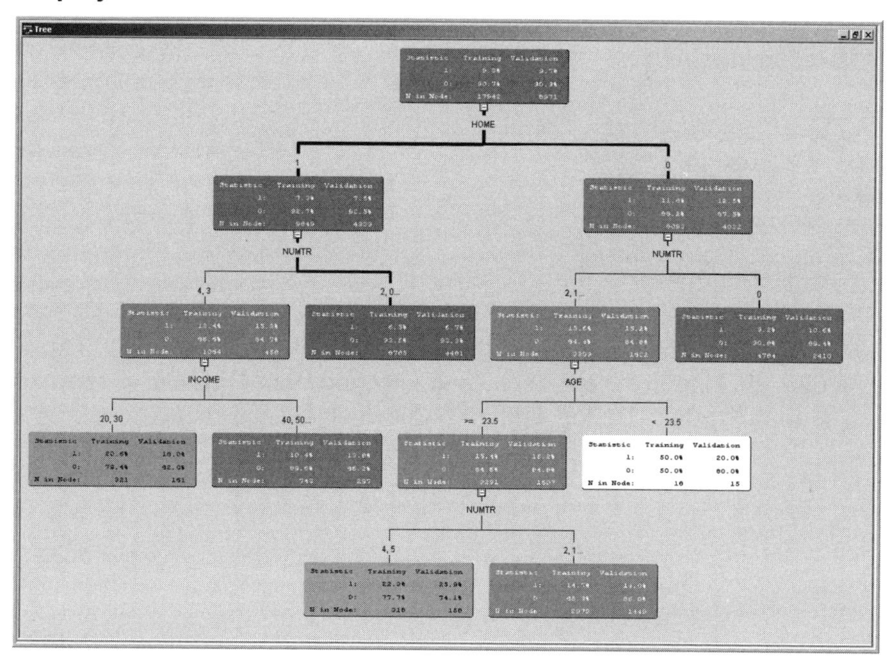

When you view the tree on your screen, the intensity of the color of the node indicates the response rate of the customers included in the node. By passing the mouse pointer over any node on the display, the statistics for that node will become more visible. In Display 4.14 the far right leaf has a white background (least intensity of color), which indicates the highest response rate. The customers in this node do not own a home (HOME = 0), they own one or two credit cards (NUMTR = 1, 2), and they are 23.5 years old or younger (AGE < 23.5).

The Cumulative Lift chart is shown in the **Score Rankings Overlay** pane in the top left-hand section of the **Results** window, as shown Display 4.15A.

Display 4.15A

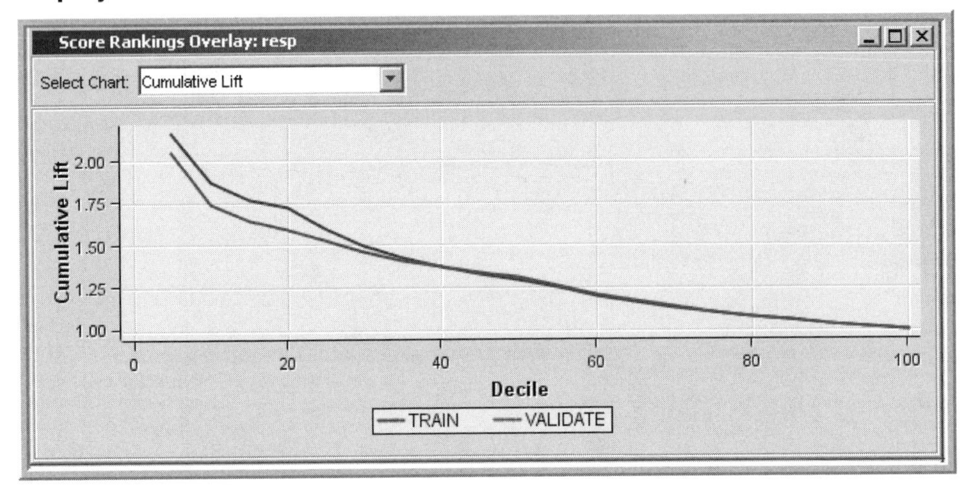

When you click on the arrow in the box next to the label **Select Chart**, you get a drop-down menu with a list of charts to choose from, as shown in Display 4.15B.

Display 4.15B

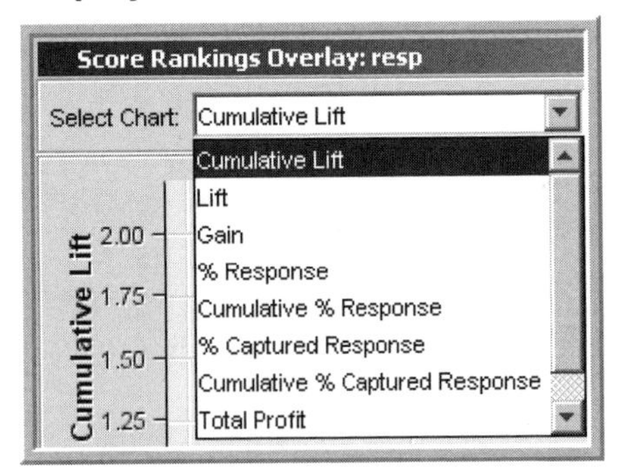

An alternative way of making charts is through the **Data Options Dialog** window. This window can be opened by right-clicking in the chart area, and selecting **Data Options**, as shown in Display 4.15C. The **Data Options Dialog** window is shown in Display 4.16.

Display 4.15C

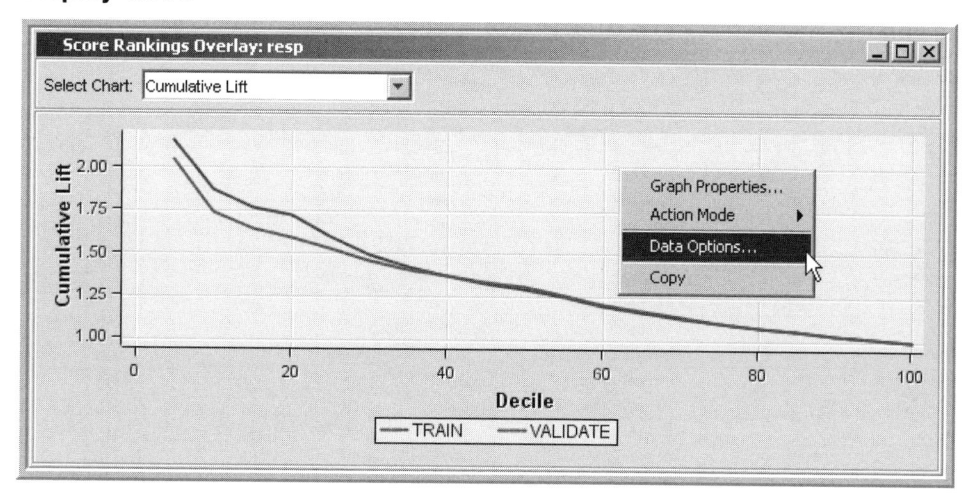

Display 4.16

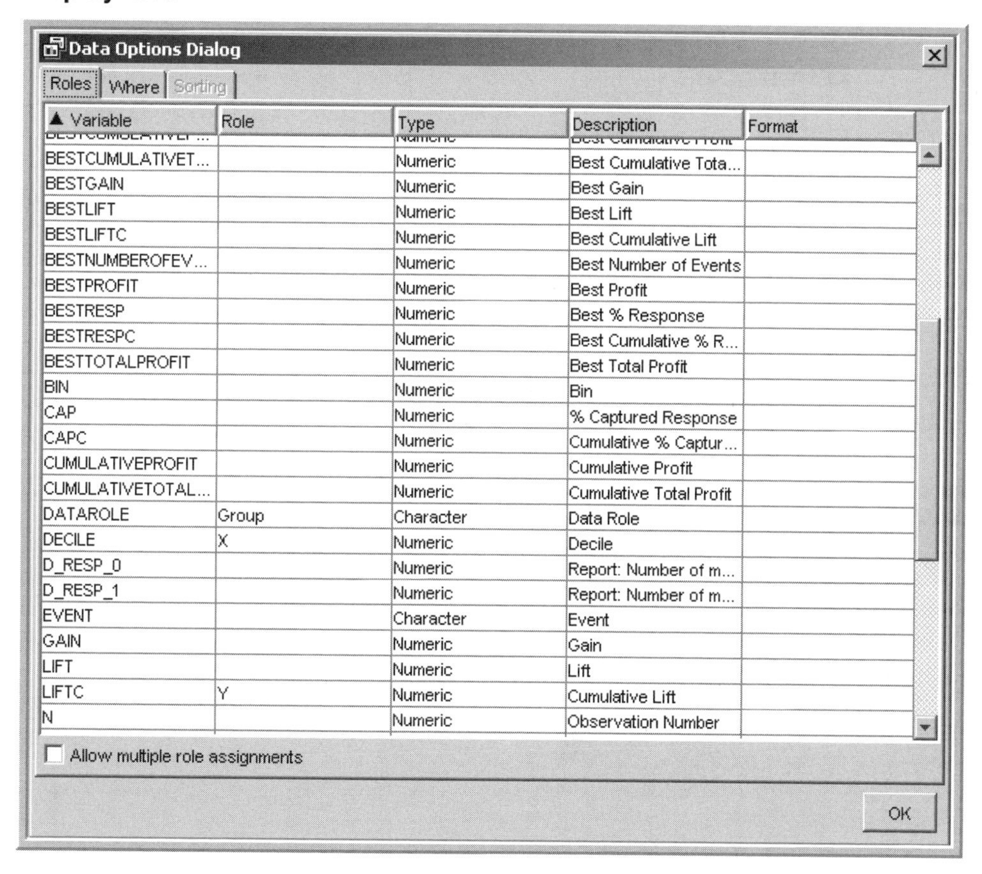

Variable	Role	Type	Description	Format
BESTCUMULATIVET...		Numeric	Best Cumulative Tota...	
BESTGAIN		Numeric	Best Gain	
BESTLIFT		Numeric	Best Lift	
BESTLIFTC		Numeric	Best Cumulative Lift	
BESTNUMBEROFEV...		Numeric	Best Number of Events	
BESTPROFIT		Numeric	Best Profit	
BESTRESP		Numeric	Best % Response	
BESTRESPC		Numeric	Best Cumulative % R...	
BESTTOTALPROFIT		Numeric	Best Total Profit	
BIN		Numeric	Bin	
CAP		Numeric	% Captured Response	
CAPC		Numeric	Cumulative % Captur...	
CUMULATIVEPROFIT		Numeric	Cumulative Profit	
CUMULATIVETOTAL...		Numeric	Cumulative Total Profit	
DATAROLE	Group	Character	Data Role	
DECILE	X	Numeric	Decile	
D_RESP_0		Numeric	Report: Number of m...	
D_RESP_1		Numeric	Report: Number of m...	
EVENT		Character	Event	
GAIN		Numeric	Gain	
LIFT		Numeric	Lift	
LIFTC	Y	Numeric	Cumulative Lift	
N		Numeric	Observation Number	

In Display 4.16, there is an X in the **Role** column of the variable DECILE, and Y in the **Role** column of the variable LIFTC. This means that the decile ranking is shown on the X-axis (horizontal axis) and the cumulative lift is shown on the Y-axis. With these settings you get the lift chart shown in the **Score Rankings** pane in Display 4.13. When you select different variables for the **X** and **Y** axis, you can get different charts. To make this selection, you have to click on the **Role** column corresponding to a variable, and select the role as shown in Display 4.17.

Display 4.17

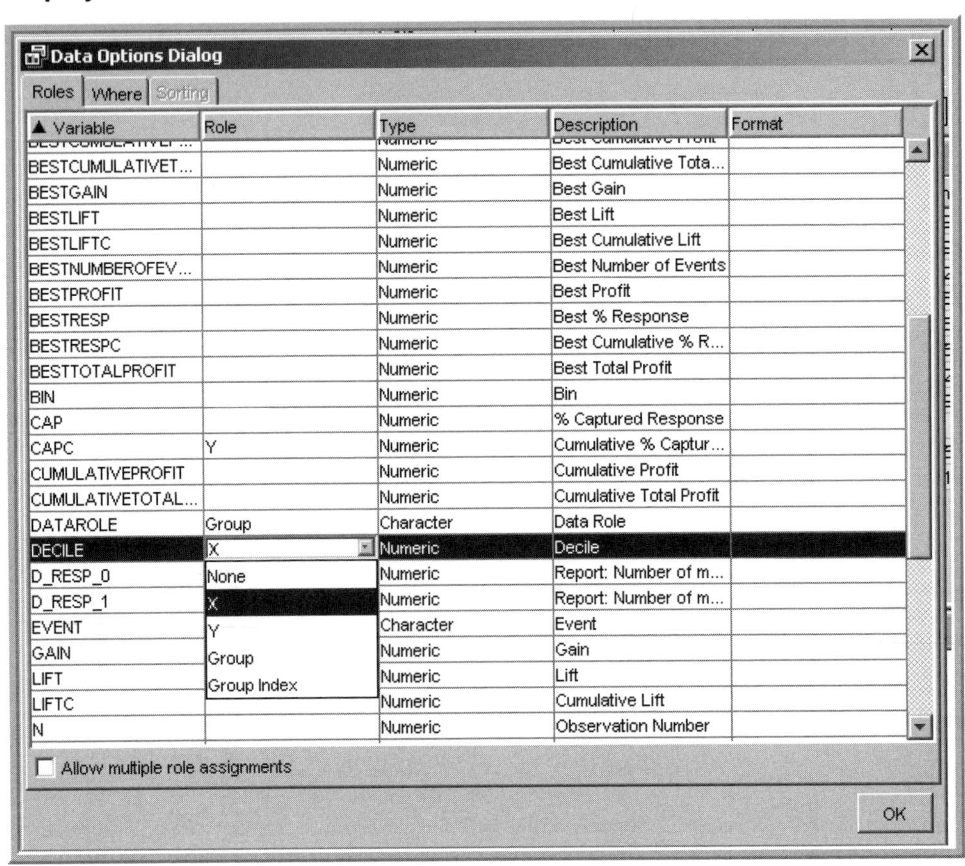

For example, by setting the **Role** of the variable CAPC to **Y** and the **Role** of **DECILE** to **X**, I get a **Cumulative % Captured Response** chart as shown in Display 4.18.

Display 4.18

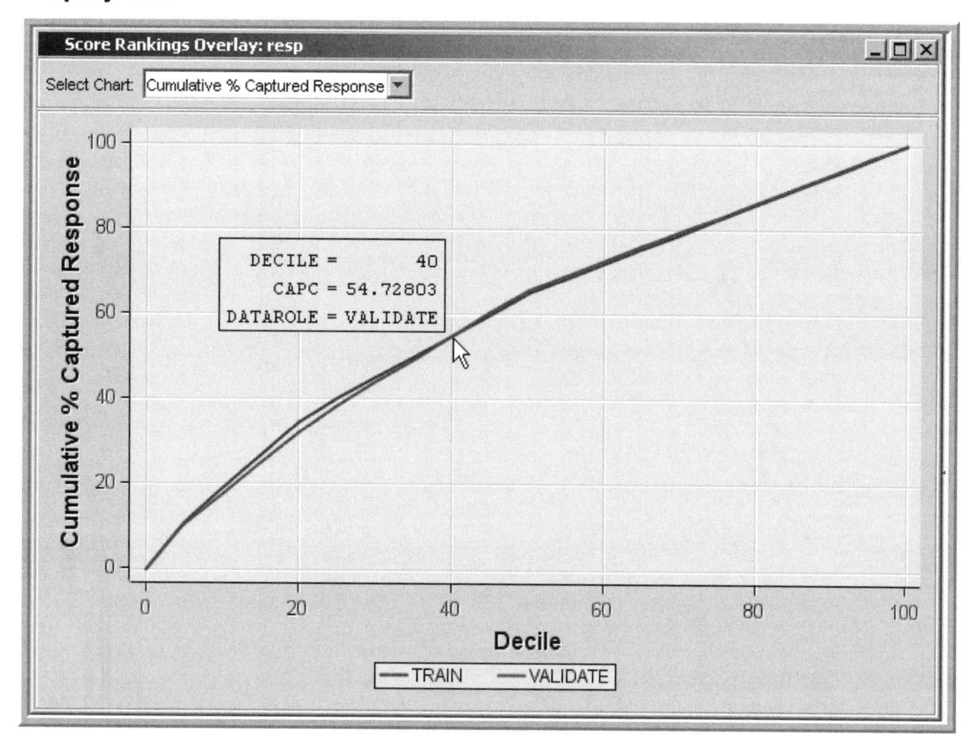

According to this chart, if I target the top 40% of the customers ranked by the predicted probability of response, then I will capture 54.7% percent of all responders. These capture rates are not impressive, but they are sufficient for demonstrating the tool.

As pointed out earlier, you can obtain the Cumulative Captured % Response chart directly by selecting it from the drop-down menu shown in Display 4.15B.

To get a view of the pruning process displayed by the **Iteration Plot**, select **View→Model→Iteration Plot** in the **Results** window, as shown in Display 4.19.

Display 4.19

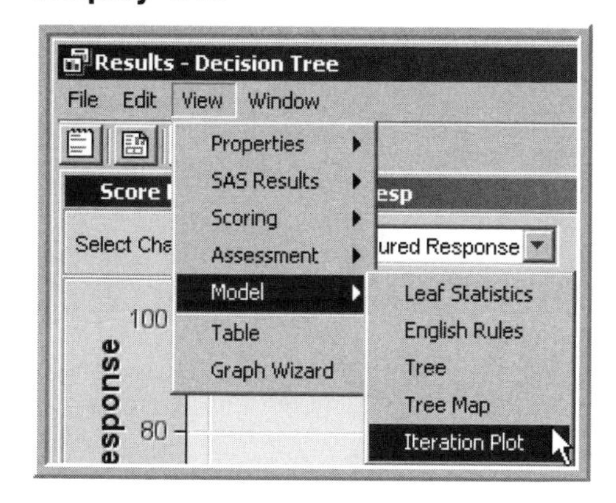

The plot generated by choosing the settings in Display 4.19 is shown in Display 4.20.

Display 4.20

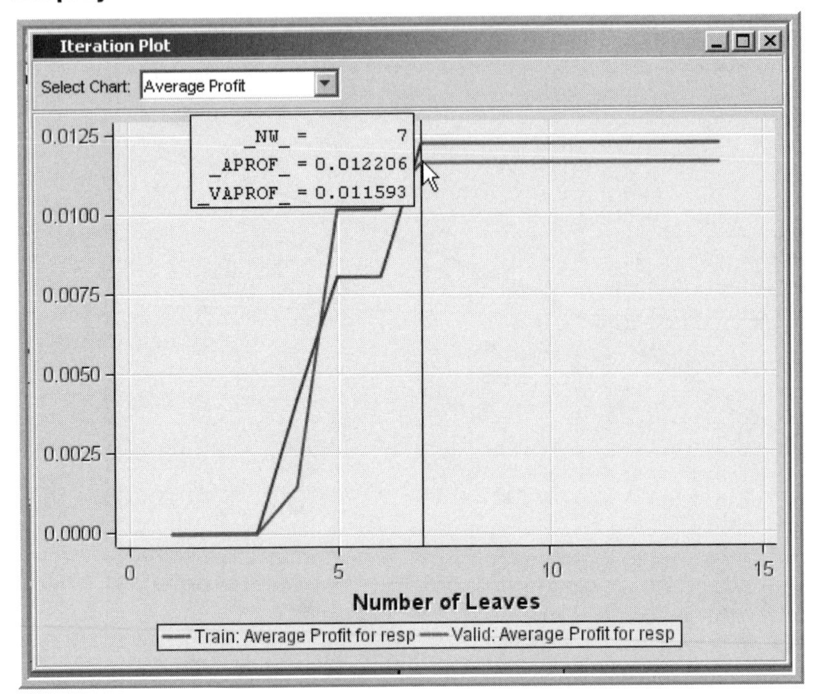

Display 4.20 shows the average profit for trees of different sizes. On the horizontal axis, the number of leaves is shown indicating the size of the tree. On the vertical axis, the average profit of the tree is shown. The calculation of the average profit is illustrated in Section 4.2.5 and in Step 2 of Section 4.3.5.

The graph shown in Display 4.20 depicts the pruning process. An initial tree with 14 leaves is grown with the training data set, and a sub-tree of only 7 leaves is selected using the validation data set. The selected tree is shown in Display 4.14.

The data behind these charts can be retrieved by clicking on the Table icon from the toolbar in the **Results** window. The data is stored in a SAS data set in the project directory. Output 4.1 is printed from the saved data set, and it shows the data underlying the graph shown in Display 4.20. The SAS code for generating Output 4.1 is given in Display 4.21.

Output 4.1

Number of Leaves	Train: Average Profit for resp	Valid: Average Profit for resp
1	0.000000	0.000000
2	0.000000	0.000000
3	0.000000	0.000000
4	0.004180	0.001449
5	0.008026	0.010144
6	0.008026	0.010144
7	0.012206	0.011593
8	0.012206	0.011593
9	0.012206	0.011593
10	0.012206	0.011593
11	0.012206	0.011593
12	0.012206	0.011593
13	0.012206	0.011593
14	0.012206	0.011593

Display 4.21

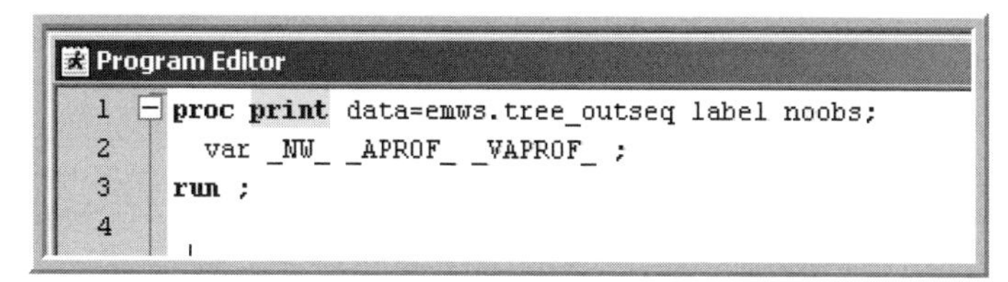

Output 4.1 and the graph in Display 4.20 show that the average profit reached its maximum at seven leaves for both the validation data and training data. This may not always be the case. The **Decision Tree** node selects the right-sized tree based on validation profits only.

Display 4.22 shows a partial list of the score code.

Display 4.22

```
******               ASSIGN OBSERVATION TO NODE              ******;
 DROP _ARB_P_;
 DROP _ARB_PPATH_; _ARB_PPATH_ = 1;

********** LEAF     1, NODE     8 ***************;
_ARB_P_      = 0;
_ARB_PPATH_  = 1;

 DROP _BRANCH_;
_BRANCH_ = -1;
_ARBFMT_12 = PUT( HOME , BEST12.);
 %DMNORMIP( _ARBFMT_12);
  IF _ARBFMT_12 IN ('1' ) THEN DO;
   _BRANCH_ =    1;
  END;
IF _BRANCH_ LT 0 THEN DO;
   IF MISSING( HOME  ) THEN _BRANCH_ = 1;
  ELSE IF _ARBFMT_12 NOTIN (
     '1' ,'0'
     ) THEN _BRANCH_ = 1;
END;
IF _BRANCH_ GT 0 THEN DO;

  _BRANCH_ = -1;
  _ARBFMT_18 = PUT( NUMTR , BEST12.);
   %DMNORMIP( _ARBFMT_18);
   IF _ARBFMT_18 IN ('4' ,'3' ,'7' ) THEN DO;
    _BRANCH_ =    1;
   END;

  IF _BRANCH_ GT 0 THEN DO;

    _BRANCH_ = -1;
    _ARBFMT_13 = PUT( INCOME , BEST12.);
     %DMNORMIP( _ARBFMT_13);
     IF _ARBFMT_13 IN ('20' ,'30' ) THEN DO;
      _BRANCH_ =    1;
     END;

    IF _BRANCH_ GT 0 THEN DO;
     _NODE_   =                 8;
     _LEAF_   =                 1;
     P_resp1  =     0.20560747663551;
     P_resp0  =     0.79439252336448;
     V_resp1  =     0.18012422360248;
     V_resp0  =     0.81987577639751;
     D_RESP   = '1' ;
     EP_RESP  =     0.23364485981308;
     END;
    END;
  END;
```

The segment of the code included in Display 4.22 shows how an observation is assigned to leaf node 1. Note that P_resp1 = 0.20560747663551 is computed from the training data set and it is called the *posterior probability of response.* P_resp0 = 0.794399252336448 is also computed from the training data set. It is called the *posterior probability of non-response.* These *posterior probabilities* are the same as $\hat{P}_1$ and $\hat{P}_0$ discussed in Section 4.2.2. Any observation that meets the definition of *leaf node* 1 is assigned these *posterior probabilities.* This, in essence, is how a decision tree model predicts the *probability of response.* This code can be applied to score the target population, as long as the data records in the prospect data set are consistent with the sample used for developing the model. Display 4.23 shows another segment of the score code.

Display 4.23

```
*****  DECISION VARIABLES *******;

*** Update Posterior Probabilities;

*** Decision Processing;
label D_RESP = 'Decision: resp' ;
label EP_RESP = 'Expected Profit: resp' ;

length D_RESP $ 9;

D_RESP = ' ';
EP_RESP = .;

*** Compute Expected Consequences and Choose Decision;
_decnum = 1; drop _decnum;

D_RESP = '1' ;
EP_RESP = P_resp1 * 5 + P_resp0 * -1;
drop _sum;
_sum = P_resp1 * 0 + P_resp0 * 0;
if _sum > EP_RESP + 2.273737E-12 then do;
   EP_RESP = _sum; _decnum = 2;
   D_RESP = '0' ;
end;


*** End Decision Processing ;
```

This code shows how decisions are made with regard to assigning an observation to a target level (such as response and non-response). The decision-making process is also discussed in Section 4.2.2 and Step 1C of Section 4.3.5.

The decision process is clear from the code segments in Displays 4.22 and 4.23. In any data set an observation is assigned to a node first, as shown in Display 4.22. Then, based on the posterior probabilities (P_resp1 and P_resp0) of the assigned node and the profit matrix, expected profits are computed under Decision1 and Decision2. The expected profit under Decision1 is given by the SAS statement `EP_RESP = P_resp1*5 + p_resp0*(-1)`. Note that the numbers 5 and 1 are from the first column of the profit matrix. If you classify a true responder as a responder then you make a profit of $5, and if you classify a non-responder as a responder, the profit is –$1.0. Hence, EP_RESP represents the consequences of classifying an observation as a responder. The alternative decision is to classify an observation as a non-responder, and this is given by the SAS statement `_sum = P_resp1*0 + P_resp0* 0`. The two numbers 0 and 0 are the second column of the profit matrix. The code shows that if _sum > EP_RESP, then the decision is to assign the target level 0 to the observation. The variable that indicates the decision is D_RESP, which takes the value 0 if the observation is assigned to the target level 0, or 1 if the observation is assigned to the target level 1. Since observations that are assigned to the same node will all have the same posterior probabilities, they are all assigned the same target level.

4.4.1 Testing Model Performance with a Test Data Set

I used the training data set for developing the node definitions and posterior probabilities, and I used the validation data set for pruning the tree to find the optimal tree. Now I need an independent data set on which to evaluate the model. The test data set is set aside for this assessment. I will connect the **Model Comparison** node to the **Decision Tree** node and run it. The **Results** window of the **Model Comparison** node shows the lift charts for all three data sets. These charts are shown in Display 4.24.

Display 4.24

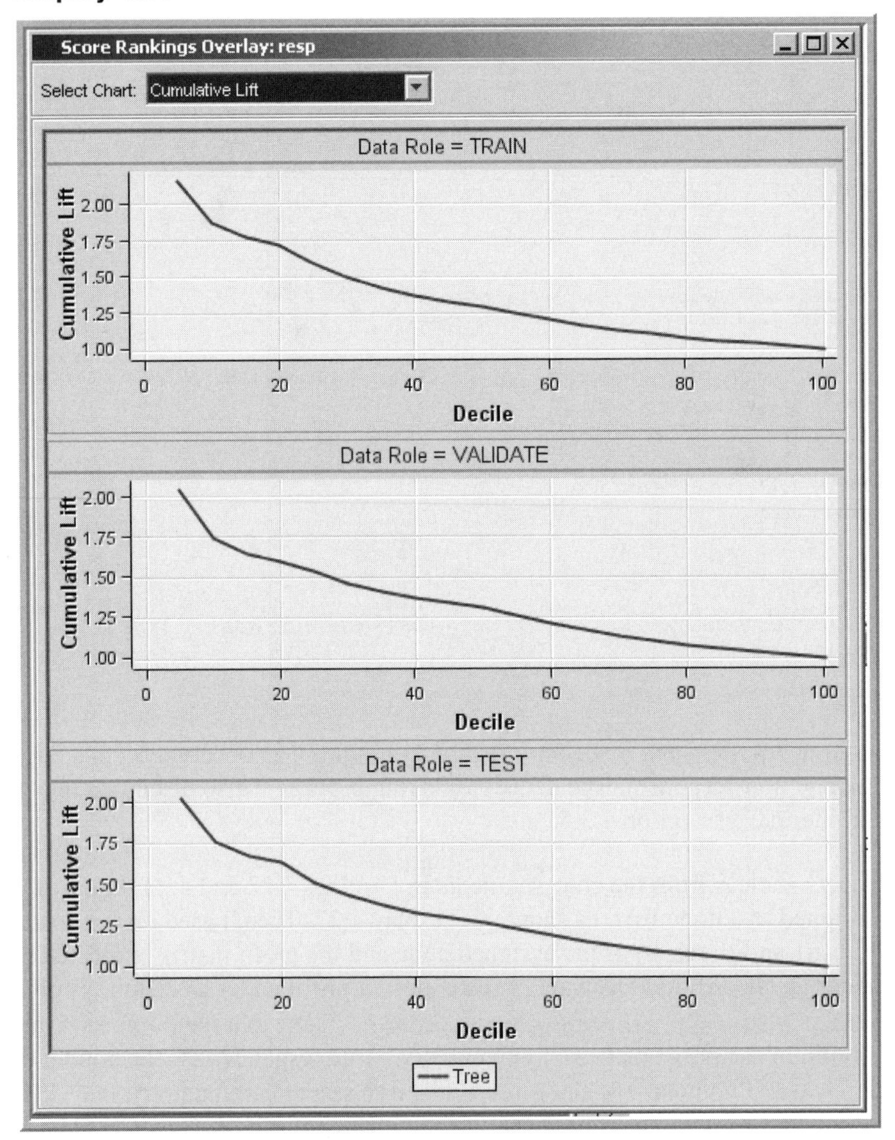

The performance of the model on the test data is very similar to its performance on the training and validation data sets. This comparison shows that the model is robust.

4.4.2 Applying the Decision Tree Model to Score a Data Set

This section shows the application of the decision tree model developed earlier to score a prospect data set. The records in the prospect data set do not have the value of the target variable. The target variable has two levels, response and no response. I will use the model to predict the probability of response for each record in the prospect data set.

I use the **Score** node to apply the model and make the predictions. I add an **Input Data** node for the prospect data and a **Score** node in the process flow as shown in Display 4.25.

Display 4.25

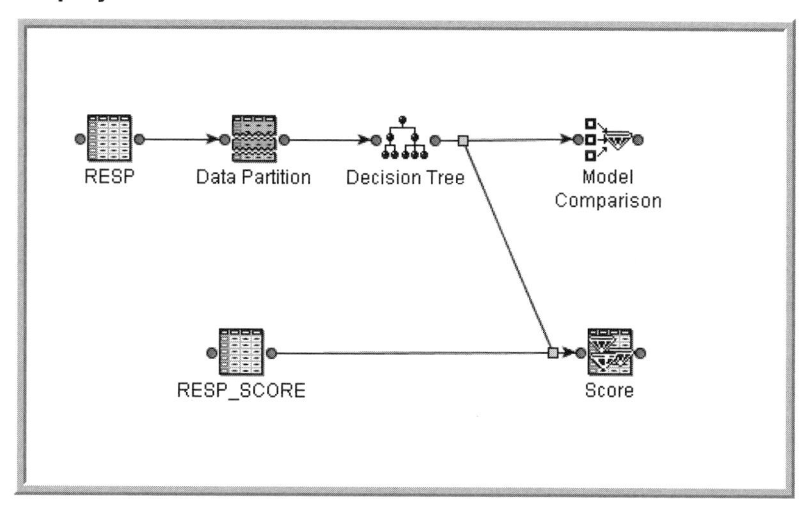

In Display 4.25, the data set that is being scored is named RESP_SCORE, and its role is set to Score. When the **Score** node is run, it applies the **Decision Tree** model to the data set RESP_SCORE, generates predictions of the target levels and their probabilities, and creates a new data set by appending the predictions to the prospect data set. Output 4.2 shows the appended variables in the new data set.

Output 4.2

Score Output Variables

NAME	FUNCTION	CREATOR	LABEL
D_RESP	DECISION	Tree	Decision: resp
EM_CLASSIFICATION	CLASSIFICATION	Score	Prediction for resp
EM_DECISION	DECISION	Score	Recommended Decision for resp
EM_EVENTPROBABILITY	PREDICT	Score	Probability for level 1 of resp
EM_PROBABILITY	PREDICT	Score	Probability of Classification
EM_PROFIT	ASSESS	Score	Expected Profit for resp
EM_SEGMENT	TRANSFORM	Score	Segment
EP_RESP	ASSESS	Tree	Expected Profit: resp
I_resp	CLASSIFICATION	Tree	Into: resp
P_resp0	PREDICT	Tree	Predicted: resp=0
P_resp1	PREDICT	Tree	Predicted: resp=1
U_resp	CLASSIFICATION	Tree	Unnormalized Into: resp
V_resp0	PREDICT	Tree	Validated: resp=0
V_resp1	PREDICT	Tree	Validated: resp=1
LEAF	TRANSFORM		Leaf
NODE	TRANSFORM	Tree	Node
WARN	ASSESS	Tree	Warnings

The predicted probabilities are P_resp1 and P_resp0. These are the *posterior probabilities* created in the training process. They are part of the decision tree model. The **Score** node gives the predicted values of the target levels based on profit maximization. These predicted values are stored as the variable EM_Decision. Display 4.27 shows the frequency distribution of the predicted target levels based on profit maximization.

Display 4.27

DECISIONS BASED ON PROFIT MAXIMIZATION

The FREQ Procedure

Frequency	Table of D_RESP by EM_DECISION		
	EM_DECISION(Recommended Decision for resp)		
D_RESP(Decision: resp)	0	1	Total
Decision 0: Profit under decision 1 < Profit under decision 0	28785	0	28785
Decision 1: Profit under decision 1 >= profit under decision 0	0	1119	1119
Total	28785	1119	29904

Display 4.27 shows that, under profit maximization, 1119 out of the total 29904 customers are designated as responders. The score code gives an alternative prediction of the target levels based on error minimization. If predicted probability of non-response is greater than the predicted probability of response for any record, then the variable EM_Classification is given the value 0 for that record. If the predicted probability of response is greater than the predicted probability of non-response for a given record, then the variable EM_Classification for that record is given the value 1. Thus, the variable EM_Classification indicates whether a record is designated as a responder or a non-responder record. Display 4.28 shows that, in the data set, the variable EM_Classification takes the value 0 for 29867 records, and it takes the value 1 for 37 records. Thus, according to the decisions based on error minimization, only 37 customers (records) are designated as responders, while 29867 customers are designated as non-responders.

The frequency distribution of EM_Classification is shown in Display 4.28.

Display 4.28

DECISIONS BASED ON MAXIMUM POSTERIOR PROBABILITY

The FREQ Procedure

Frequency	Table of CL by EM_CLASSIFICATION		
	EM_CLASSIFICATION(Prediction for resp)		
CL	0	1	Total
Decision 0: P_resp1 < P_resp0	29867	0	29867
Decision 1: P_resp1 >= P_resp0	0	37	37
Total	29867	37	29904

The program that generated Displays 4.27 and 4.28 is shown in Display 4.29.

Display 4.29

```
proc format ;
 value clfmt
1 = 'Decision 1: P_resp1 >= P_resp0'
0 = 'Decision 0: P_resp1 <  P_resp0 ' ;
 value $pfmt
'1' = 'Decision 1: Profit under decision 1 >= profit under decision 0'
'0' = 'Decision 0: Profit under decision 1 <  Profit under decision 0 '
;
run ;
ods html file='C:\TheBook\EM_5.2\Nov2006\Ch4_Dec1.html' ;
proc freq data=emws.score_SCORE ;
  table D_resp*EM_Decision / nopercent norow nocol  ;
  Title 'DECISIONS BASED ON PROFIT MAXIMIZATION' ;
 format d_resp $pfmt. ;
run ;
ods html close ;

data temp ;
  set emws.score_score ;
 CL =  p_resp1 ge p_resp0 ;
run ;
ods html file ='C:\TheBook\EM_5.2\Nov2006\Ch4_Dec2.html' ;
proc freq data=temp ;
  table CL*EM_Classification / nopercent norow nocol ;
  format CL clfmt. ;
  Title 'DECISIONS BASED ON MAXIMUM POSTERIOR PROBABILITY' ;
run ;
ods html close ;
```

4.5 Developing a Regression Tree Model to Predict Risk

In this section, I develop a decision tree model to predict *risk.* In general, *risk* is defined as the expected amount of loss to the company. An auto insurance company incurs losses resulting from such events as accidents or thefts. An auto insurance company is often interested in predicting the expected *loss frequency* of its customers or prospects. The insurance company can set the premium for a customer according to the predicted *loss frequency. Loss frequency* is measured as the number of *losses* per car year, where car year is defined as the duration of the insurance policy multiplied by the number of cars insured under the policy. If the insurance policy has been in effect for four months and one car is covered, then the number of car years is 0.3333. If there is one loss during this four months, then the *loss frequency* is 1/ (4/12) = 3. If, during these four months, two cars are covered under the same policy, then the number of car years is 0.6666; and if there is one loss, the *loss frequency* is 1.5.

The target variable in this model is the *loss frequency*, and is continuous. When the target variable is continuous, the tree constructed is called a *regression tree.* A *regression tree*, like a decision tree, partitions the records (customers) into segments or *nodes*. The *regression tree model* consists of the *leaf nodes*. While posterior *probabilities* are calculated for each *leaf node* in a *decision tree model, the expected value* or *mean* of the target variable is calculated for each leaf node in a *regression tree model*. In the current example, the tree calculates the expected frequency of losses for each *leaf node* or *segment*. These target means are calculated from the training data.

The process flow for modeling risk is shown in Display 4.30.

Display 4.30

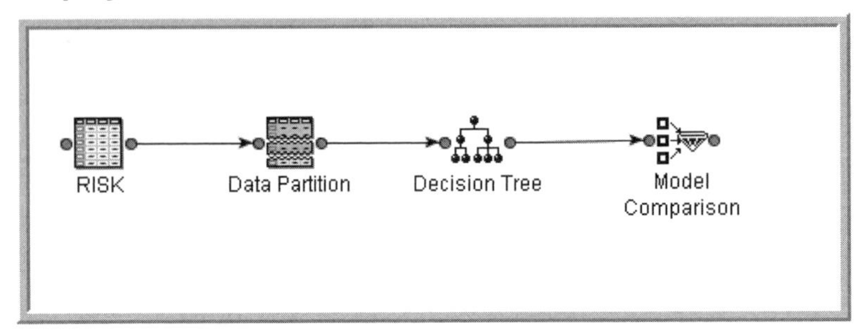

The property settings of the **Decision Tree** node are shown in Display 4.31

In this example, the modeling data set is partitioned such that 60% of the records are allocated for training, 30% for validation, and 10% for testing. The selection method used for this partitioning is simple random sampling.

Display 4.31

Property	Value
Node ID	Tree
Imported Data	...
Exported Data	...
Variables	...
Interactive	...
Splitting Rule	
Criterion	ProbF
Significance Level	0.2
Missing Values	Use in search
Use Input Once	No
Maximum Branch	2
Maximum Depth	6
Minimum Categorical Size	5
Node	
Leaf Size	5
Number of Rules	5
Number of Surrogate Rules	0
Split Size	
Split Search	
Exhaustive	5000
Node Sample	5000
Subtree	
Method	Assessment
Number of Leaves	1
Assessment Measure	Average Square Error
Assessment Fraction	0.25
P-Value Adjustment	
Bonferroni Adjustment	Yes
Time of Kass Adjustment	Before
Inputs	No
Number of Inputs	1
Split Adjustment	Yes
Output Variables	
Variable Selection	Yes
Leaf Variable	Yes
Leaf Role	Segment

The main difference between trees with binary targets and trees with continuous targets is in the test statistics used for evaluating the splits and the calculation of the worth of the trees. For the continuous target, the test statistic has an F-distribution. The model assessment is done by using the Error Sums of Squares. The **Splitting Criterion** property is **ProbF**, which is the default for interval targets. The **Assessment Measure** property for assessment is set to **Average Square Error**. Section 4.3.1.4 discusses the **ProbF** criterion for calculating the worth of a split. Section 4.3.8 discusses how Average Square Error is calculated.

Displays 4.32A through 4.32C show the tree model developed by the **Decision Tree** node.

Display 4.32A

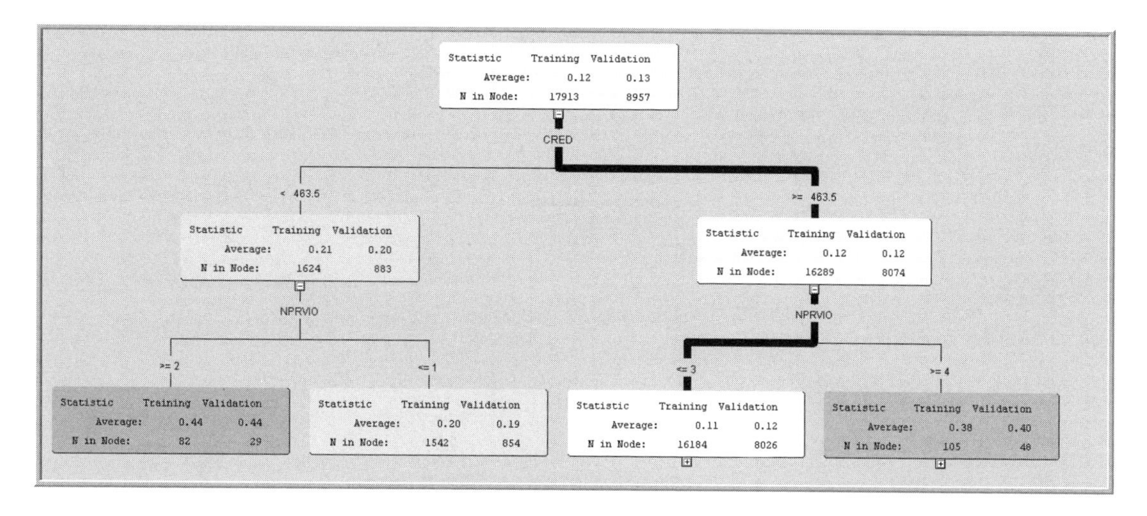

Display 4.32B

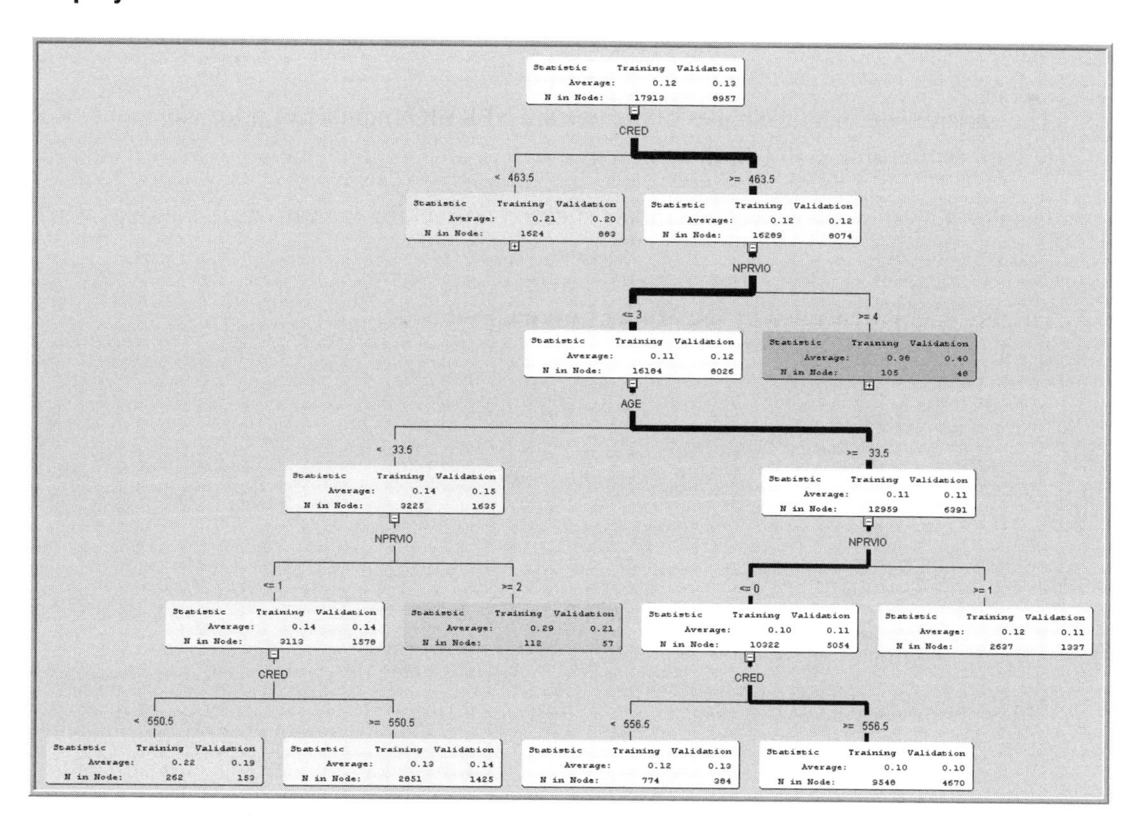

Display 4.32C

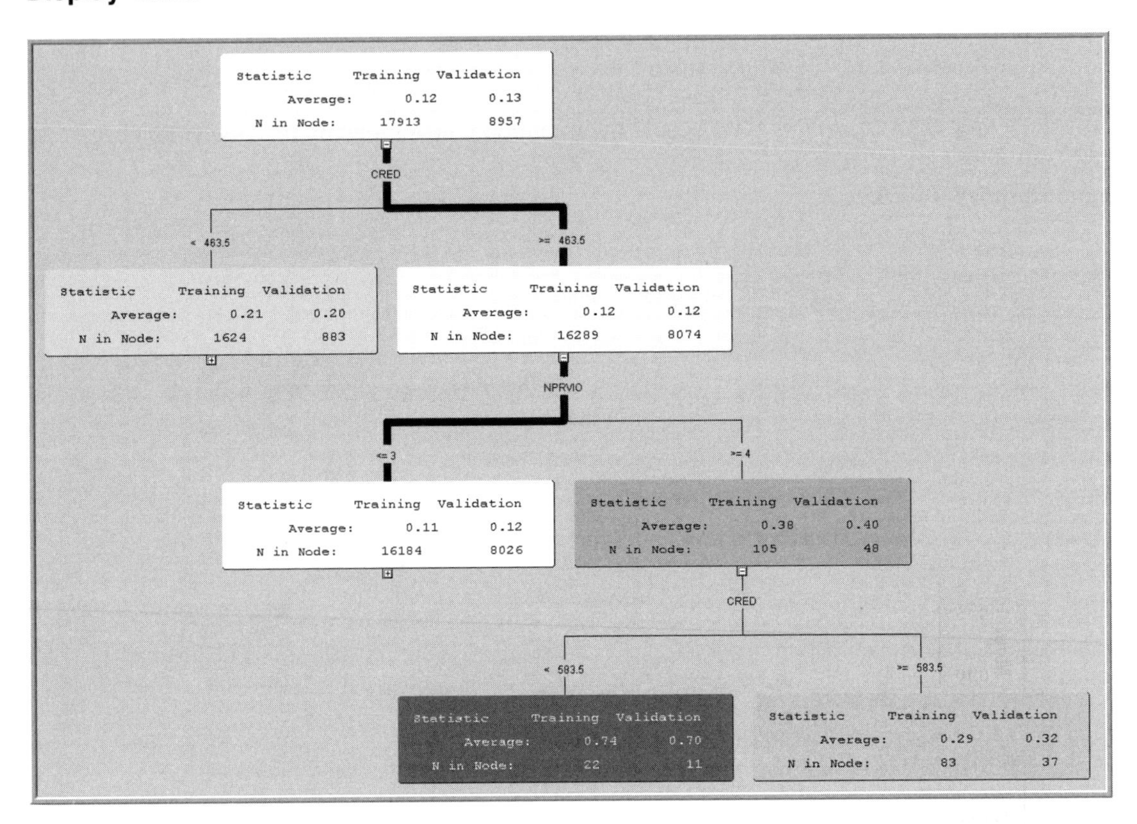

The variables defining the nodes of the tree are NPRVIO (number of prior violations), CRED (Credit score), and AGE (Age).

Display 4.33 shows how well the model succeeds in ranking the various risk groups when applied to a test data set.

Display 4.33 is generated by the **Model Comparison** node.

Display 4.33

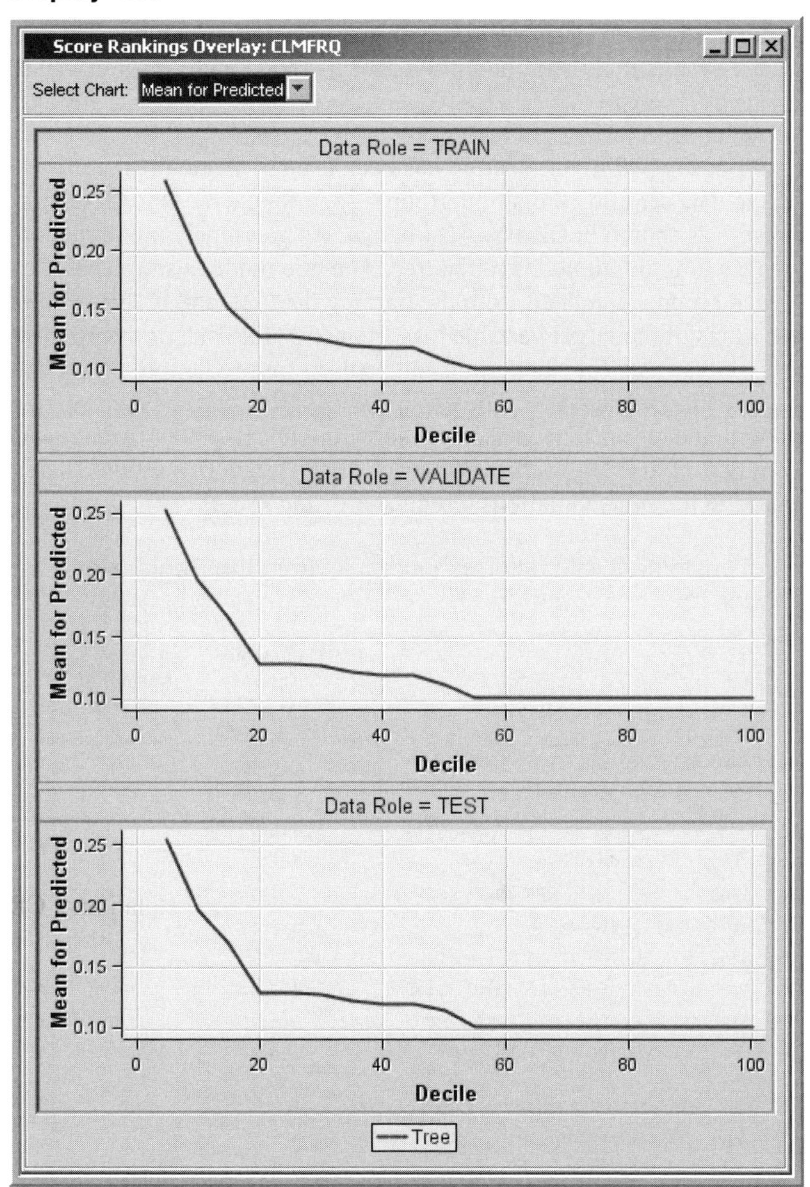

The title of the **Results** window shown in Display 4.33 is **Score Rankings Overlay: CLMFRQ**. CLMFRQ, which measures *loss frequency*, is the target, and is measured as a continuous variable.

The deciles computed from ordered values of the predicted *loss frequency* are shown on the horizontal axis, with those customers with the highest loss frequency in the first decile and those with the lowest losses in the tenth decile. The vertical axis shows the mean of actual *loss frequency* for each decile.

4.5.1 Summary of the Regression Tree Model to Predict Risk

Although the regression tree shown in Displays 4.32A through 4.32C is developed from simulated data, it provides a realistic illustration of how this method can be used to identify the profiles of high-risk groups.

The regression tree model developed in this section has been used to group the records (in this case, the customers) in the data set into 10 disjoint groups—one for each of the 10 leaf nodes or terminal nodes on the tree—as shown in Display 4.34 below. As previously mentioned, these groups are called the leaves or terminal nodes of the tree. The tree model also reports the mean of the target variable for each group, calculated from the training data set, and in this example, these means are used as predictions of the target variable loss frequency for each of the 10 groups. Since the mean of the target variable for each group can be used for prediction, I have referred to it as *Expected loss frequency* in Display 4.34. The expected loss frequency of a group reflects the level of risk associated with the group. The groups are shown in Display 4.34 in decreasing order of their risk levels. Display 4.34 also shows the profiles of the different risk groups, that is, their characteristics with regard to previous violations, credit score, and age.

Display 4.34 Profiles of customers with different risk ranks from the Regression Tree model

Risk Rank	Leaf ID	Node ID	Input ranges	Mean of the Target (Expected loss frequency)	Number of records
1	9	14	463.5 ≤ CRED < 583.5, 4 ≤ NPRVIO ≤ 7	0.74324	22
2	2	5	2 ≤ NPRVIO ≥ 7, CRED < 463.5	0.43836	82
3	5	21	2 ≤ NPRVIO ≤ 3, 463.5 ≤ CRED, AGE < 33.5	0.28802	21
4	10	15	583.5 ≤ CRED, 4 ≤ NPRVIO ≤ 7	0.28551	83
5	3	24	463.5 ≤ CRED < 550.5, NPRVIO ≤ 1, AGE < 33.5	0.21925	262
6	1	4	NPRVIO ≤1, CRED < 463.5	0.19754	1542
7	4	25	550.5 ≤ CRED , NPRVIO ≤ 1, AGE < 33.5	0.12926	2851
8	6	28	463.5 ≤ CRED < 556.5, NPRVIO = 0, 33.5 ≤ AGE	0.12334	774
9	8	23	1 ≤ NPRVIO ≤ 3, 33.5 ≤ AGE, 463.5 ≤ CRED	0.11973	2637
10	7	29	556.5 ≤ CRED, NPRVIO = 0, 33.5 ≤ AGE	0.10243	9548
				Total records	*17822*
			NPRVIO = Number of previous violations		
			CRED = Credit Score		
			AGE = Age of the insured		

In an effort to prevent unfair or discriminatory practices by insurance companies, many of the state agencies that oversee the insurance industry have decreed that certain variables cannot be used for risk rating or rate setting or both. The variables that appear in Display 4.34 are not necessarily legally permitted variables. The table is given only for illustrating the regression tree model.

4.6 Summary

- *Growing a decision tree* means arriving at certain rules for segmenting a given data set, stated in terms of ranges of values for inputs, and assigning a target class[3] (such as response or non-response) to each segment that is created.
- The *training data set* is used for developing the rules for the segmentation, for calculating the *posterior probabilities* of each segment (or the target mean of each segment when the target variable is continuous), and for assigning each segment to a target class when the target is categorical.
- Node definitions, target means, posterior probabilities, and the assignment of the target levels to nodes do not change when the tree is applied to the validation, or prospect, data set.
- Enterprise Miner offers several criteria for splitting a node; you can select the criterion suitable for the type of model being developed by setting the **Splitting Rule Criterion** property to an appropriate value.
- In general, there are a number of sub-trees within the initial tree that need to be examined during the tree development. The *training data* set is used to create this initial tree, while the *validation data set* is used to select the best sub-tree by applying an assessment measure that you choose. Several measures of assessment are available in Enterprise Miner, and you can select one of them by setting the **Sub-tree Assessment Measure** property to a certain value. Selecting the best sub-tree is often referred to as *pruning*.
- Enterprise Miner offers different methods for selecting a sub-tree: instead of applying an assessment measure to select a sub-tree, you can also select the tree with the maximum number of leaves, or the tree with a specified number of leaves (regardless of whether or not the tree has maximum profit, or minimum cost, or meets any other criterion of assessment).
- A *decision tree model* consists finally of only the *leaf* nodes (also known as the terminal nodes) of the selected tree, the attributes of which can be used for scoring prospect data, etc.
- The definition of the leaf nodes, the target classes assigned to the nodes and their posterior probabilities can all be seen in the SAS code (score code) generated by Enterprise Miner. The SAS code can be exported if the scoring is to be applied to external data.
- The two models developed in this chapter, one for a binary target and one for a continuous target, serve as examples of how to use the **Decision Tree** node: how to interpret the results given in the **Output** window and how to interpret the various types of graphs and tables associated with each of them.
- The regression tree model developed for predicting risk is used to rank the customers according to their degree of risk. Each risk group is then profiled and represented as an exclusive segment of the tree model.

---

[3] The terms *target class* and *target level* are used synonymously.

4.6.1 Extensions

The **Decision Tree** node can be used for variable selection by setting the **Variable Selection** property to **Yes**. The selected variables can be passed to another modeling tool, such as the **Regression** node. In addition, a categorical input that indicates the leaf to which the observation is assigned can also be created by setting the **Leaf Variable** property to **Yes**. These extensions are not demonstrated in this chapter, but you are encouraged to experiment with them.

4.7 Appendix to Chapter 4

4.7.1 Pearson's Chi-Square Test

As an illustration of the Chi-Square test, consider a simple example. Suppose there are 100 cases in the parent node (τ), of which 10 are responders and 90 are non-responders. Suppose this node is split into two child nodes, A and B. Let A have 50 cases, and let B have 50 cases. Also, suppose that there are 9 responders in child node A, and there is 1 responder in child node B. The parent node has 10 responders and 90 non-responders.

Here is a 2 x 2 contingency table representing the split of parent node τ :

	Child node A	**Child node B**
Responders	9	1
Non-responders	41	49

The null hypothesis is that the child nodes A and B are not different from their parent nodes in terms of class composition (i.e., their mix of *responders* and *non-responders*). Under the null hypothesis, each child node is similar to the parent node, and contains 10% *responders* and 90% *non-responders*.

Expected frequencies under the null hypothesis are as follows:

	Child node A	**Child node B**
Responders	5	5
Non-responders	45	45

$$\chi^2 = \sum \frac{(O-E)^2}{E} = \frac{(4)^2}{5} + \frac{(-4)^2}{45} + \frac{(-4)^2}{5} + \frac{(4)^2}{45} = 7.1$$

A Chi-Square value of 7.1 implies the following values for p-value and for logworth: p-value = 0.0076655, and *logworth* = $\log_{10}(p\text{-value}) = 2.11546$.

The *p*-value is the probability of the Chi-Square statistic taking a value of 7.1 or higher when there is no difference in the child nodes. In this case the *p*-value is very small, 0.0076655. This indicates in this case the child nodes do differ significantly with respect to their composition of responders and non-responders.

4.7.2 Adjusting the Predicted Probabilities for Over-sampling

When modeling rare events, it is common practice to include all the records having the rare event, but only a fraction of the non-event records in the modeling sample. This practice is often referred to as over-sampling, case-control sampling, or biased sampling.

For example, in response modeling, if the response rate is very low, you can include in the modeling sample all the responders available and only a randomly selected fraction of non-responders. The bias introduced by such over-sampling is corrected by adjusting the predicted probabilities with prior probabilities. Display 2.18 in Chapter 2 shows how to enter the prior probabilities.

Adjustment of the predicted probabilities can be illustrated by a simple example. Suppose that, based on a biased sample the *unadjusted* probabilities of response and no-response calculated by the decision tree process for a particular node T are $\hat{P}_{1T}$ and $\hat{P}_{0T}$. If the prior probabilities, that is, the proportion of responders and non-responders in the entire population, are known to be π_1 and π_0, and if the proportion of responders and the non-responders in the biased sample are ρ_1 and ρ_0, respectively, then the *adjusted* predictions of the probabilities are given by

$$P_{1T} = \frac{\hat{P}_{1T}(\pi_1 / \rho_1)}{\hat{P}_{0T}(\pi_0 / \rho_0) + \hat{P}_{1T}(\pi_1 / \rho_1)}, \text{ and}$$

$$P_{0T} = \frac{\hat{P}_{0T}(\pi_0 / \rho_0)}{\hat{P}_{0T}(\pi_0 / \rho_0) + \hat{P}_{1T}(\pi_1 / \rho_1)},$$

where P_{1T} is the adjusted probability of response, and P_{0T} is the adjusted probability of no-response. It can be easily verified that $P_{1T} + P_{0T} = 1$. If there are n_{1T} responders and n_{oT} non-responders in node T prior to adjustment, then the adjusted number of responders and non-responders will be $\left(\pi_1 / \rho_1\right) n_{1T}$ and $\left(\pi_0 / \rho_0\right) n_{0T}$, respectively.

The effect of adjustment on the number of responders and non-responders, and on the expected profits for node T can be seen from a numerical example. Let the proportions of response and non-response for the entire population be $\pi_1 = 0.05$ and $\pi_0 = 0.95$ while the proportions for the biased sample are $\rho_1 = 0.5$ and $\rho_0 = 0.5$. If the predicted proportions for node T are $\hat{P}_{1T} = 0.4$ and $\hat{P}_{0T} = 0.6$, and if the predicted number of responders and non-responders are $n_{1T} = 400$ and $n_{0T} = 600$, then the adjusted probability of response (P_{1T}) is equal to 0.0339, and the adjusted probability of no-response (P_{0T}) equals 0.9661. In addition, the adjusted number of responders is 40, and the adjusted number of non-responders is 1140. One more measure that might need adjustment in the case of a sample biased due to over-sampling is expected profits. To show expected profits under alternative decisions with adjusted and unadjusted probabilities, I will use the profit matrix given in Table 4.8 below.

Table 4.8

	Decision1	**Decision2**
Actual target level/class		
1	$10	-$2
0	-$1	$1

4.7.3 Expected Profits Using Unadjusted Probabilities

Using the unadjusted predicted probabilities in my example, under Decision1 the expected profits are ($10*0.4) + (–$1*0.6) = $3.4. Under Decision2 the expected profits are (–$2*0.4) + ($1*0.6) = –$0.2.

4.7.4 Expected Profits Using Adjusted Probabilities

Using the adjusted probabilities, under Decision1 the expected profits are ($10*0.0339) + (–$1*0.9661) = –$0.6271. Under Decision2 the expected profits are (–$2*0.0339) + ($1*0.9661) = $0.8983.

Neural Network Models to Predict Response and Risk

5.1 Introduction

I now turn to using SAS Enterprise Miner to derive and interpret neural network models. This chapter develops two neural network models using simulated data. The first model is a response model with a binary target, and the second is a risk model with an ordinal target.

In this chapter:

- A general example of a *multilayer perceptron* (MLP) is presented.
- Calculation of the outputs of the units in each layer of the network is described by means of the underlying formulas.
- A general description of how the **Neural Network** node estimates the weights in the case of a binary target is given.
- Using simulated data, a neural network model is developed to predict response, and the results are discussed. The model is evaluated using a test data set.

- The response model is used to score a prospect or score a data set.
- The score code generated by the **Neural Network** node is examined to show how the probabilities in the target layer are computed, how these probabilities (also called posterior probabilities) are updated, and how the decision of assigning a target level (class) to each customer or record in the prospect data set is made.
- A risk model is developed using the **Neural Network** node to predict an ordinal variable. The theoretical underpinnings behind the chosen target layer activation function are discussed in detail.
- The SAS code generated by the **Neural Network** node is examined step-by-step to make sure that the probabilities calculated in the target layer activation function are consistent with the theoretical specification.
- Lift charts generated by the **Neural Network** node based on the probability of the highest level of the target variable are examined. An alternative way of calculating the lift charts based on expected loss frequency is presented to demonstrate the various data sets created by the **Neural Network** node and to show how to access them to make custom reports.
- A data set is scored using the risk model that was developed, and the score code is examined to verify that what is actually done by the **Neural Network** node is consistent with what is expected of the risk model.
- Alternative architectural specifications, which produce alternative neural network models, are examined, using a binary target as an example.

This chapter develops two neural network models using simulated data. The first model is a *response model* with a binary target, and the second is a *risk model* with an ordinal target.

5.1.1 Target Variables for the Models

The target variable for the response model is $RESP$, which takes the value of 0 if there is no response and the value of 1 if there is a response. The neural network produces a model to predict the probability of response based on the values of a set of input variables for a given customer. The output of the **Neural Network** node gives probabilities $\Pr(RESP = 0 \mid X)$ and $\Pr(RESP = 1 \mid X)$, where X is a given set of inputs.

The target variable for the risk model is a discrete version of *loss frequency,* defined as the number of losses or accidents per car-year, where car-year is defined as the duration of the insurance policy multiplied by the number of cars insured under the policy. If the policy has been in effect for four months and one car is covered under the policy, then the number of car-years is 4/12 at the end of the fourth month. If one accident occurred during this four-month period, then the loss frequency is approximately $1/(4/12) = 3$. If the policy has been in effect for 12 months covering one car, and if one accident occurred during that 12-month period, then the loss frequency is $1/(12/12) = 1$. If the policy has been in effect for only four months, and if two cars are covered under the same policy, then the number of car-years is (4/12) x 2. If one accident occurred during the four-month period, then the loss frequency is (1 / (8/12) = 1.5. Defined this way, *loss frequency* is a continuous variable. However, the target variable *LOSSFRQ*, which I will use in the model developed here, is a discrete version of *loss frequency* defined as follows:

$$LOSSFRQ = 0 \textit{ if loss frequency } = 0,$$
$$LOSSFRQ = 1 \textit{ if } 0 < \textit{loss frequency} < 1.5, \textit{ and}$$
$$LOSSFRQ = 2 \textit{ if loss frequency } \geq 1.5.$$

The three levels 0, 1, and 2 of the variable *LOSSFRQ* represent low, medium, and high risk, respectively.

The output of the **Neural Network** node gives:

$$\Pr(LOSSFRQ = 0 | X) = \Pr(loss\ frequency = 0 | X),$$
$$\Pr(LOSSFRQ = 1 | X) = \Pr(0 < loss\ frequency < 1.5 | X),$$
$$\Pr(LOSSFRQ = 2 | X) = \Pr(loss\ frequency \geq 1.5 | X).$$

5.1.2 Neural Network Node Details

I now specify the properties of the **Neural Network** node such that the target variable *LOSSFRQ* is treated as an ordinal variable and such that the model developed is of the *cumulative logits* type, also referred to as the *proportional odds* model. A detailed explanation of the proportional odds model is given in Section 5.5.1.1.

Neural network models use mathematical functions to map inputs into outputs. When the target is categorical, the outputs of a neural network are the probabilities of the target levels. The neural network model for risk (in the current example) gives formulas to calculate the probability of the variable *LOSSFRQ* taking each of the values 0, 1, or 2, whereas the neural network model for response provides formulas to calculate the probability of response and non-response. Both of these models use the complex nonlinear transformations available in the **Neural Network** node of Enterprise Miner.

In general, you can fit logistic or proportional odds models to your data directly by first transforming the variables and then selecting the inputs, combining and further transforming the inputs, and finally estimating the model with the original combined and transformed inputs. On the other hand, as I will show in this chapter, these combinations and transformations of the inputs can be done within the neural network framework. A neural network model can be thought of as a complex nonlinear model where the tasks of variable transformation, composite variable creation, and model estimation (estimation of weights) are done simultaneously in such a way that a specified error function is minimized. This does not rule out transforming some or all of the variables before running the neural network models, if you choose to do so.

In the following pages, I will make clear how the transformations of the inputs are performed and how they are combined through different "layers" of the neural network, as well as how the final model that uses these transformed inputs is specified.

A neural network model is represented by a number of layers, each layer containing computing elements known as *units* or *neurons*. Each unit in a layer takes inputs from the preceding layer and computes outputs. The outputs from the neurons in one layer become inputs to the next layer in the sequence of layers.

The first layer in the sequence is the *input layer*, and the last layer is the *output,* or *target*, layer. Between the input and output layers there can be a number of *hidden* layers. The units in a hidden layer are called *hidden units*. The hidden units perform *intermediate calculations* and pass the results to the next layer.

The calculations performed by the hidden units involve combining the inputs they receive from the previous layer and performing a mathematical transformation on the combined values. In

Enterprise Miner, the formulas used by the hidden units for combining the inputs are called *hidden layer combination functions*. The formulas used by the hidden units for transforming the combined values are called *hidden layer activation functions*. The values generated by the combination and activation operations are called the *outputs*. As noted above, outputs of one layer become inputs to the next layer.

The units in the target layer or output layer also perform the two operations of combination and activation. The formulas used by the units in the target layer to combine the inputs are called *Target Layer Combination Functions*, and the formulas used for transforming the combined values are called *Target Layer Activation Functions*. Interpretation of the outputs produced by the target layer depends on the target layer activation function used. Hence, the specification of the target layer activation function cannot be arbitrary. It should reflect the business question and the theory behind the answer you are seeking. For example, if the target variable is response, which is binary, then you should select **Logistic** as the Target Layer Activation Function. This selection will ensure that the outputs of the target layer are the probabilities of response and non-response.

The combination and activation functions in the hidden layers and in the target layer are key elements of the *architecture* of a neural network. Enterprise Miner offers a wide range of choices for these functions. Hence, the number of combinations of hidden layer combination function, hidden layer activation function, target layer combination function, and target layer activation function to choose from is quite large, and each produces a different neural network model.

5.2 A General Example of a Neural Network Model

To help you understand the models presented later, I now provide a general description of a neural network with two *hidden layers*[1] and an *output layer.*

Suppose a data set consists of n records, and each record contains, along with a person's response to a direct mail campaign, some measures of demographic, economic, and other characteristics of the person. These measures are often referred to as *explanatory variables* or *inputs*. Suppose there are p inputs $x_{i1}, x_{i2}, \ldots x_{ip}$ for the ith person. In a neural network model, the input layer passes these inputs to the next layer. In the input layer, there is one input unit for each input. The units in the input layer do not make any computations other than optionally standardizing the inputs. The inputs are passed from the input layer to the next layer. In this illustration, the next layer is the first hidden layer. The units in the first hidden layer produce *intermediate outputs*. Suppose there are three hidden units in this layer. Let H_{i11} denote the intermediate output produced by the first unit of the first hidden layer for the ith record. In general, H_{ikj} stands for the output of the jth unit in the kth hidden layer for the ith record in the data set.

Similarly, the intermediate outputs produced by the second and third units are H_{i12} and H_{i13}. These intermediate outputs are passed to the second hidden layer. Suppose the second hidden layer also contains three units. The outputs of these three hidden units are H_{i21}, H_{i22}, and H_{i23}. These intermediate outputs become inputs to the *output* or *target layer*, which is the last layer in the network. The output produced by this layer is the predicted value of the target, which is the probability of response (and non-response) in the case of the response model (a binary target), and the probability of the number of losses occurring in the case of the risk model (an ordinal

---

[1] Although the general discussion here considers two hidden layers, the two neural network models developed later in this book are based on a single hidden layer.

target with three levels). In the case of a continuous target, such as value of deposits in a bank or household savings, the output of the final layer is the expected value of the target.

Below I show how the units in each layer calculate their outputs.

The network just described is shown in Display 5.0A.

Display 5.0A

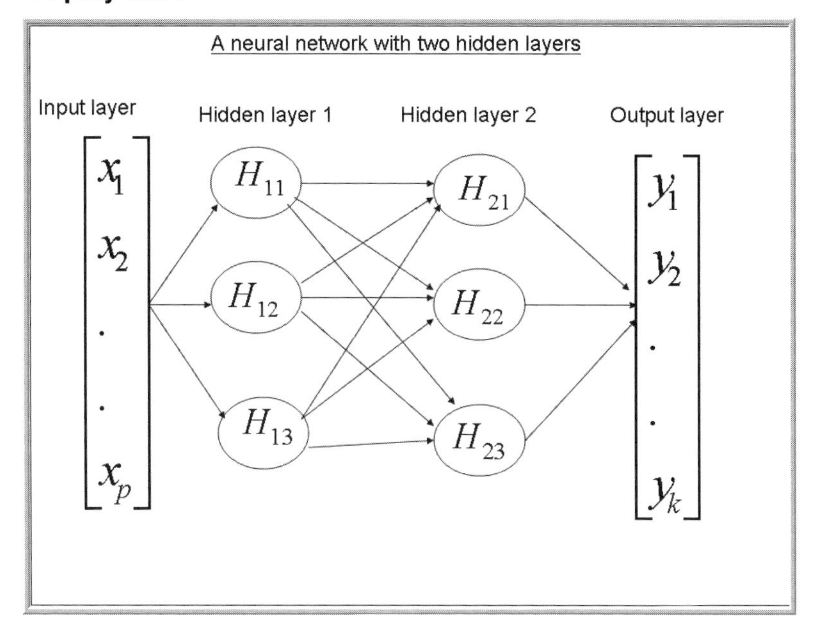

5.2.1 Input Layer

This layer passes the inputs to the next layer in the network either without transforming them or, in the case of interval-scaled inputs, after standardizing the inputs. Categorical inputs are converted to dummy variables. For each record, there are p inputs: $x_1, x_2, ... x_p$.

5.2.2 Hidden Layers

There are three hidden units in each of the two hidden layers in this example. Each hidden unit produces an intermediate output as described below.

Hidden Layer 1: Unit 1

A weighted sum of the inputs for the ith record for the first unit in the first hidden layer is calculated as

$$\eta_{i11} = w_{011} + w_{111}x_{i1} + w_{211}x_{i2} + ... + w_{p11}x_{ip}. \tag{5.1}$$

where $w_{111}, w_{211}, ...w_{p11}$ are the *weights* to be estimated by the iterative algorithm to be described later, one weight for each of the *p* inputs. w_{011} is called *bias.* η_{i11} is the weighted sum for the ith person[2] or record in the data set. Equation 5.1 is called the *combination function.* Since the combination function given in Equation 5.1 is a linear function of the inputs, it is called a *linear combination function.* Other types of combination function are discussed in Section 5.7.

A nonlinear transformation of the weighted sum would yield the output from unit 1 as

$$H_{i11} = \tanh(\eta_{i11}) = \frac{\exp(\eta_{i11}) - \exp(-\eta_{i11})}{\exp(\eta_{i11}) + \exp(-\eta_{i11})} \tag{5.2}$$

The effect of this transformation is to map values of η_{i11}, which may range from $-\infty$ to $+\infty$, into the narrower range of –1 to +1. The transformation shown in Equation 5.2 is referred to as a *hyperbolic tangent activation function.*

To illustrate this I will take a simple example with only two inputs, AGE and CREDIT, both interval-scaled. The input layer first standardizes these variables and passes them to the units in hidden layer 1. Suppose at some point in the iteration the weighted sum presented in Equation 5.1 is $\eta_{i11} = 0.01 + 0.4 * AGE + 0.7 * CRED,$ where *AGE* is the age and *CRED* is the credit score of the person reported in the ith record. The value of η_{i11} (the combination function) may range from –5 to +5. Substituting the value of η_{i11} in Equation 5.2, I get the value of the activation function H_{i11} for the ith record. If I plot these values for all the records in the data set, I get the chart shown in Display 5.0B.

[2] The terms *observation, record, case*, and *person* are used interchangeably.

Display 5.0B

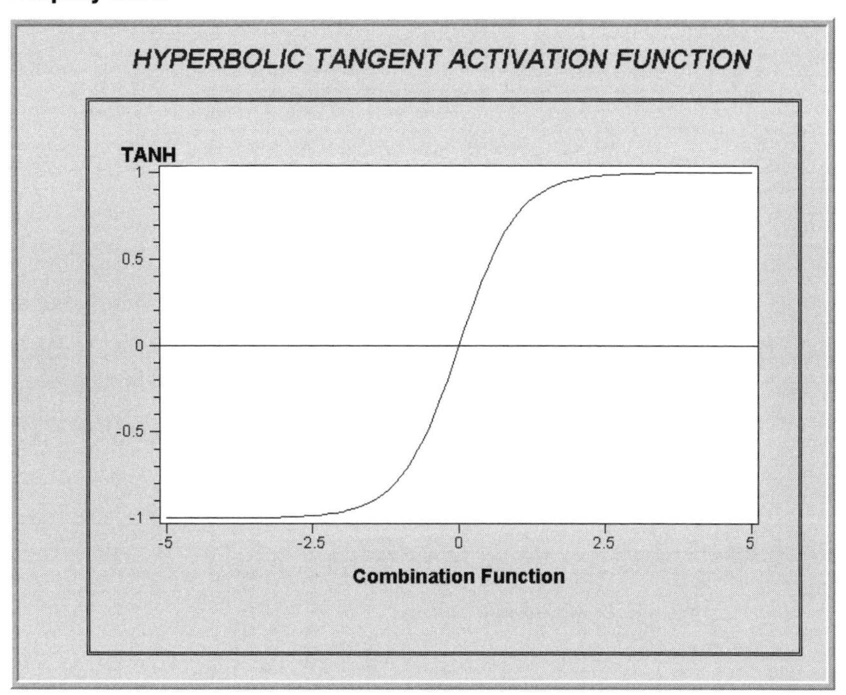

In Enterprise Miner a variety of activation functions are available for use in the hidden layers. These include *Arc Tangent*, *Elliot*, *Hyperbolic Tangent*, *Logistic*, *Gauss*, *Sine,* and *Cosine* in addition to several other types of activation functions discussed in Section 5.7.

The four activation functions: Arc Tangent, Elliot, Hyperbolic Tangent, and Logistic, are called Sigmoid functions; they are *s*-shaped and give values that are bounded within the range of 0 to 1 or –1 to 1.

The Hyperbolic Tangent function shown in Display 5.0B has values bounded between –1 and 1. Displays 5.0C, 5.0D, and 5.0E show Logistic, Arc Tangent, and Elliot functions, respectively.

Display 5.0C

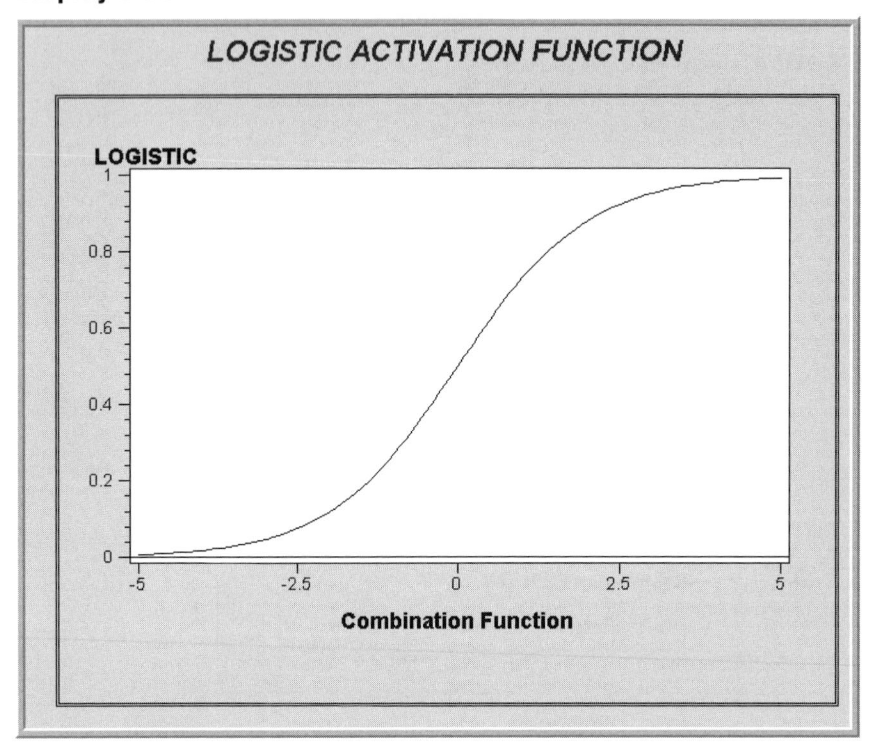

Display 5.0D

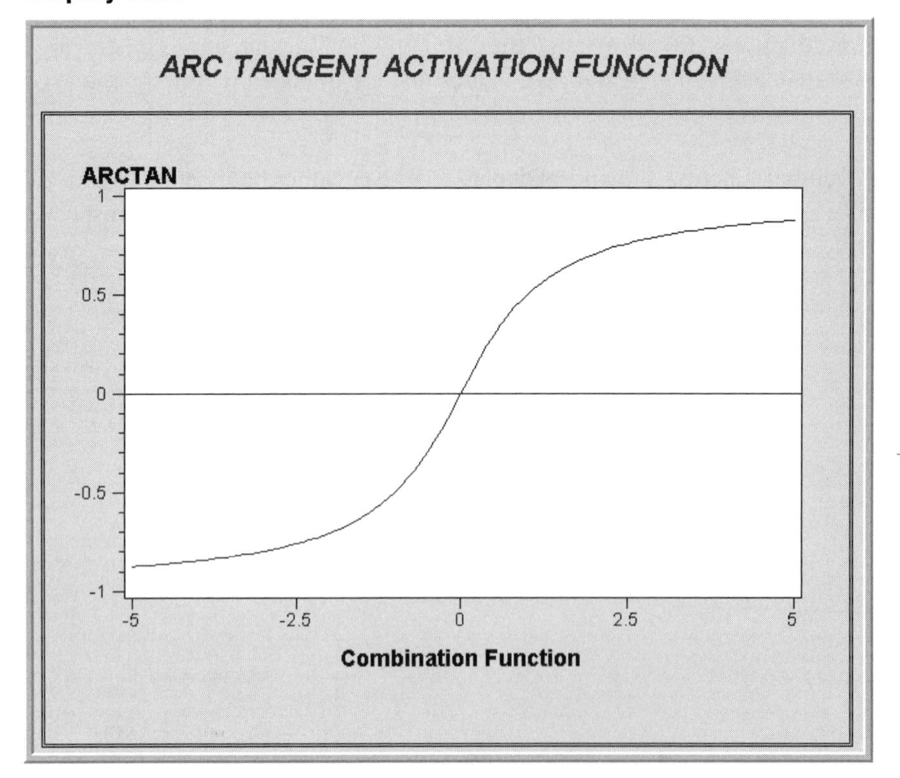

Display 5.0E

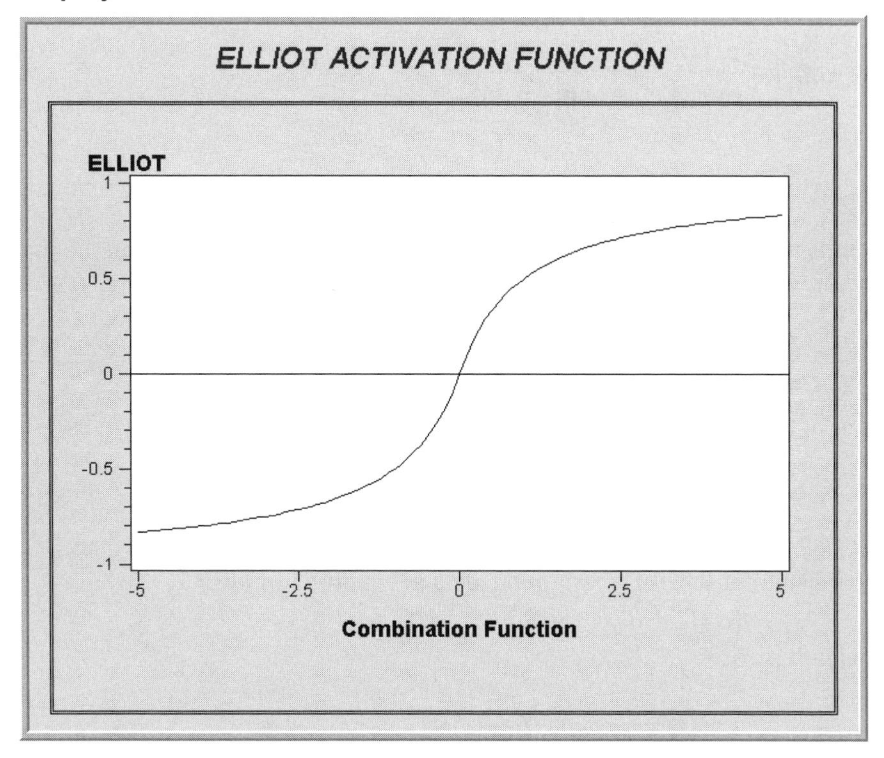

In general, w_{mkj} is the weight of the mth input in the jth neuron (unit) of the kth hidden layer, and w_{0kj} is the corresponding bias. The weighted sum and the output calculated by the jth neuron in the kth hidden layer are η_{ikj} and H_{ikj}, respectively, where i stands for the ith record in the data set.

Hidden Layer 1: Unit 2

The calculation of the output of unit 2 proceeds in the same way as in unit 1, but with a *different set* of weights and bias. The weighted sum of the inputs for the ith person or record in the data is

$$\eta_{i12} = w_{012} + w_{112}x_{i1} + w_{212}x_{i2} + \ldots w_{p12}x_{ip}. \tag{5.3}$$

The output of unit 2 is

$$H_{i12} = \tanh(\eta_{i12}) = \frac{\exp(\eta_{i12}) - \exp(-\eta_{i12})}{\exp(\eta_{i12}) + \exp(-\eta_{i12})} \tag{5.4}$$

Hidden Layer 1: Unit 3

The calculations in unit 3 proceed exactly as in units 1 and 2, but with a *different set* of weights and bias. The weighted sum of the inputs for i^{th} person in the data set is

$$\eta_{i13} = w_{013} + w_{113}x_{i1} + w_{213}x_{i2} + \ldots w_{p13}x_{ip}, \tag{5.5}$$

and the output of this unit is

$$H_{i13} = \tanh(\eta_{i13}) = \frac{\exp(\eta_{i13}) - \exp(-\eta_{i13})}{\exp(\eta_{i13}) + \exp(-\eta_{i13})} \qquad (5.6)$$

Hidden Layer 2

There are three units in this layer, each producing an intermediate output based on the inputs it receives from the previous layer.

Hidden Layer 2: Unit 1

The weighted sum of the inputs calculated by this unit is

$$\eta_{i21} = w_{021} + w_{121}H_{i11} + w_{221}H_{i12} + w_{321}H_{i13}, \qquad (5.7)$$

η_{i21} is the weighted sum for the ith record in the data set in unit 1 of layer 2.

The output of this unit is

$$H_{i21} = \tanh(\eta_{i21}) = \frac{\exp(\eta_{i21}) - \exp(-\eta_{i21})}{\exp(\eta_{i21}) + \exp(-\eta_{i21})} \qquad (5.8)$$

Hidden Layer 2 : Unit 2

The weighted sum of the inputs for the ith person or record in the data is

$$\eta_{i22} = w_{022} + w_{122}H_{i11} + w_{222}H_{i12} + w_{322}H_{i13}, \qquad (5.9)$$

and the output of this unit is

$$H_{i22} = \tanh(\eta_{i22}) = \frac{\exp(\eta_{i22}) - \exp(-\eta_{i22})}{\exp(\eta_{i22}) + \exp(-\eta_{i22})} \qquad (5.10)$$

Hidden Layer 2: Unit 3

The weighted sum of the inputs for i^{th} person in the data set is

$$\eta_{i23} = w_{023} + w_{123}H_{i11} + w_{223}H_{i12} + w_{323}H_{i13}, \qquad (5.11)$$

and the output of this unit is

$$H_{i23} = \tanh(\eta_{i23}) = \frac{\exp(\eta_{i23}) - \exp(-\eta_{i23})}{\exp(\eta_{i23}) + \exp(-\eta_{i23})} \qquad (5.12)$$

5.2.3 Output Layer or Target Layer

This layer produces the final output of the neural network—the predicted values of the targets. In the case of a response model, the output of this layer will be the predicted probability of response and non-response, and the output layer will have two units. In the case of a categorical target with more than two levels, there will be more than two output units. When the target is continuous, such as a bank account balance, the output is the expected value. Display 5.0A shows *k* outputs, and they are: $y_1, y_2, \ldots\ldots, y_k$. For the response model, k = 2, and for the loss frequency (risk) model, k = 3.

The inputs into this layer are H_{i21}, H_{i22}, and H_{i23}, which are the intermediate outputs produced by the preceding layer. These intermediate outputs can be thought of as the nonlinear transformations and combinations of the original inputs performed through successive layers of the network, as described in Equations 5.1 through 5.13. H_{i21}, H_{i22}, and H_{i23} can be considered *synthetic variables* constructed from the inputs.

In the output layer these transformed inputs (H_{i21}, H_{i22}, and H_{i23}) are used to construct predictive equations of the target. First, for each output unit, a linear combination of H_{i21}, H_{i22}, and H_{i23} called a linear predictor (or combination function) is defined and an activation function is specified to give the desired predictive equation.

In the case of response model, the first output unit gives the probability of response represented by y_1, while the second output unit gives the probability of no response represented by y_2. In the case of the loss frequency model, there are three outputs: $y_1, y_2,$ and y_3. The first output unit gives the probability that LOSSFRQ is 0 (y_1), the second output unit gives the probability that LOSSFRQ is 1 (y_2), and the third output gives the probability that LOSSFRQ is (y_3).

The linear combination of the inputs from the previous layer is calculated as

$$\eta_{i31} = w_{031} + w_{131}H_{i21} + w_{231}H_{i22} + w_{331}H_{i23}. \tag{5.13}$$

Equation 5.13 is the linear predictor (or linear combination function), because it is linear in terms of the weights.

5.2.4 Activation Function of the Output Layer

The *activation function* is a formula for calculating the target values from the inputs coming from the final hidden layer. The outputs of the final hidden layer are first combined as shown in Equation 5.13. The resulting combination is transformed using the activation function to give the final target values. The activation function is chosen according to the type of target and what you are interested in predicting. In the case of a binary target I use a logistic activation function in the final output or target layer wherein the ith person's probability of response (π_i) is calculated as

$$\pi_i = \frac{\exp(\eta_{i31})}{1+\exp(\eta_{i31})} \tag{5.14}$$

Equation 5.14 is called the *logistic activation function*, and is based on a logistic distribution. By substituting Equations 5.1 through 5.13 into Equation 5.14, I can easily verify that it is possible to write the output of this node as an explicit nonlinear function of the weights and the inputs. This can be denoted by

$$\pi_i(W, X_i). \tag{5.15}$$

In Equation 5.15, W is the vector whose elements are the weights shown in Equations 5.1, 5.3, 5.5, 5.7, 5.9, 5.11, and 5.13, and X_i is the vector of inputs $x_{i1}, x_{i2}, \ldots x_{ip}$ for the ith person in the data set.

The neural network described above is called a multilayer perceptron or MLP. In general, a network which uses linear combination functions and sigmoid activation functions in the hidden layers is called a multilayer perceptron. In Enterprise Miner, if you set the **Architecture** property to **MLP**, then you get an MLP with only one hidden layer.

5.3 Estimation of Weights in a Neural Network Model

The weights are estimated iteratively using the *training data set* in such a way that the error function specified by the user is minimized. In the case of a response model, I use the following Bernoulli error function:

$$E = -2\sum_{i=1}^{n}\left\{ y_i \ln\frac{\pi(W, X_i)}{y_i} + (1 - y_i)\ln\frac{1 - \pi(W, X_i)}{1 - y_i} \right\} \tag{5.16}$$

where π is the estimated probability of response. It is a function of the vector of weights, W, and the vector of explanatory variables for the ith person, X_i. The variable y_i is the observed response of the ith person, in which $y_i = 1$ if the ith person responded to direct mail and $y_i = 0$ if the ith person did not respond. If $y_i = 1$ for the ith observation, then its contribution to the error function (Eq. 5.16) is $-2\log\pi(W, X_i)$; if $y_i = 0$, then the contribution is $-2\log(1 - \pi(W, X_i))$.

The total number of observations in the *training data set* is n. The error function is summed over all of these observations. The inputs for the ith record are known and the values for the weights W are chosen so as to minimize the *error function*. A number of iterative methods of finding the optimum weights are discussed in Bishop.

In general, calculation of optimum weights is done in two steps:

Step 1: Finding the error-minimizing weights from the training data set

In the first step, an iterative procedure is used with the training data set to find a set of weights that minimize the error function given in Equation 5.16. In the first iteration, a set of initial weights is used, and the error function in Equation 5.16 is evaluated. In the second iteration, the weights are changed by a small amount in such a way that the error is reduced. This process continues until the error cannot be further reduced, or until the specified maximum number of iterations is reached. At each iteration, a set of weights is generated. If it takes 100 iterations to find the error-minimizing weights for the training data set, then 100 sets of weights are generated and saved. Thus, the training process generates 100 models in this example. The next task is to select the best model out of these 100 models. This is done using the *validation data set.*

Step 2: Finding the optimum weights from the validation data set.

In this step, the user-supplied **Model Selection Criterion** is applied to select one of the 100 sets of weights generated in Step 1. This selection is done using the validation data set. Suppose I set the **Model Selection Criterion** property to **Average Error**. Then the average error is calculated for each of the 100 models using the validation data set. If the validation data set has m observations, then the error calculated from the weights generated at the k_{th} iteration is

$$E_{(k)} = -2\sum_{i=1}^{m}\left\{ y_i \ln\frac{\pi(W_{(k)}, X_i)}{y_i} + (1-y_i)\ln\frac{1-\pi(W_{(k)}, X_i)}{1-y_i} \right\} \tag{5.17}$$

where $y_i = 1$ for a responder, and 0 for a non-responder. The *average error* at the k_{th} iteration is $\frac{E_{(k)}}{2m}$. The set of weights with the smallest average error is used in the final predictive equation. Alternative model selection criteria can be chosen by changing the **Model Selection Criterion** property.

As pointed out earlier, Equations 5.1, 5.3, 5.5, 5.7, 5.9, 5.11, and 5.13 are called *combination functions*, and Equations 5.2, 5.4, 5.6, 5.8, 5.10, 5.12, and 5.14 are called *activation functions*.

By substituting Equations 5.1 through 5.13 into Equation 5.14, I can write the neural network model as an explicit function of the *inputs*. This is an arbitrary nonlinear function, and its complexity (and hence the degree of nonlinearity) depends on network features such as the number of hidden layers included, the number of units in each hidden layer, and the combination and activation functions in each hidden unit.

In the neural network model described above, the final layer inputs (H_{i21}, H_{i22}, and H_{i23}) can be thought of as final transformations of the original inputs made by the calculations in different layers of the network.

In each hidden unit, I used a linear combination function to calculate the weighted sum of the inputs, and a hyperbolic tangent activation function to calculate the intermediate output for each hidden unit. In the output layer, I specified a linear combination function and a logistic activation function. The error function I specified is called the Bernoulli error function. A neural network with two hidden layers and one output layer is called a three-layered network. In characterizing the network, the input layer is not counted, since units in the input layer do not perform any computations except standardizing the inputs.

5.4 A Neural Network Model to Predict Response

This topic discusses the neural network model developed to predict the response to a planned direct mail campaign. The campaign's purpose was to solicit customers for a hypothetical insurance company. A two-layered network with one hidden layer was chosen. Three units are included in the hidden layer. In the hidden layer the *combination function* chosen is linear, and the *activation function* is hyperbolic tangent. In the output layer, a *logistic activation function* and *Bernoulli error function* are used. The logistic activation function results in a logistic regression type model with nonlinear transformation of the inputs as shown in Equation 5.14 in Section 5.2.4. Models of this type are in general estimated by minimizing the Bernoulli error functions shown in Equation 5.16. Minimization of the Bernoulli error function is equivalent to maximizing the likelihood function.

Display 5.1 shows the process flow for the response model. The first node in the process flow diagram is the **Input Data** node, which makes the SAS data set available for modeling. The next node is the **Data Partition** node, which creates the **Train**, **Validate**, and **Test** data sets. The training data set is used for preliminary model fitting. The validation data set is used for selecting the optimum weights. The **Model Selection Criterion** property is set to **Average Error**.

As pointed out earlier, the estimation of the weights is done by minimizing the error function. This minimization is done by an iterative procedure. Each iteration yields a set of weights. Each set of weights defines a model. If I set the **Model Selection Criterion** property to **Average Error**, the algorithm selects the set of weights that results in the smallest error, where the error is calculated from the validation data set.

Since both the **Train** and **Validate** data sets are used for parameter estimation and parameter selection, respectively, an additional holdout data set is required for an independent assessment of the model. The **Test** data set is set aside for this purpose.

Display 5.1

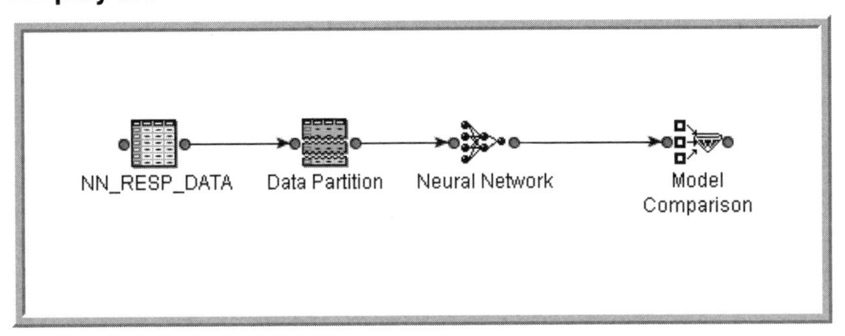

5.4.1 Setting the Neural Network Node Properties

Here is a summary of the neural network specifications for this application:

- one hidden layer with three neurons
- *linear combination functions* for both the hidden and output units
- *hyperbolic tangent activation functions* for the hidden units
- *logistic activation functions* for the *output units*
- the *Bernoulli error function*
- the model assessment criterion is **Average Error**.

These settings are shown in Display 5.2.

Display 5.2

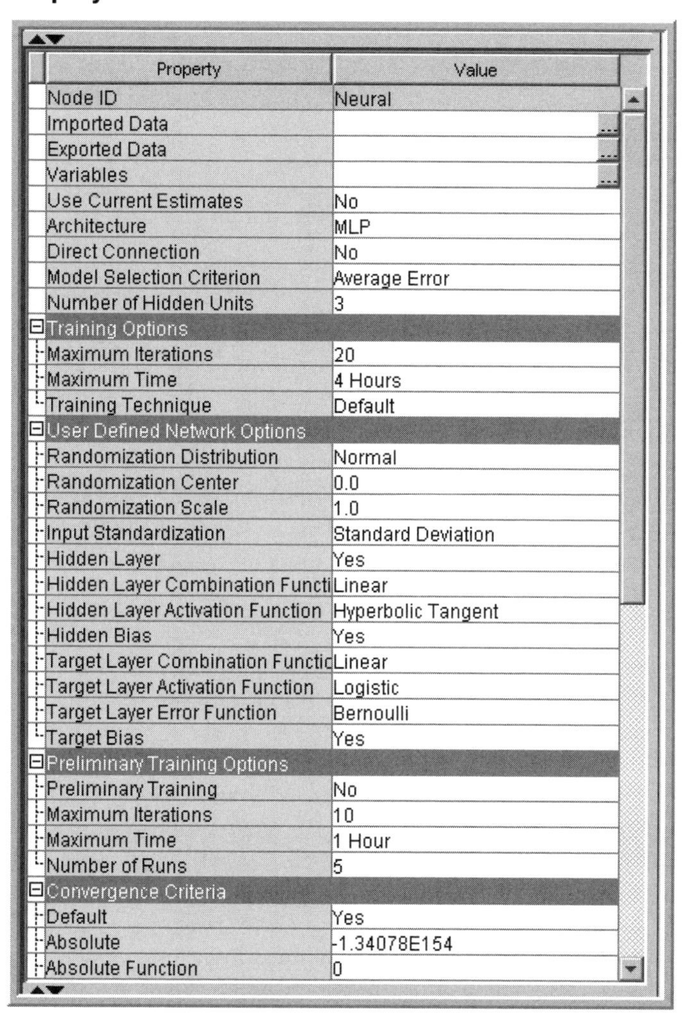

Property	Value
Node ID	Neural
Imported Data	
Exported Data	
Variables	
Use Current Estimates	No
Architecture	MLP
Direct Connection	No
Model Selection Criterion	Average Error
Number of Hidden Units	3
Training Options	
Maximum Iterations	20
Maximum Time	4 Hours
Training Technique	Default
User Defined Network Options	
Randomization Distribution	Normal
Randomization Center	0.0
Randomization Scale	1.0
Input Standardization	Standard Deviation
Hidden Layer	Yes
Hidden Layer Combination Functi	Linear
Hidden Layer Activation Function	Hyperbolic Tangent
Hidden Bias	Yes
Target Layer Combination Functic	Linear
Target Layer Activation Function	Logistic
Target Layer Error Function	Bernoulli
Target Bias	Yes
Preliminary Training Options	
Preliminary Training	No
Maximum Iterations	10
Maximum Time	1 Hour
Number of Runs	5
Convergence Criteria	
Default	Yes
Absolute	-1.34078E154
Absolute Function	0

After running the **Neural Network** node, you can open the **Results** window. The **Results** window, shown in Display 5.3, contains four windows: **Score Rankings Overlay**, **Iteration Plot**, **Fit Statistics**, and **Output**.

The **Score Rankings Overlay** window in Displays 5.3 and 5.3A shows the cumulative lift for the training and validation data sets.

Display 5.3

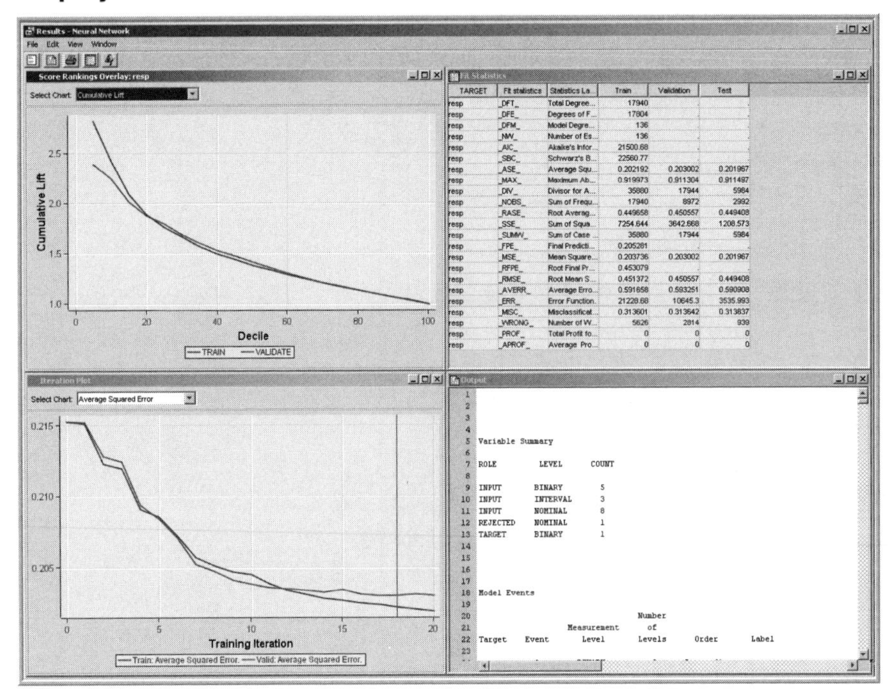

Display 5.3A

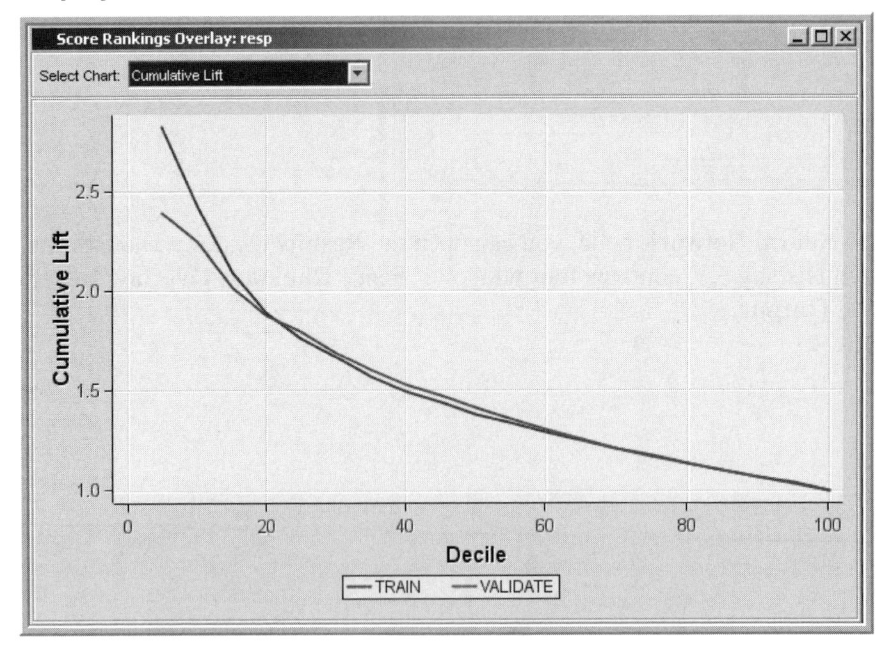

Display 5.3B shows the list of charts that can be displayed in the **Score Rankings Overlay** window.

Display 5.3B

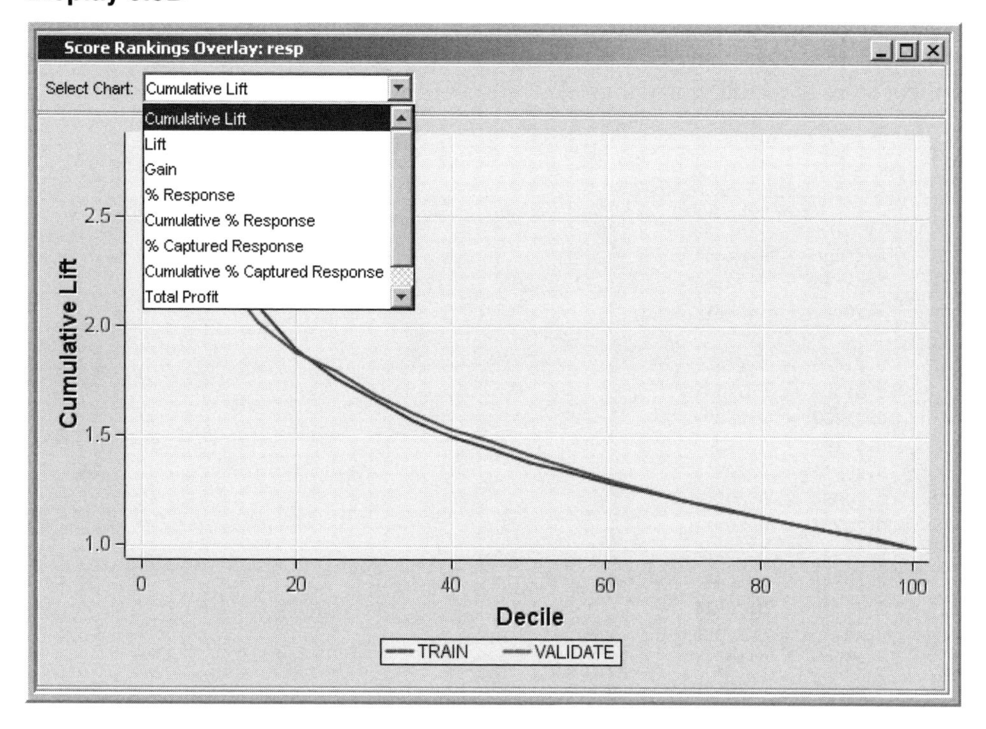

Display 5.3C displays the iteration plot with Average Squared Error at each iteration for the training and validation data sets. The estimation process required 20 iterations (this is the limit I set in the properties panel). The weights from the 18th iteration were selected. Around the 18th iteration, the Average Squared Error flattened out in the validation data set, although it continued to decline in the training data set.

Display 5.3C

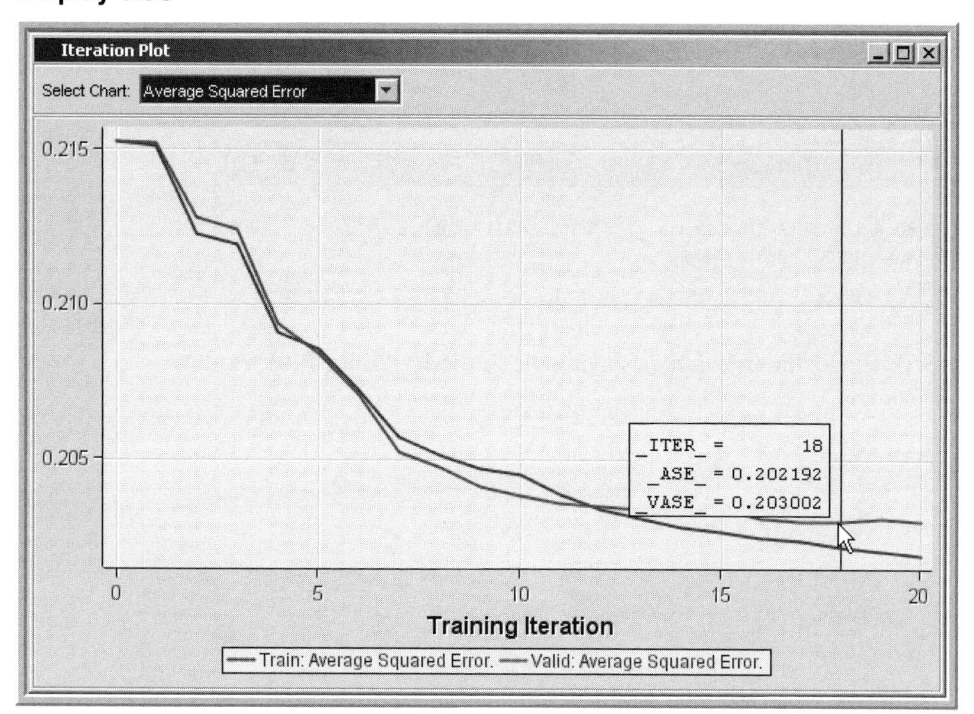

You can save the table corresponding to the plots shown in Display 5.3C by selecting the Tables icon and then selecting **File → Save As**. Table 5.1 shows the three variables _ITER_ (iteration number), _ASE_ (Average Squared Error from the training data set), and _VASE_ (Average Squared Error from the validation data set).

Table 5.1

Training Iteration	Train: Average Squared Error.	Valid: Average Squared Error.
0	0.21526	0.21527
1	0.21509	0.21521
2	0.21227	0.21279
3	0.21193	0.21242
4	0.20911	0.20936
5	0.20854	0.20840
6	0.20726	0.20711
7	0.20572	0.20525
8	0.20516	0.20473
9	0.20470	0.20415
10	0.20450	0.20383
11	0.20387	0.20361
12	0.20342	0.20352
13	0.20310	0.20340
14	0.20284	0.20324
15	0.20270	0.20347
16	0.20248	0.20312
17	0.20238	0.20303
18	0.20219	0.20300
19	0.20208	0.20312
20	0.20194	0.20302

Display 5.3D shows the SAS code used for printing Table 5.1.

Display 5.3D

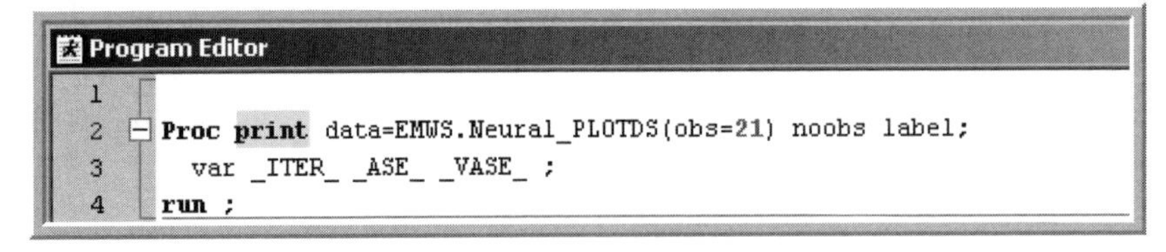

Display 5.3E shows the list of charts available in the **Iteration Plot** window.

Display 5.3E

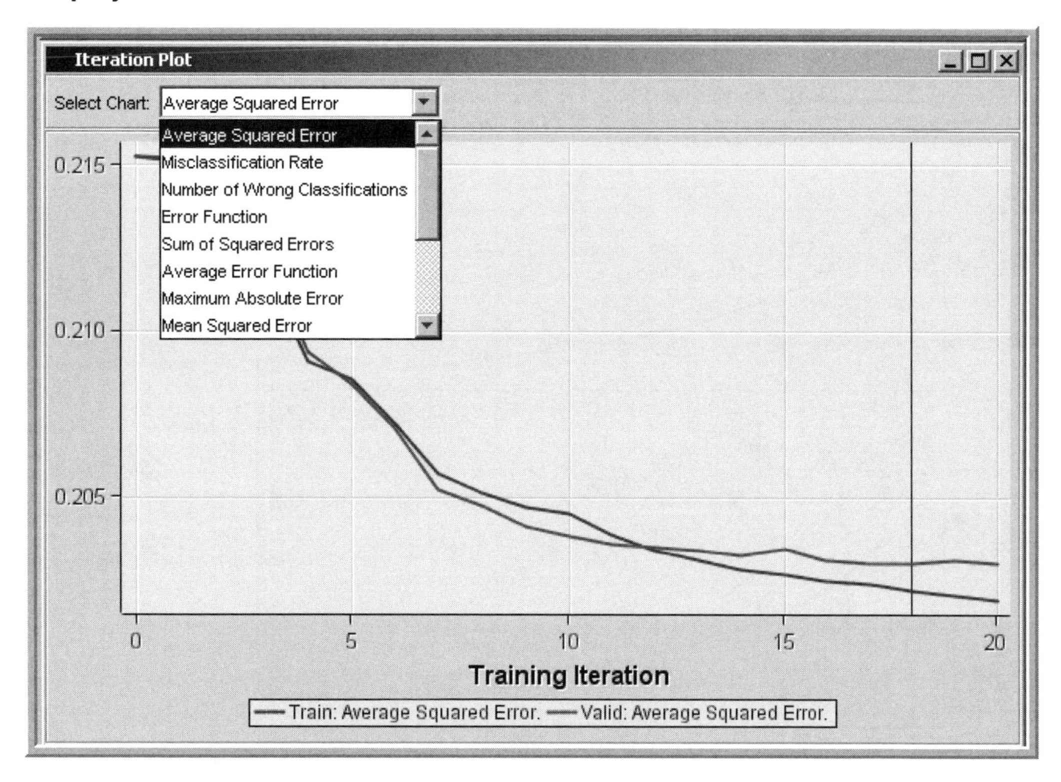

5.4.2 Assessing the Predictive Performance of the Estimated Model

In order to assess the predictive performance of the neural network model, you can run the **Model Comparison** node. After running the **Model Comparison** node, select **Results** from the pop-up menu. The **Results** window appears. In the **Results** window, the **Score Rankings Overlay** shows the lift charts for the training, validation, and test data sets. These are shown in Display 5.4.

Display 5.4

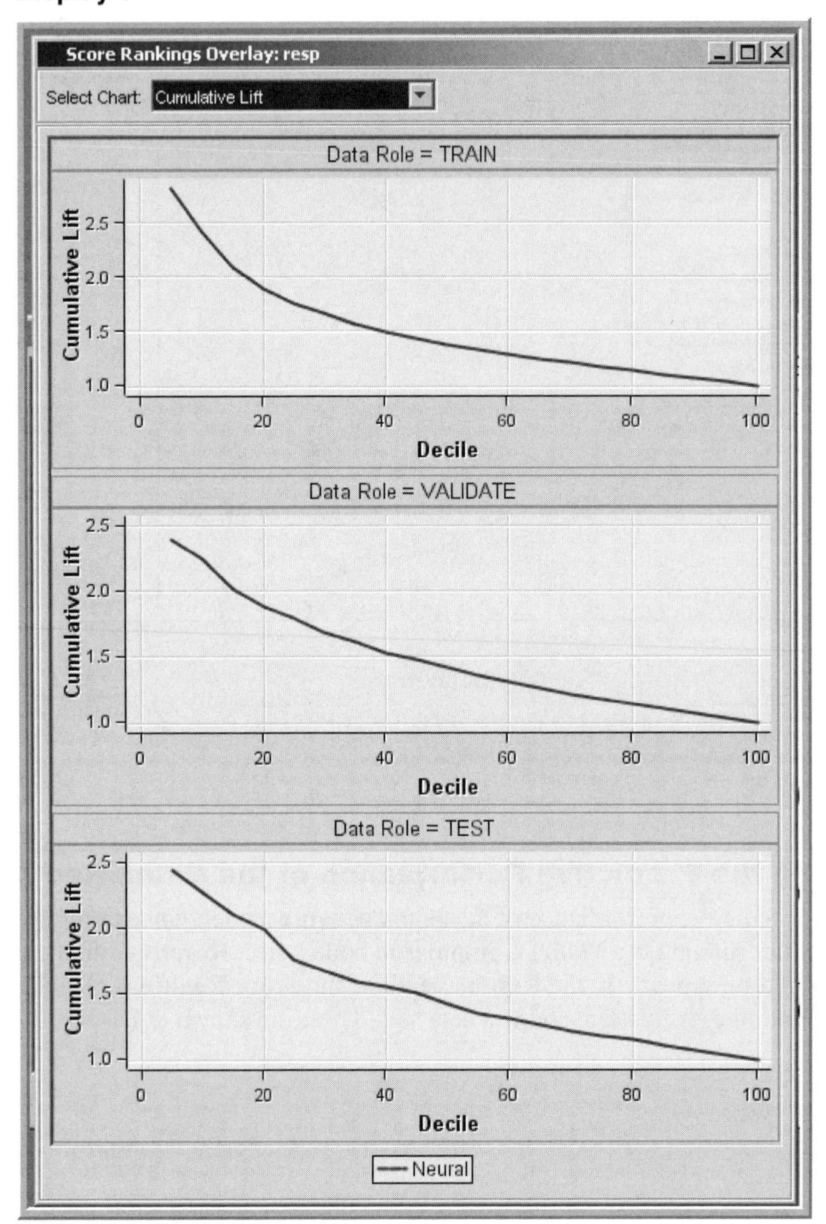

The list of available charts can be seen by clicking the arrow on the **Select Chart** box as shown in Display 5.4A.

Display 5.4A

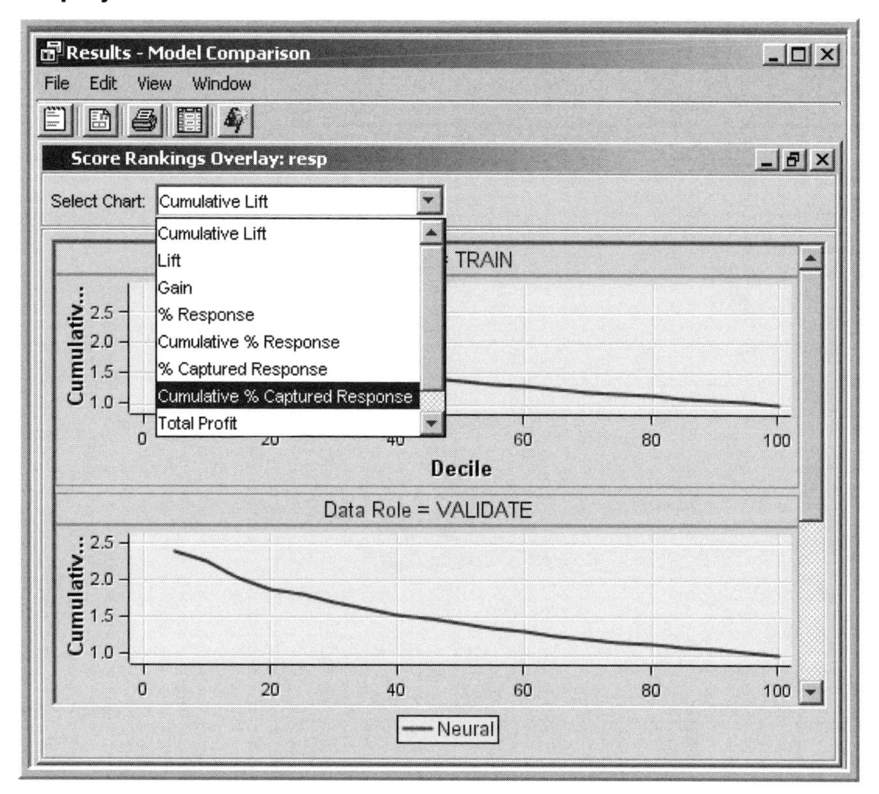

By selecting **Cumulative % Captured Response**, you get the charts of the capture rates as shown in Display 5.5.

Display 5.5

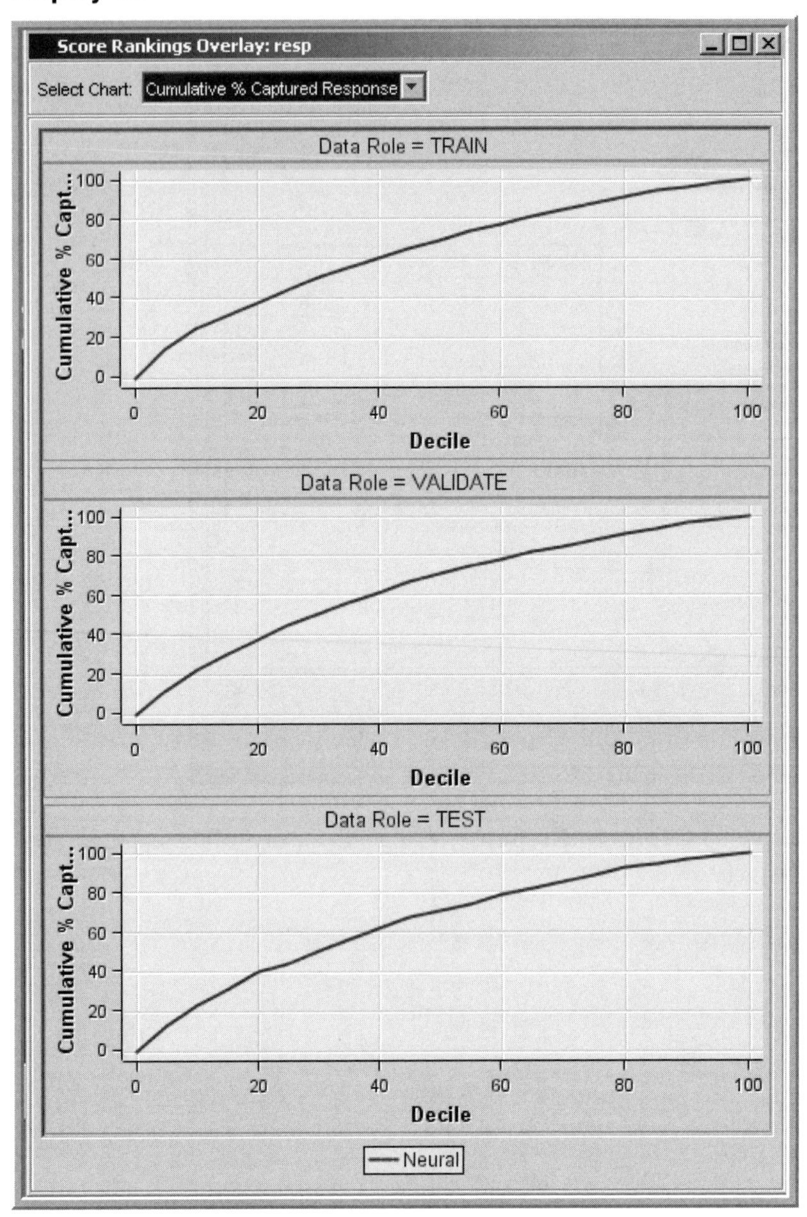

You can save the data corresponding to the charts shown in Displays 5.4 and 5.5 by selecting the Table icon in the **Results** – **Model Comparison** window shown in Display 5.6.

Display 5.6

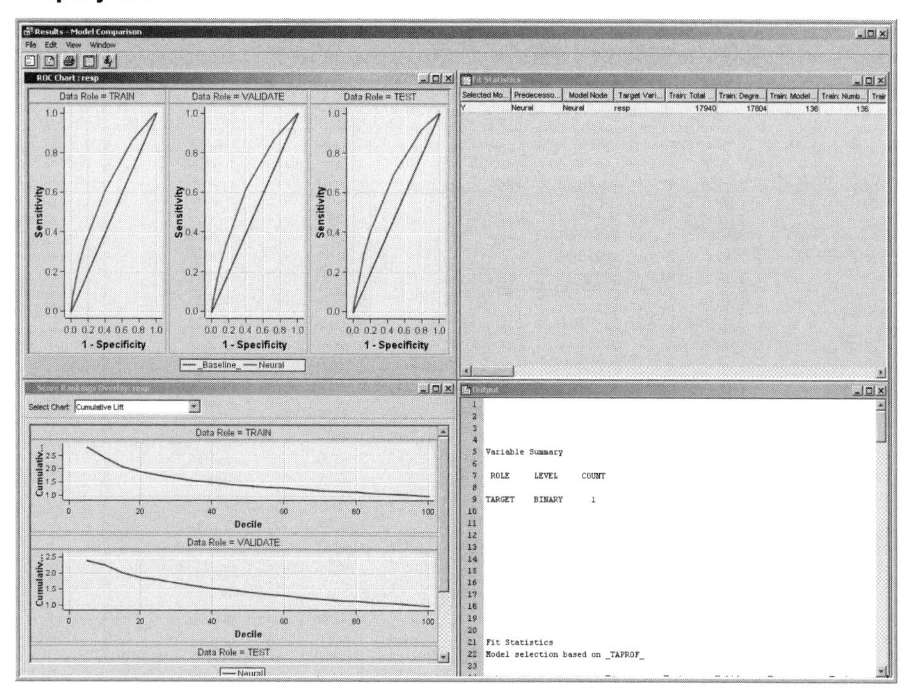

Enterprise Miner saves the Score Rankings table as EMWS.MdlComp_EMRANK. Tables 5.2, 5.3, and 5.4 are created from the saved data set using the simple SAS code shown in Display 5.6A.

Display 5.6A

Program Editor

```
1   options center ;
2   proc print data=EMWS.MdlComp_EMRANK label noobs ;
3    where upcase(datarole) = "TRAIN" and bin ne . ;
4    var bin decile resp respc lift liftc cap capc ;
5   title "Lift and Capture Rates: TRAINING Data set" ;
6   run ;
7   proc print data=EMWS.MdlComp_EMRANK label noobs ;
8    where upcase(datarole) = "VALIDATE" and bin ne . ;
9    var bin decile resp respc lift liftc cap capc ;
10  title "Lift and Capture Rates: Validation Data set" ;
11  run ;
12  proc print data=EMWS.MdlComp_EMRANK label noobs ;
13   where upcase(datarole) = "TEST" and bin ne . ;
14   var bin decile resp respc lift liftc cap capc ;
15  title "Lift and Capture Rates: TEST Data set" ;
16  run ;
```

Table 5.2

Lift and Capture Rates: TRAINING Data set

Bin	Decile	% Response	Cumulative % Response	Lift	Cumulative Lift	% Captured Response	Cumulative % Captured Response
1	5	8.46783	8.46783	2.82261	2.82261	14.1130	14.113
2	10	6.06825	7.26804	2.02275	2.42268	10.1138	24.227
3	15	4.28724	6.27444	1.42908	2.09148	7.1454	31.372
4	20	3.87131	5.67366	1.29044	1.89122	6.4522	37.824
5	25	3.72200	5.28333	1.24067	1.76111	6.2033	44.028
6	30	3.59042	5.00118	1.19681	1.66706	5.9840	50.012
7	35	2.93642	4.70621	0.97881	1.56874	4.8940	54.906
8	40	2.97519	4.48983	0.99173	1.49661	4.9586	59.864
9	45	2.95442	4.31923	0.98481	1.43974	4.9240	64.788
10	50	2.53822	4.14113	0.84607	1.38038	4.2304	69.019
11	55	2.66006	4.00649	0.88669	1.33550	4.4334	73.452
12	60	2.33105	3.86687	0.77702	1.28896	3.8851	77.337
13	65	2.20448	3.73899	0.73483	1.24633	3.6741	81.011
14	70	2.09342	3.62145	0.69781	1.20715	3.4890	84.501
15	75	1.98365	3.51226	0.66122	1.17075	3.3061	87.807
16	80	1.85258	3.40853	0.61753	1.13618	3.0876	90.894
17	85	1.65553	3.30542	0.55184	1.10181	2.7592	93.653
18	90	1.37636	3.19825	0.45879	1.06608	2.2939	95.947
19	95	1.39708	3.10345	0.46569	1.03448	2.3285	98.276
20	100	1.03448	3.00000	0.34483	1.00000	1.7241	100.000

Table 5.3

Lift and Capture Rates: Validation Data set

Bin	Decile	% Response	Cumulative % Response	Lift	Cumulative Lift	% Captured Response	Cumulative % Captured Response
1	5	7.16418	7.16418	2.38806	2.38806	11.9403	11.940
2	10	6.31130	6.73774	2.10377	2.24591	10.5188	22.459
3	15	4.64819	6.04122	1.54940	2.01374	7.7470	30.206
4	20	4.39232	5.62900	1.46411	1.87633	7.3205	37.527
5	25	4.41365	5.38593	1.47122	1.79531	7.3561	44.883
6	30	3.43284	5.06041	1.14428	1.68680	5.7214	50.604
7	35	3.36663	4.81844	1.12221	1.60615	5.6110	56.215
8	40	2.92335	4.58156	0.97445	1.52719	4.8722	61.087
9	45	3.09168	4.41602	1.03056	1.47201	5.1528	66.240
10	50	2.66525	4.24094	0.88842	1.41365	4.4421	70.682
11	55	2.36674	4.07056	0.78891	1.35685	3.9446	74.627
12	60	2.15352	3.91080	0.71784	1.30360	3.5892	78.216
13	65	2.04735	3.76746	0.68245	1.25582	3.4122	81.628
14	70	1.81193	3.62778	0.60398	1.20926	3.0199	84.648
15	75	1.85501	3.50959	0.61834	1.16986	3.0917	87.740
16	80	1.76972	3.40085	0.58991	1.13362	2.9495	90.689
17	85	1.70576	3.30114	0.56859	1.10038	2.8429	93.532
18	90	1.47122	3.19948	0.49041	1.06649	2.4520	95.984
19	95	1.27932	3.09842	0.42644	1.03281	2.1322	98.117
20	100	1.13006	3.00000	0.37669	1.00000	1.8834	100.000

Table 5.4

Lift and Capture Rates: TEST Data set

Bin	Decile	% Response	Cumulative % Response	Lift	Cumulative Lift	% Captured Response	Cumulative % Captured Response
1	5	7.28435	7.28435	2.42812	2.42812	12.1406	12.141
2	10	6.28078	6.78256	2.09359	2.26085	10.4680	22.609
3	15	5.28471	6.28328	1.76157	2.09443	8.8079	31.416
4	20	4.98403	5.95847	1.66134	1.98616	8.3067	39.723
5	25	2.42812	5.25240	0.80937	1.75080	4.0469	43.770
6	30	3.83387	5.01597	1.27796	1.67199	6.3898	50.160
7	35	3.32268	4.77408	1.10756	1.59136	5.5378	55.698
8	40	3.83387	4.65655	1.27796	1.55218	6.3898	62.087
9	45	3.19489	4.49414	1.06496	1.49805	5.3248	67.412
10	50	2.04473	4.24920	0.68158	1.41640	3.4079	70.820
11	55	1.98083	4.04299	0.66028	1.34766	3.3014	74.121
12	60	2.74317	3.93467	0.91439	1.31156	4.5720	78.693
13	65	2.04916	3.78963	0.68305	1.26321	3.4153	82.109
14	70	1.72524	3.64217	0.57508	1.21406	2.8754	84.984
15	75	1.91693	3.52716	0.63898	1.17572	3.1949	88.179
16	80	2.30032	3.45048	0.76677	1.15016	3.8339	92.013
17	85	1.27796	3.32268	0.42599	1.10756	2.1299	94.143
18	90	1.15016	3.20199	0.38339	1.06733	1.9169	96.060
19	95	1.34185	3.10409	0.44728	1.03470	2.2364	98.296
20	100	1.02236	3.00000	0.34079	1.00000	1.7039	100.000

The lift and capture rates calculated from the test data set (shown in Table 5.4) should be used for evaluating the models or comparing the models because the test data set is not used in training or fine-tuning the model.

To calculate the lift and capture rates, Enterprise Miner first calculates the predicted probability of response for each record in the test data. Then it sorts the records in descending order of the predicted probabilities (also called the scores) and divides the data set into 20 groups of equal size. In Table 5.4, the column called **Bin** shows the ranking of these groups. If the model is accurate, the table should show the highest actual response rate in the first bin, the second highest in the next bin, and so on. From the column called **%Response**, it is clear that the average response rate for observations in the first bin is 7.284%. The average response rate for the entire test data set is 3%. Hence the lift, which is the ratio of the response rate in bin 1 to the overall response rate, is 2.428. The lift for each bin is calculated in the same way. The first row of the column **Cumulative %Response** rate shows the response rate for the first bin. The second row shows the response rate for bins 1 and 2 combined, and so on.

The capture rate of a bin shows the percentage of likely responders that it is reasonable to expect to be captured in the bin. From the column called **% Captured Response**, it can be seen that the 12.141% of all responders are in bin 1.

From the **Cumulative % Captured Response** column of Table 5.3, it can be seen that by sending mail to customers in the first four bins, or the top 20% of the target population, it is reasonable to expect to capture 39.7% of all potential responders from the target population. This assumes that the modeling sample represents the target population.

5.4.3 Receiver Operating Characteristic Charts

Display 5.6B, taken from the **Results** window of the **Model Comparison** node, displays receiver operating characteristic (ROC) curves for the training, validation, and test data sets. An ROC curve shows the values of the *true positive fraction* and the *false positive fraction* at different *cut-off values*, which can be denoted by P_c. In the case of the response model, if the estimated probability of response for a customer record were above a cut-off value P_c, then you would classify the customer as a responder; otherwise, you would classify the customer as a non-responder.

Display 5.6B

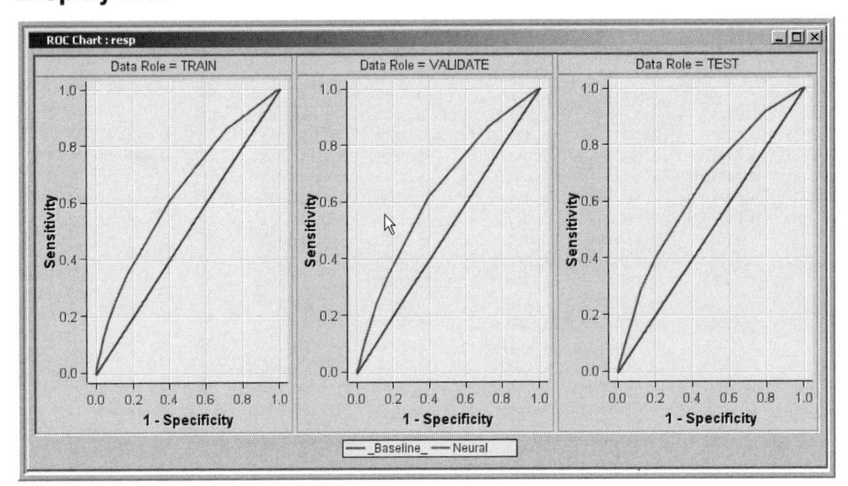

In the ROC chart, the *true positive fraction* is shown on the vertical axis, and the *false positive fraction* is shown on the horizontal axis for each cut-off value (P_c). *True positive fraction* is the proportion of responders correctly classified as responders. *False positive fraction* is the proportion of non-responders incorrectly classified as responders. The true positive fraction is also called *sensitivity,* and *specificity* is the proportion of non-responders correctly classified as non-responders. Hence, the false positive fraction is 1–*specificity*. An ROC curve reflects the tradeoff between *sensitivity* and *specificity*.

The straight diagonal lines in Display 5.6B that are labeled Baseline are the ROC charts of a model that assigns customers at random to the responder group and the non-responder group, and hence has no predictive power. On these lines, *sensitivity* = 1–*specificity* at all cut-off points. The larger the area between the ROC curve of the model being evaluated and the diagonal line, the better the model. The area under the ROC curve is a measure of the predictive accuracy of the model, and can be used for comparing different models.

Table 5.4A shows *sensitivity* and *1–specificity* at various cut-off points in the validation data. Display 5.6C shows the SAS code that generated Table 5.4A.

Table 5.4A

ROC Table: Validation Data

cutoff	sensitivity	oneMinus Specificity
0.13696	0.00071	0.00000
0.11143	0.00213	0.00114
0.10001	0.00746	0.00390
0.08987	0.02736	0.01104
0.07947	0.05899	0.02111
0.06945	0.11158	0.04287
0.05945	0.16205	0.06934
0.04938	0.24271	0.10669
0.03932	0.33795	0.17230
0.02931	0.61834	0.39769
0.01930	0.86674	0.72702
0.00931	0.99609	0.98522
0.00483	1.00000	1.00000

From Table 5.4A it can be seen that at a cut-off probability (P_c) of 0.01930, for example, the sensitivity is 0.86674. That is, at this cut-off point, you will correctly classify 86.7% of responders as responders, but you will also *incorrectly* classify 72.7% of non-responders as responders, since 1–*specificity* at this point is 0.72702. If instead you chose a much higher cut-off point of $P_c = 0.11143$, you would classify 0.213% of true responders as responders and 0.114% of non-responders as responders. In this case, by increasing the cut-off probability beyond which you would classify an individual as a responder, you would be reducing the fraction of false-positive decisions made, while at the same time also reducing the fraction of true-positive decisions made. These pairings of a *true positive fraction* with a *false positive fraction* are plotted as the ROC curve for the VALIDATE case in Display 5.6B.

For further information on ROC curves, see the textbook by A.A. Afifi, Virginia Clark, and Susanne May (2004).[3]

Display 5.6C

```
options center ;
proc print data=emws.mdlcomp_emroc noobs;
    var cutoff sensitivity oneminusspecificity ;
    where upcase(datarole) = 'VALIDATE' and upcase(model)='NEURAL' ;
    title "ROC Table:  Validation Data" ;
run;
```

5.4.4 How Did the Neural Network Node Pick the Optimum Weights for This Model?

In Section 5.3, I described how the optimum weights are found in a neural network model. There, I described the two-step procedure of estimating and selecting the weights. In this section, I show the results of these two steps with reference to the neural network model discussed in Sections 5.4.1 and 5.4.2.

[3] A.A. Afifi, V. A. Clark, and S. May, *Computer Aided Multivariate Analysis*, 4th ed. (London: Chapman & Hall/CRC Press, 2004).

The weights such as those shown in Equations 5.1, 5.3, 5.5, 5.7, 5.9, 5.11, and 5.13 are shown in the **Results** window of the **Neural Network** node. The estimated weights created at each iteration can be seen by performing the following steps:

1. Open the **Results** window.
2. Select **View** from the menu bar.
3. Select **Model**.
4. Select **Weights-History**.

These steps are shown in Displays 5.7 and 5.7A.

Display 5.7

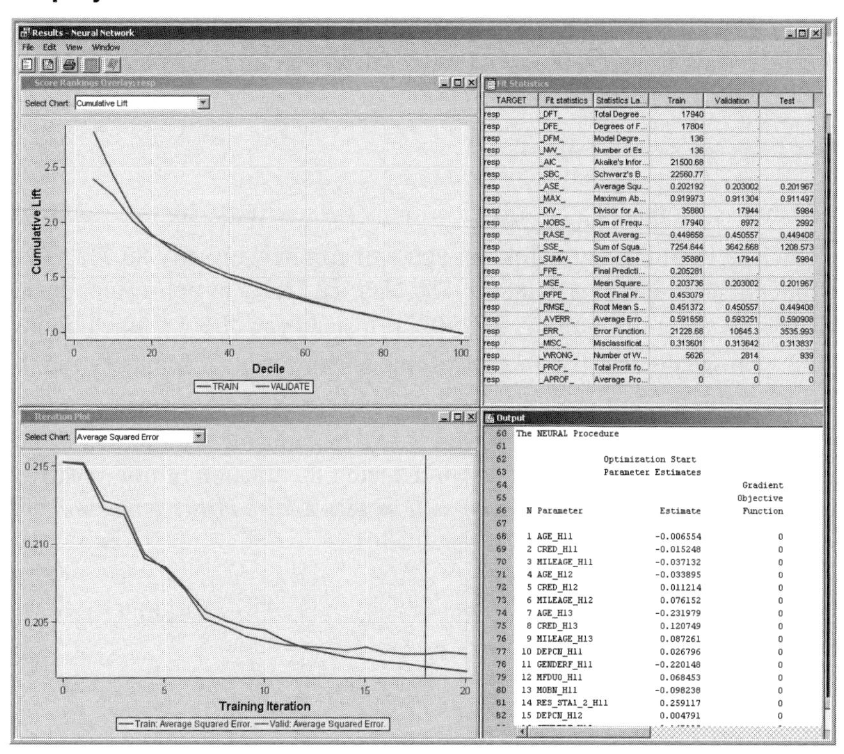

Display 5.7A

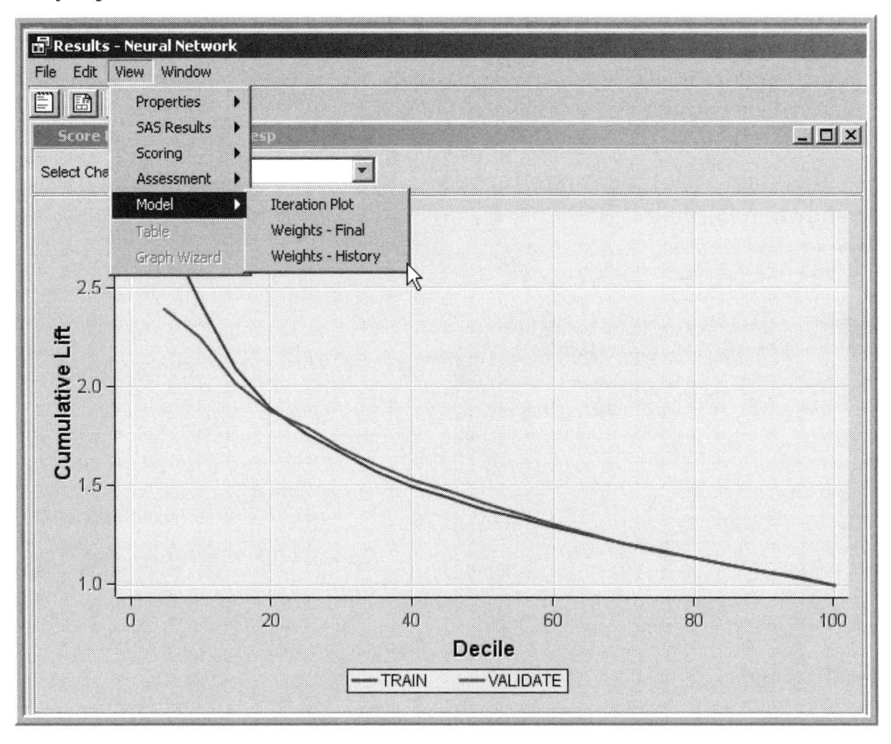

Display 5.7B shows a partial view of the **Weights-History** window.

Display 5.7B

Weights - History

ITER	AGE -> H11	CRED -> H11	MILEAGE -> H11	AGE -> H12	CRED -> H12	MILEAGE -> H12	AGE -> H13	CRED -> H13	MILEAGE ->...
0	-0.00655	-0.01525	-0.03713	-0.03389	0.011214	0.076152	-0.23198	0.120749	0.087261
1	-0.00655	-0.01525	-0.03713	-0.03389	0.011214	0.076152	-0.23198	0.120749	0.087261
2	-0.00639	-0.01514	-0.03726	-0.01432	0.024458	0.060054	-0.21665	0.133041	0.072556
3	-0.00918	-0.01542	-0.03535	0.005418	0.028405	0.046823	-0.20324	0.136935	0.061694
4	-0.04466	-0.01631	-0.01159	0.247052	0.059343	-0.10574	-0.04393	0.170102	-0.0646
5	-0.04942	-0.01162	-0.0071	0.287137	0.051616	-0.13885	-0.01457	0.168202	-0.09234
6	-0.03637	-0.00152	-0.01274	0.220467	0.017068	-0.11037	-0.06089	0.146524	-0.06911
7	-0.02428	0.021814	-0.01603	0.174104	0.012491	-0.09421	-0.07697	0.147599	-0.07079
8	-0.02199	0.018874	-0.02104	0.1528	0.054569	-0.04856	-0.07576	0.17279	-0.05087
9	-0.01602	0.034694	-0.01965	0.111818	0.053061	-0.03113	-0.07532	0.169474	-0.06197
10	-0.01423	0.075511	-0.00357	0.065508	-0.00239	-0.06489	-0.03922	0.116303	-0.12792
11	-0.03222	0.06229	-0.01148	0.100904	0.022698	-0.03694	-0.00278	0.119795	-0.08649
12	-0.05012	0.054068	-0.01268	0.110186	0.059501	-0.03955	0.038873	0.111461	-0.05503
13	-0.05906	0.066871	-0.00585	0.065915	0.063226	-0.05796	0.069335	0.072811	-0.05478
14	-0.06266	0.065735	0.000069	0.010912	0.098405	-0.09118	0.038698	0.044454	-0.0384
15	-0.10262	0.05492	-0.00637	0.011591	0.126087	-0.08468	0.020465	0.010397	0.008725
16	-0.11778	0.055197	-0.00606	0.029418	0.130263	-0.0839	0.02726	0.025289	-0.00596
17	-0.13382	0.056688	-0.00195	0.040487	0.121518	-0.09933	0.015674	0.027964	-0.02599
18	-0.18049	0.06755	-0.00429	0.038434	0.123747	-0.10437	-0.01431	0.018264	-0.02861
19	-0.23399	0.082986	-0.01638	0.028343	0.130387	-0.09921	-0.04786	0.044168	-0.02466
20	-0.25697	0.141751	-0.03412	-0.00663	0.095957	-0.09129	-0.05244	0.084811	-0.03792

The second column in Display 5.7B shows the weight of the variable AGE in hidden unit 1 at each of the 20 iterations. The fifth column shows the weight of AGE in hidden unit 2 at each iteration. The eighth column shows the weight of AGE in the third hidden unit. Similarly, you can trace through the weights of other variables. The Weights-History table can be saved as a SAS data set.

The final weights (iteration 18) of the variables AGE, CRED, and MILEAGE into each of the hidden units are shown in Display 5.7C.

Display 5.7C

LABEL	FROM	TO	WEIGHT
AGE -> H11	AGE	H11	-0.18049
CRED -> H11	CRED	H11	0.06755
MILEAGE -> H11	MILEAGE	H11	-0.00429
AGE -> H12	AGE	H12	0.03843
CRED -> H12	CRED	H12	0.12375
MILEAGE -> H12	MILEAGE	H12	-0.10437
AGE -> H13	AGE	H13	-0.01431
CRED -> H13	CRED	H13	0.01826
MILEAGE -> H13	MILEAGE	H13	-0.02861

Outputs of the hidden units become inputs to the target layer. In the target layer, these inputs are combined using the weights estimated by the **Neural Network** node. The weights of the three hidden units and the bias coefficient in the target layer are shown for each iteration in Display 5.7D.

Display 5.7D

ITER	H11 -> resp1	H12 -> resp1	H13 -> resp1	BIAS -> resp1
0	0.00000	0.00000	0.00000	-0.78334
1	-0.00056	-0.06081	-0.05492	-0.78245
2	0.09556	-0.56601	-0.46552	-1.03018
3	0.10574	-0.63507	-0.51696	-1.06052
4	0.15817	-0.77680	-0.56418	-1.20221
5	0.14018	-0.75626	-0.55507	-1.17912
6	0.10931	-0.59275	-0.45323	-1.12531
7	0.18830	-0.69503	-0.56389	-1.21613
8	0.10463	-0.60154	-0.53486	-1.13736
9	0.10151	-0.65818	-0.58927	-1.12936
10	0.16339	-0.70226	-0.78295	-1.05582
11	0.12824	-0.72107	-0.72586	-1.05553
12	0.16906	-0.68377	-0.76414	-1.00694
13	0.21300	-0.71257	-0.84584	-0.95796
14	0.26815	-0.73487	-0.86254	-0.94828
15	0.27137	-0.74066	-0.89318	-0.92753
16	0.25630	-0.70146	-0.89208	-0.92028
17	0.29055	-0.71203	-0.87218	-0.93389
18	0.34749	-0.68139	-0.90797	-0.91040
19	0.41966	-0.65697	-0.94482	-0.88618
20	0.49722	-0.65745	-0.96656	-0.85943

Display 5.7E shows the final weights (iteration 18) of the three hidden units, and the bias coefficient in the target layer.

Display 5.7E

LABEL	FROM	TO	WEIGHT
H11 -> resp1	H11	resp1	0.34749
H12 -> resp1	H12	resp1	-0.68139
H13 -> resp1	H13	resp1	-0.90797
BIAS -> resp1	BIAS	resp1	-0.91040

In the model I have developed, the weights generated at the 18th iteration are the optimal weights, because the Average Squared Error computed from the validation data set reaches its *minimum*. This is shown in Display 5.8.

Display 5.8

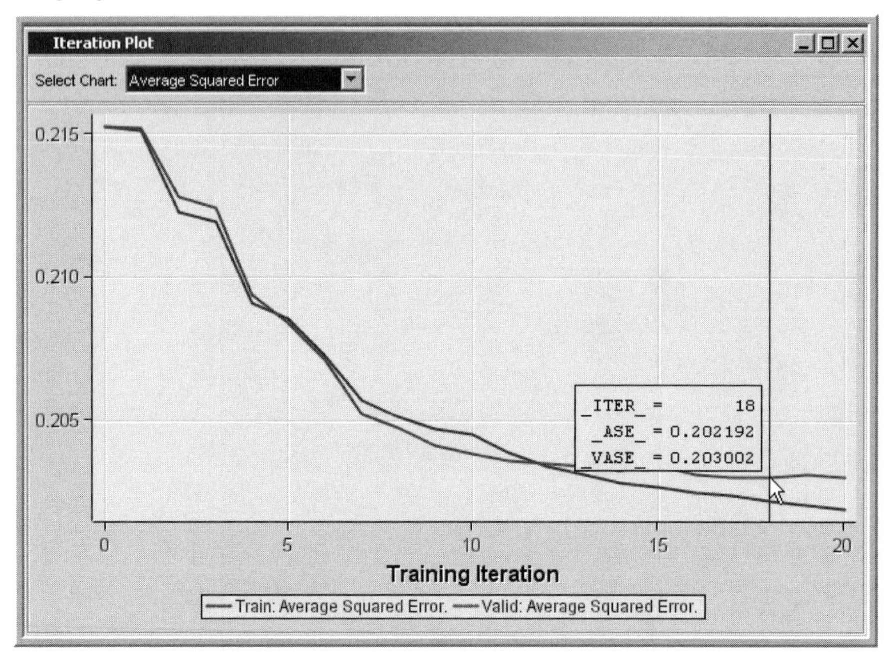

5.4.5 Scoring a Data Set Using the Neural Network Model

The SAS code generated by **Neural Network** node can be used to score a data set within Enterprise Miner or outside. Here, I will score a data set inside Enterprise Miner.

The process flow diagram with a scoring data set is shown in Display 5.9.

Display 5.9

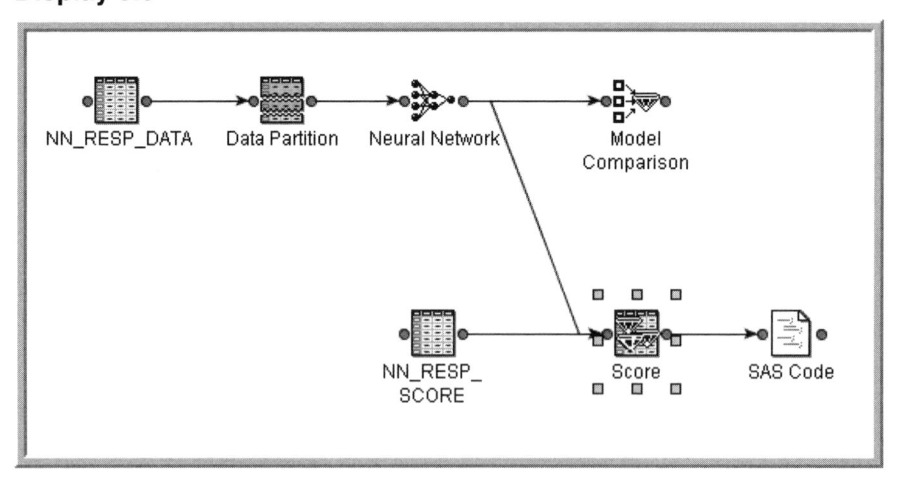

The **Role** property of the data set to be scored should be set to **Score** as shown in Display 5.10.

Display 5.10

Property	Value
Node ID	Ids2
Imported Data	...
Exported Data	...
⊟Settings	
Variables	...
Decisions	...
Output Type	View
Role	Score
Rerun	No
⊟Data	
Data Selection	Data Source
New Table	
⊟Metadata	
Data Source	NN_RESP_SCORE ...
Table	NN_RESP_SCORE
Library	THEBOOK
Description	
Number of Observations	6017
Number of Columns	18
External Data	No
Sampling Rate	.
Segment	
Scope	Local
⊟Status	
Time of Creation	1/13/07 1:07 PM
Run Id	5d2304a1-99d9-4751-85eb-0808c
Last Error	
Last Status	Complete
Needs Updating	No
Needs to Run	No
Time of Last Run	1/13/07 1:12 PM
Run Duration	0 Hr. 0 Min. 14.88 Sec.
Grid Host	

The score code applies the SAS code generated by the **Neural Network** node to the Score data set NN_RESP_SCORE shown in Display 5.9.

For each record, the probability of response, the probability of non-response, and the expected profit of each record is calculated and appended to the scored data set.

Display 5.11 shows the segment of the score code where the probabilities of response and non-response are calculated. The coefficients of H11, H12, and H13 in Display 5.11 are the weights in the final output layer. These are same as the coefficients shown in Display 5.7E.

Display 5.11

```
P_resp1  =     0.347491113970005 * H11  +    -0.68139124506479 * H12
        +     -0.9079702335076 * H13 ;
P_resp1  =     -0.91040266030932 + P_resp1 ;
P_resp0  = 0;
_MAX_ = MAX (P_resp1 , P_resp0 );
_SUM_ = 0.;
P_resp1  = EXP(P_resp1  - _MAX_);
_SUM_ = _SUM_ + P_resp1 ;
P_resp0  = EXP(P_resp0  - _MAX_);
_SUM_ = _SUM_ + P_resp0 ;
P_resp1  = P_resp1  / _SUM_;
P_resp0  = P_resp0  / _SUM_;
```

The code segment given in Display 5.11 calculates the probability of response using the formula $P\_resp1_i = \dfrac{\exp(\eta_{i21})}{1+\exp(\eta_{i21})}$. This formula is the same as Equation 5.14 in Section 5.2.4, re-written to be consistent with the SAS code shown in Display 5.11. The subscript i is added to emphasize that this is a record-level calculation. In the code shown in Display 5.11, the probability of non-response is calculated as $p\_resp0 = \dfrac{1}{1+\exp(\eta_{i21})}$.

The probabilities calculated above are modified by the *prior probabilities* I entered before running the **Neural Network** node. These probabilities are shown in Display 5.12.

Display 5.12

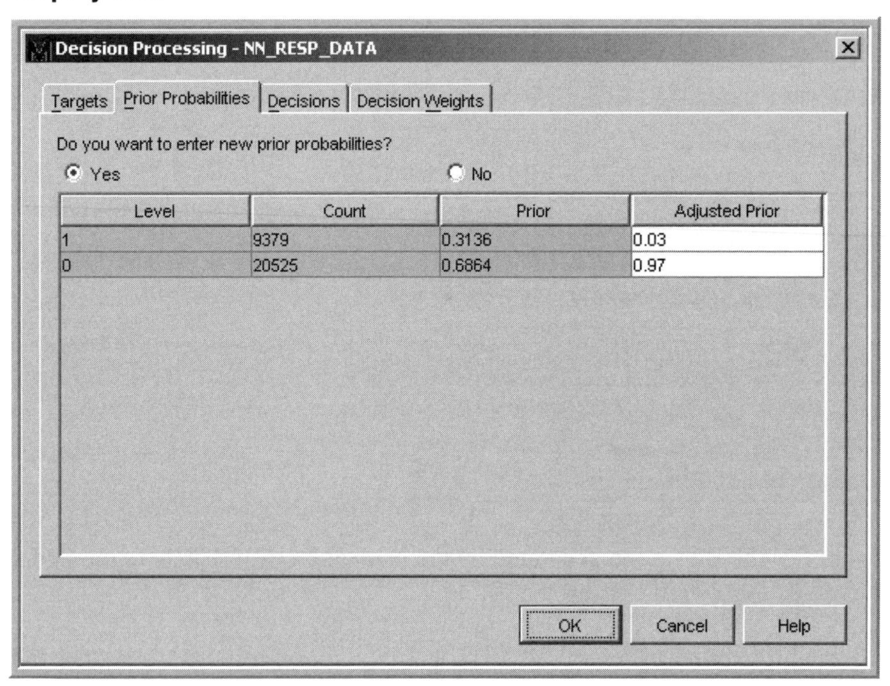

The *prior probabilities* can be entered when the data source is created. *Prior probabilities* are entered because the responders are overrepresented in the modeling sample, which is extracted from a larger sample. In the larger sample, the proportion of responders is only 3%. In the modeling sample, the proportion of responders is 31.36%. Hence, the probabilities should be adjusted before expected profits are computed. The SAS code generated by the **Neural Network** node and passed on to the **Score** node includes statements for making this adjustment. Display 5.13 shows these statements.

Display 5.13

```
*** Update Posterior Probabilities;
P_resp1 = P_resp1 * 0.03 / 0.31368937998772;
P_resp0 = P_resp0 * 0.97 / 0.68631062001227;
drop _sum; _sum = P_resp1 + P_resp0 ;
if _sum > 0 then do;
   P_resp1 = P_resp1 / _sum;
   P_resp0 = P_resp0 / _sum;
end;
```

In this example, I used the following profit matrix.

Display 5.14

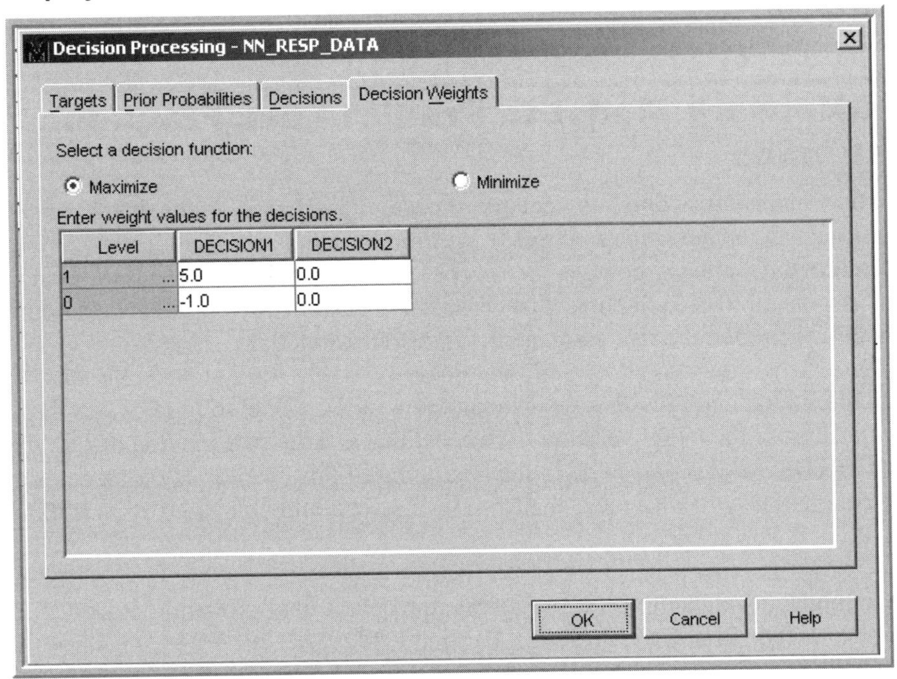

Given the above profit matrix, calculation of expected profit under the alternative decisions of classifying an individual as responder or non-responder proceeds as follows. Using the neural network model, the scoring algorithm first calculates the individual's probability of response and non-response. Suppose the calculated probability of response for an individual is 0.3, and probability of non-response is 0.7. The expected profit if the individual is classified as responder is 0.3x\$5 + 0.7x (–\$1.0) = \$0.8. The expected profit if the individual is classified as non-responder is 0.3x (\$0) + 0.7x (\$0) = \$0. Hence classifying the individual as responder (Decision1) yields a higher profit than if the individual is classified as non-responder (Decision2). An additional field is added to the record in the scored data set indicating the decision to classify the individual as a responder.

These calculations are shown in the score code segment shown in Display 5.15.

Display 5.15

```
*** Compute Expected Consequences and Choose Decision;
_decnum = 1; drop _decnum;

D_RESP = '1' ;
EP_RESP = P_resp1 * 5 + P_resp0 * -1;
drop _sum;
_sum = P_resp1 * 0 + P_resp0 * 0;
if _sum > EP_RESP + 2.273737E-12 then do;
   EP_RESP = _sum; _decnum = 2;
   D_RESP = '0' ;
end;


*** End Decision Processing ;
```

5.4.6 Saving the SAS Code

In order to use the model to score a data set outside Enterprise Miner, you must save the score code for the model that has been estimated.

5.5 A Neural Network Model to Predict Loss Frequency in Auto Insurance

The premium that an insurance company charges a customer is based on the degree of risk of monetary loss to which the customer exposes the insurance company. The higher the risk, the higher the premium the company charges. For proper rate setting, it is essential to predict the degree of risk associated with each current or prospective customer. Neural networks can be used to develop models to predict the risk associated with each individual.

In the example presented here, I will develop a neural network model to predict *loss frequency* at the customer level. I use the target variable *LOSSFRQ* that is a discrete version of *loss frequency.* The definition of *LOSSFRQ* is presented in the beginning of this chapter. If the target is a discrete form of a continuous variable with more than two levels, it should be treated as an ordinal variable, as I did in this example.

The goal of the model developed here is to estimate the conditional probabilities as $\Pr(LOSSFRQ = 0 \mid X)$, $\Pr(LOSSFRQ = 1 \mid X)$, and $\Pr(LOSSFRQ = 2 \mid X)$, where X is a set of inputs or explanatory variables.

I have already discussed the general framework for neural network models in Section 5.2. There I gave an example of a neural network model with two hidden layers. I also pointed out that the outputs of the final hidden layer can be considered as complex transformations of the original inputs. These are given in Equations 5.1 through 5.12.

In the example presented below, there is only one hidden layer, and it has three units. The outputs of the hidden layer are H_{11}, H_{12}, and H_{13}. Since these are nothing but transformations of the original inputs, I can write the desired conditional probabilities as $\Pr(LOSSFRQ = 0 \mid H_{11}, H_{12}, and\ H_{13})$, $\Pr(LOSSFRQ = 1 \mid H_{11}, H_{12}, and\ H_{13})$, and $\Pr(LOSSFRQ = 2 \mid H_{11}, H_{12}, and\ H_{13})$.

5.5.1 Loss Frequency As an Ordinal Categorical Target

5.5.1.1 Target Activation Function Property

If you set the **Target Layer Activation Function** property of the **Neural Network** node to **Logistic**, Enterprise Miner gives you a proportional odds or logistic regression with a cumulative logit link. In addition, if you also set the **Target Layer Combination Function** property to **Linear**, then combination functions for different levels of the target will have equal slopes, but different intercepts. In general, in the case of ordinal targets, when you set the **Target Layer Activation Function** property to **Logistic**, the bias is the only coefficient of the combination function that will be different for different target levels.

A review of the theory behind the proportional odds or logistic regression with cumulative logit link [4] is useful to get an appreciation of the results produced by the **Neural Network** node. I present the theory using the auto insurance loss frequency example.

According to this theory, the observable variable LOSSFRQ is determined by an underlying unobservable variable y^*, also called the latent variable. The observable variable *LOSSFRQ* takes the values 0, 1, and 2 and the unobservable variable y^* is defined by the relationship $y_i^* = w'H_i + U_i$, where $w' = (w_0, w_1, w_2, w_3)$ and $H_i = \begin{pmatrix} 1 \\ H_{11} \\ H_{12} \\ H_{13} \end{pmatrix}_i$ for the ith observation. U_i represents random error terms that make up for the fact that the relationship between the y_i^* values and $w'H_i$ values does not hold exactly for each customer. U_i follows a probability distribution function I will call F below. As I pointed out above, H_{11}, H_{12}, and H_{13} are the outputs of the last hidden layer, and they are functions of the inputs. Hence, the latent variable y^* is determined by the inputs.

The relation between the underlying variable y^* and the target variable LOSSFRQ is defined as

- If $y^* > 0$ (a threshold value), then *LOSSFRQ* = 2,
- If $y^* > -C$ then *LOSSFRQ* ≥ 1, where C is a positive constant,
- If $y^* \leq -C$ then L*OSSFRQ* = 0.

To simplify notation I will denote *LOSSFRQ* by y. For the ith individual record, the variable *LOSSFRQ* is denoted by y_i.

Therefore,

$$\begin{aligned} P(y_i = 2 \mid H_i) &= P(y_i^* > 0 \mid H_i) \\ &= P(w'H_i + U_i > 0) \\ &= F(U_i > -w'H_i) \\ &= 1 - F(-w'H_i) = F(w'H_i), \end{aligned}$$

$$\begin{aligned} P(y_i \geq 1 \mid H_i) &= P(y_i = 1 \mid H_i) + P(y_i = 2 | H_i) = P(y_i^* > -C \mid H_i) \\ &= P(w'H_i + U_i > -C) \\ &= F(U_i > -w'H_i - C) \\ &= 1 - F(-w'H_i - C) = F(w'H_i + C), \text{ and} \end{aligned}$$

$$P(y_i = 0 \mid H_i) = 1 - P(y_i > 0 \mid H_i)$$

---

[4] G.S. Maddala, "Limited-Dependent and Qualitative Variables in Econometrics," in *Econometric Society Monographs* (New York: Cambridge University Press, 1986).

Hence, the individual probabilities can be derived as follows:

$$P(y_i = 2 \mid H_i) = P(U_i \le w'H_i) = F(w'H_i),$$
$$P(y_i = 1 \mid H_i) = P(w'H_i < U_i \le w'H_i + C) = F(w'H_i + C) - F(w'H_i), \text{ and}$$
$$P(y_i = 0 \mid H_i) = 1 - P(y_i = 1 \mid H_i) - P(y_i = 2 \mid H_i) = 1 - F(w'H_i + C)$$

The above equations can be written as

$$P(y_i = 2 \mid H_i) = F(w'H_i)$$
$$P(y_i = 1 \mid H_i) + P(Y_i = 2) = F(w'H_i + C), \text{ and}$$
$$P(y_i = 0 \mid H_i) = 1 - P(y_i = 1 \mid H_i) - P(y_i = 2 \mid H_i)$$

Since I set the **Target Layer Activation Function** property of the **Neural Network** node to **Logistic**, $F()$ is a logistic distribution function. If I substitute the Logistic distribution function in the above equations, I have

$$P(y_i = 2 \mid H_i) = F(w'H_i) = \frac{1}{1 + e^{-(w'H_i)}} \quad (5.18)$$

$$P(y_i = 1 \mid H_i) + P(y_i = 2 \mid H_i) = F(w'H_i + C) = \frac{1}{1 + e^{-(w'H_i + C)}} \quad (5.19)$$

$$P(y_i = 0 \mid H_i) = 1 - P(y_i = 1 \mid H_i) - P(y_i = 2 \mid H_i) \quad (5.20)$$

These equations can be written as *cumulative logits*, as follows.

$$\ln\left(\frac{\mathrm{P}(y_i = 2 \mid H_i)}{1 - \mathrm{P}(y_i = 2 \mid H_i)}\right) = w_0 + w_1 H_{11i} + w_2 H_{12i} + w_3 H_{13i} \quad (5.21)$$

$$\ln\left(\frac{\mathrm{P}(y_i = 2 \mid H_i) + \mathrm{P}(y = 1 \mid H_i)}{1 - (\mathrm{P}(y_i = 2 \mid H_i) + \mathrm{P}(y = 1 \mid H_i))}\right) = (w_0 + C) + w_1 H_{11i} + w_2 H_{12i} + w_3 H_{13i} \quad (5.22)$$

$$\mathrm{P}(y_i = 0 \mid H_i) + p(y_i = 1 \mid H_i) + \mathrm{P}(y_i = 2 \mid H_i) = 1 \quad (5.23)$$

If I then take the ratio of the expressions in Equation 5.22 and Equation 5.21, I get Equation 5.24.

$$\frac{\left\{(P(y_i = 2/H_i) + P(y_i = 1/H_i)) \big/ (1 - (P(y_i = 2/H_i) + P(y_i = 1/H_i)))\right\}}{\left\{P(y_i = 2/H_i) \big/ (1 - P(y_i = 2/H_i))\right\}}$$

$$= \exp(C) \text{, a constant.} \quad (5.24)$$

This shows that the odds ratios for y = 1 or 2, and for y = 2 differ by a constant multiple, $\exp(C)$. It is for this reason that this type of model is called a *proportional odds* model.

Equations 5.21 through 5.23 are estimated by the **Neural Network** node, and solved for the probabilities, as shown in Display 5.16.

Display 5.16

```
P_LOSSFRQ2  =      1.56258835780253 * H11  +      -1.5801138416407 * H12
       +    -1.56794112123126 * H13 ;
P_LOSSFRQ1  =      1.56258835780253 * H11  +      -1.5801138416407 * H12
       +    -1.56794112123126 * H13 ;
P_LOSSFRQ0  =      1.56258835780253 * H11  +      -1.5801138416407 * H12
       +    -1.56794112123126 * H13 ;
P_LOSSFRQ2  =     -2.94734581823348 + P_LOSSFRQ2 ;
P_LOSSFRQ1  =     -0.20578811716334 + P_LOSSFRQ1 ;
DROP _EXP_BAR;
_EXP_BAR=50;
P_LOSSFRQ2  = 1.0 / (1.0 + EXP(MIN( - P_LOSSFRQ2 , _EXP_BAR)));
P_LOSSFRQ1  = 1.0 / (1.0 + EXP(MIN( - P_LOSSFRQ1 , _EXP_BAR)));
P_LOSSFRQ0  = 1. -  P_LOSSFRQ1 ;
P_LOSSFRQ1  = P_LOSSFRQ1  - P_LOSSFRQ2 ;
```

Display 5.16 is a segment of the SAS code generated by the **Neural Network** node. It can be seen that $w_0 = -2.947345$ and $w_0 + C = -0.205788$. Therefore, $C = 2.741557$.

5.5.1.2 Target Layer Error Function Property

I set the **Target Layer Error Function** property to **MBernoulli**. Equation 5.16 is a Bernoulli error function for a binary target. If the target variable (y) takes the values 1 and 0, then the Bernoulli error function would be

$$E = -2\sum_{i=1}^{n}\left\{ y_i \ln\frac{\pi(W,X_i)}{y_i} + (1-y_i)\ln\frac{1-\pi(W,X_i)}{1-y_i} \right\}$$

If the target has *k* levels, you can create *k* dummy variables $y_1, y_2, y_3, \ldots, y_k$ for each record. In this example, the observed loss frequency has the three levels: 0, 1, and 2. Hence, I can create three dummy variables: y_1, y_2, and y_3. If the observed loss frequency[5] is 0 for the ith observation, then $y_{i1} = 1$, and both yi2 and yi3 will be 0. Similarly, if the observed loss frequency is 1, then $y_{i2} = 1$, and both yi1 and yi3 will equal 0. If the observed loss frequency is 2, then $y_{i3} = 1$, and the other two dummy variables will take on values of 0. Using these indicator variables, it is possible to generalize the error function to *k* levels as

$$E = -2\sum_{i=1}^{n}\sum_{j=1}^{k}\left\{ y_{ij} \ln\frac{\pi(W_j,X_i)}{y_{ij}} + (1-y_{ij})\ln\frac{1-\pi(W_j,X_i)}{1-y_{ij}} \right\}$$

[5] This is the target variable *LOSSFRQ*.

where n is the number of observations in the training data set and k is the number of target levels.

The weights W_j for the jth level of the target variable are chosen so as to minimize the error function. The minimization is achieved by means of an iterative process. As mentioned before, at each iteration a set of weights is generated, each set defining a candidate model. The iteration stops when the error is minimized or the threshold number of iterations is reached. The next task in the model development process is to select one of these models. This is done by applying the **Model Selection Criterion** property to the validation data set.

5.5.1.3 Model Selection Criterion Property

I set the **Model Selection Criterion** property to **Profit/Loss**. I also supplied the following profit matrix.

Display 5.17

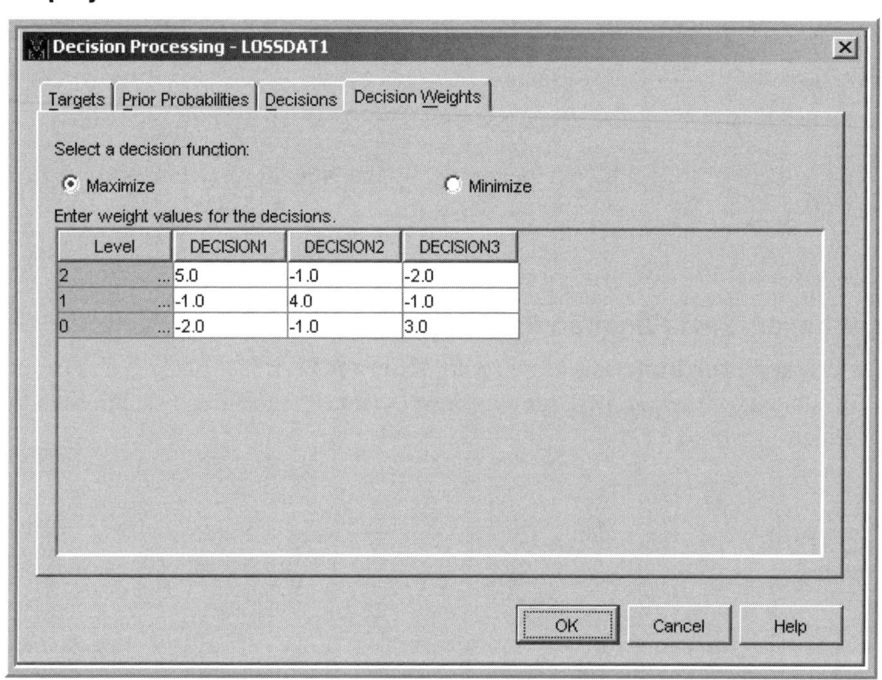

Level	DECISION1	DECISION2	DECISION3
2	5.0	-1.0	-2.0
1	-1.0	4.0	-1.0
0	-2.0	-1.0	3.0

The selected model details can be seen in the **Results** window.

5.5.1.4 Score Ranks in the Results Window

The **Results** window is shown in Display 5.18.

Display 5.18

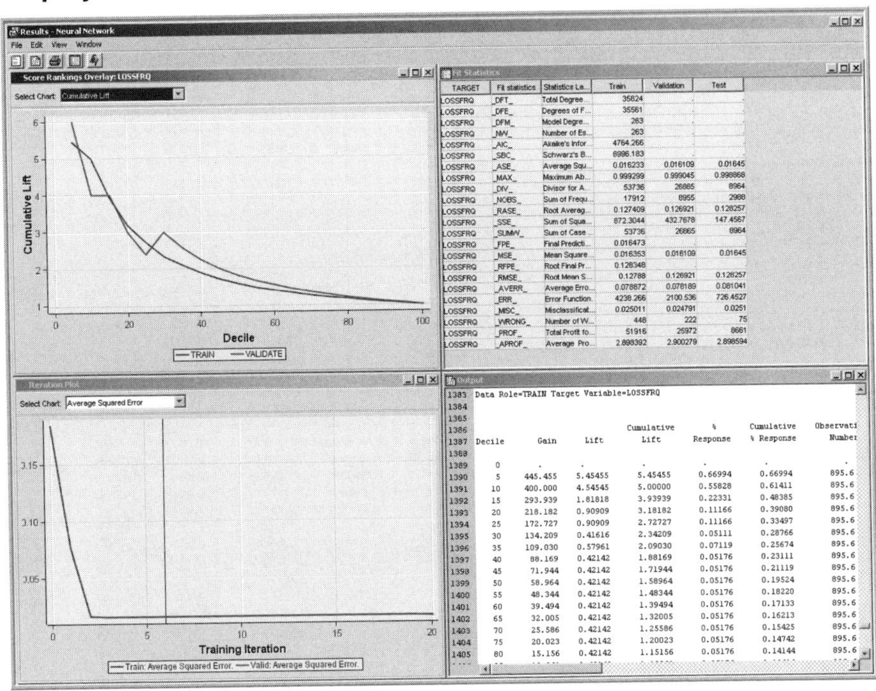

The score rankings are shown in the **Score Rankings** window. To see how lift and cumulative lift are presented for a model with an ordinal target, you can open the table corresponding to the graph by clicking on the **Score Rankings** window, and then clicking on the Table icon on the task bar. The data table behind the **Score Rankings** graph is shown in Displays 5.19A and 5.19B.

Display 5.19A

Score Rankings Overlay: LOSSFRQ - EMWS1.Neural_EMRANK

Target Vari...	Data Role	Event	Decile	Cumulative ...	Best Cumul...	Base Cumul...	Bin	Observation...	Number of E...	Gain	Lift	Cumulative Lift
LOSSFRQ	TRAIN		0	0	0	0	.	.	.	.	.	.
LOSSFRQ	TRAIN	2	5	27.27273	100	5	1	895.6	6	445.4545	5.454545	5.454545
LOSSFRQ	TRAIN	2	10	50	100	10	2	895.6	5	400	4.545455	5
LOSSFRQ	TRAIN	2	15	59.09091	100	15	3	895.6	2	293.9394	1.818182	3.939394
LOSSFRQ	TRAIN	2	20	63.63636	100	20	4	895.6	1	218.1818	0.909091	3.181818
LOSSFRQ	TRAIN	2	25	68.18182	100	25	5	895.6	1	172.7273	0.909091	2.727273
LOSSFRQ	TRAIN	2	30	70.26261	100	30	6	895.6	0.457774	134.2087	0.416159	2.342087
LOSSFRQ	TRAIN	2	35	73.16064	100	35	7	895.6	0.637567	109.0304	0.579607	2.090304
LOSSFRQ	TRAIN	2	40	75.26774	100	40	8	895.6	0.463561	88.16935	0.421419	1.881693
LOSSFRQ	TRAIN	2	45	77.37484	100	45	9	895.6	0.463561	71.94408	0.421419	1.719441
LOSSFRQ	TRAIN	2	50	79.48193	100	50	10	895.6	0.463561	58.96386	0.421419	1.589639
LOSSFRQ	TRAIN	2	55	81.58903	100	55	11	895.6	0.463561	48.34369	0.421419	1.483437
LOSSFRQ	TRAIN	2	60	83.69612	100	60	12	895.6	0.463561	39.49354	0.421419	1.394935
LOSSFRQ	TRAIN	2	65	85.80322	100	65	13	895.6	0.463561	32.00495	0.421419	1.32005
LOSSFRQ	TRAIN	2	70	87.91031	100	70	14	895.6	0.463561	25.58616	0.421419	1.255862
LOSSFRQ	TRAIN	2	75	90.01741	100	75	15	895.6	0.463561	20.02321	0.421419	1.200232
LOSSFRQ	TRAIN	2	80	92.12451	100	80	16	895.6	0.463561	15.15563	0.421419	1.151556
LOSSFRQ	TRAIN	2	85	94.2316	100	85	17	895.6	0.463561	10.86071	0.421419	1.108607
LOSSFRQ	TRAIN	2	90	96.3387	100	90	18	895.6	0.463561	7.042997	0.421419	1.07043
LOSSFRQ	TRAIN	2	95	98.44579	100	95	19	895.6	0.463561	3.627151	0.421419	1.036272
LOSSFRQ	TRAIN	2	100	100	100	100	20	895.6	0.341925	0	0.310841	1
LOSSFRQ	VALIDATE		0	0	0	0	.	.	.	.	.	.
LOSSFRQ	VALIDATE	2	5	30	100	5	1	447.75	3	500	6	6
LOSSFRQ	VALIDATE	2	10	40	100	10	2	447.75	1	300	2	4
LOSSFRQ	VALIDATE	2	15	60	100	15	3	447.75	2	300	4	4
LOSSFRQ	VALIDATE	2	20	60	100	20	4	447.75	0	200	0	3
LOSSFRQ	VALIDATE	2	25	60	100	25	5	447.75	0	140	0	2.4
LOSSFRQ	VALIDATE	2	30	89.89583	100	30	6	447.75	2.989583	199.6528	5.979167	2.996528
LOSSFRQ	VALIDATE	2	35	90	100	35	7	447.75	0.010417	157.1429	0.020833	2.571429
LOSSFRQ	VALIDATE	2	40	90	100	40	8	447.75	0	125	0	2.25
LOSSFRQ	VALIDATE	2	45	90	100	45	9	447.75	0	100	0	2
LOSSFRQ	VALIDATE	2	50	90.79954	100	50	10	447.75	0.079954	81.59907	0.159907	1.815991
LOSSFRQ	VALIDATE	2	55	91.74535	100	55	11	447.75	0.094582	66.80973	0.189163	1.668097
LOSSFRQ	VALIDATE	2	60	92.69117	100	60	12	447.75	0.094582	54.48528	0.189163	1.544853
LOSSFRQ	VALIDATE	2	65	93.63699	100	65	13	447.75	0.094582	44.0569	0.189163	1.440569
LOSSFRQ	VALIDATE	2	70	94.58281	100	70	14	447.75	0.094582	35.11829	0.189163	1.351183
LOSSFRQ	VALIDATE	2	75	95.52862	100	75	15	447.75	0.094582	27.3715	0.189163	1.273715
LOSSFRQ	VALIDATE	2	80	96.47444	100	80	16	447.75	0.094582	20.59305	0.189163	1.205931
LOSSFRQ	VALIDATE	2	85	97.42026	100	85	17	447.75	0.094582	14.61207	0.189163	1.146121
LOSSFRQ	VALIDATE	2	90	98.36608	100	90	18	447.75	0.094582	9.295639	0.189163	1.092956
LOSSFRQ	VALIDATE	2	95	99.31189	100	95	19	447.75	0.094582	4.538834	0.189163	1.045388
LOSSFRQ	VALIDATE	2	100	100	100	100	20	447.75	0.068811	0	0.137621	1

Display 5.19B

Score Rankings Overlay: LOSSFRQ - EMWS1.Neural_EMRANK

Lift	Cumulative Lift	% Response	Cumulative ...	% Captured...	Report: Num...	Report: Num...	Report: Num...	Posterior Pr...	Posterior Pr...	Posterior Pr...	Computed P...
.	.	.	.	.	.	.	.	.	.	.	.
5.454545	5.454545	0.669942	0.669942	27.27273	0	7	888.6	0.009307	0.004204	0.11474	2.665029
4.545455	5	0.558285	0.614113	22.72727	0	0	895.6	0.003263	0.002666	0.004204	2.780036
1.818182	3.939394	0.223314	0.483847	9.090909	0	0	895.6	0.002328	0.002072	0.002667	2.85038
0.909091	3.181818	0.111657	0.390799	4.545455	0	0	895.6	0.001895	0.00175	0.002075	2.838834
0.909091	2.727273	0.111657	0.334971	4.545455	0	0	895.6	0.001627	0.001527	0.00176	2.895422
0.416159	2.342087	0.051114	0.287661	2.080793	0	0	895.6	0.001414	0.001282	0.001529	2.918243
0.579607	2.090304	0.071189	0.256737	2.898033	0	0	895.6	0.001274	0.000584	0.001473	2.91002
0.421419	1.881693	0.05176	0.231115	2.107096	0	0	895.6	0.000898	0.000584	0.001281	2.930469
0.421419	1.719441	0.05176	0.211186	2.107096	0	0	895.6	0.000898	0.000584	0.001281	2.930469
0.421419	1.589639	0.05176	0.195244	2.107096	0	0	895.6	0.000898	0.000584	0.001281	2.930469
0.421419	1.483437	0.05176	0.1822	2.107096	0	0	895.6	0.000898	0.000584	0.001281	2.930469
0.421419	1.394935	0.05176	0.17133	2.107096	0	0	895.6	0.000898	0.000584	0.001281	2.930469
0.421419	1.32005	0.05176	0.162132	2.107096	0	0	895.6	0.000898	0.000584	0.001281	2.930469
0.421419	1.255862	0.05176	0.154248	2.107096	0	0	895.6	0.000898	0.000584	0.001281	2.930469
0.421419	1.200232	0.05176	0.147416	2.107096	0	0	895.6	0.000898	0.000584	0.001281	2.930469
0.421419	1.151556	0.05176	0.141437	2.107096	0	0	895.6	0.000898	0.000584	0.001281	2.930469
0.421419	1.108607	0.05176	0.136162	2.107096	0	0	895.6	0.000898	0.000584	0.001281	2.930469
0.421419	1.07043	0.05176	0.131473	2.107096	0	0	895.6	0.000898	0.000584	0.001281	2.930469
0.421419	1.036272	0.05176	0.127278	2.107096	0	0	895.6	0.000898	0.000584	0.001281	2.930469
0.310841	1	0.038178	0.122823	1.554207	0	0	895.6	0.000808	0.000498	0.001281	2.944247
.	.	.	.	.	.	.	.	.	.	.	.
6	6	0.670017	0.670017	30	0	1	446.75	0.008742	0.004432	0.068769	2.73646
2	4	0.223339	0.446678	10	0	0	447.75	0.003392	0.002766	0.004432	2.756561
4	4	0.446678	0.446678	20	0	0	447.75	0.002386	0.002098	0.002768	2.807929
0	3	0	0.335008	0	0	0	447.75	0.001923	0.001773	0.002098	2.857063
0	2.4	0	0.268007	0	0	0	447.75	0.001651	0.001546	0.00177	2.865997
5.979167	2.996528	0.66769	0.334621	29.89583	0	0	447.75	0.001474	0.001403	0.001549	2.841918
0.020833	2.571429	0.002326	0.28715	0.104167	0	0	447.75	0.001332	0.001256	0.001417	2.918848
0	2.25	0	0.251256	0	0	0	447.75	0.001208	0.001099	0.001301	2.931193
0	2	0	0.223339	0	0	0	447.75	0.00116	0.001099	0.001227	2.917355
0.159907	1.815991	0.017857	0.202791	0.799535	0	0	447.75	0.000882	0.000582	0.001227	2.941446
0.189163	1.668097	0.021124	0.186276	0.945817	0	0	447.75	0.000835	0.000582	0.001093	2.941487
0.189163	1.544853	0.021124	0.172513	0.945817	0	0	447.75	0.000835	0.000582	0.001093	2.941487
0.189163	1.440569	0.021124	0.160868	0.945817	0	0	447.75	0.000835	0.000582	0.001093	2.941487
0.189163	1.351183	0.021124	0.150886	0.945817	0	0	447.75	0.000835	0.000582	0.001093	2.941487
0.189163	1.273715	0.021124	0.142235	0.945817	0	0	447.75	0.000835	0.000582	0.001093	2.941487
0.189163	1.205931	0.021124	0.134666	0.945817	0	0	447.75	0.000835	0.000582	0.001093	2.941487
0.189163	1.146121	0.021124	0.127987	0.945817	0	0	447.75	0.000835	0.000582	0.001093	2.941487
0.189163	1.092956	0.021124	0.12205	0.945817	0	0	447.75	0.000835	0.000582	0.001093	2.941487
0.189163	1.045388	0.021124	0.116738	0.945817	0	0	447.75	0.000835	0.000582	0.001093	2.941487
0.137621	1	0.015368	0.111669	0.688107	0	0	447.75	0.000758	0.000508	0.001093	2.95743

In Display 5.19A, column 3 has the title **Event**. This column has a value of 2 for all rows. From this, it is possible to infer that Enterprise Miner created the lift charts (shown in Display 5.18) for the target level 2, which is the highest level for the target variable *LOSSFRQ*.

When the target is ordinal, Enterprise Miner creates lift charts based on the probability of the highest level of the target variable. For each record in the test data set, Enterprise Miner computes the predicted or *posterior probability* $Pr(lossfrq = 2 \mid X_i)$ from the model. Then it sorts the data set in descending order of the predicted probability and divides the data set into 20 deciles (bins)[6]. Within each decile (bin) it calculates the proportion of cases with actual $lossfrq = 2$. The lift for a decile (bin) is the ratio of the proportion of cases with $lossfrq = 2$ in the decile (bin) to the proportion of cases with $lossfrq = 2$ in the entire data set.

To assess model performance, the insurance company will have to rank its customers by the expected value of $LOSSFRQ$, instead of the highest level of the target variable. To demonstrate how to construct a lift table based on expected value of the target, I first constructed the expected value for each record in the data set using the formula given below.

$$E(lossfrq \mid X_i) = Pr(lossfrq = 0 \mid X_i) * 0 + Pr(lossfrq = 1 \mid X_i) * 1 + Pr(lossfrq = 2 \mid X_i) * 2$$

[6] Enterprise Miner uses the terms *bin* and *decile*. The term *demi-decile* is sometimes used when the data set is divided into 20 bins.

Then I sorted the data set in descending order of $E(lossfrq \mid X_i)$, and divided the data set into 20 bins (demi-deciles). Within each bin, I calculated the mean of the *actual* or observed value of the target variable $lossfrq$. The ratio of the mean of $actual\ lossfrq$ in a demi-decile to the overall mean of actual $lossfrq$ for the data set is the lift of the demi-decile.

These calculations were done in the **SAS Code** node, which is connected to the **Neural Network** node as shown Display 5.20.

Display 5.20

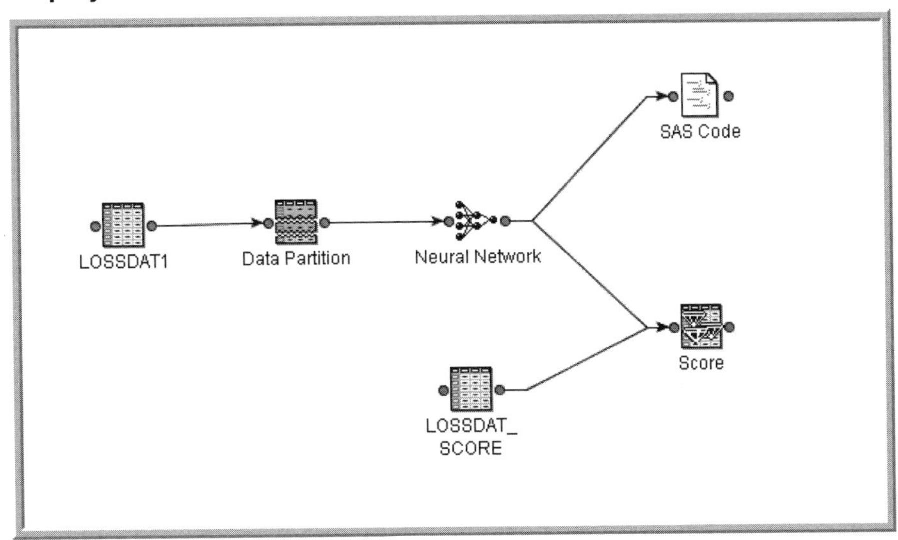

The SAS program used in the **SAS Code** node is shown in Displays 5.33, 5.34, and 5.35 in the appendix to this chapter.

The lift tables based on $E(lossfrq \mid X_i)$ are shown in Tables 5.5, 5.6, and 5.7.

Table 5.5

Loss Frequency: TRAIN

Decile	Lossfrq Mean	Lossfrq Cumulative Mean	Cumulative Lift
5	0.086034	0.086034	3.28578
10	0.059152	0.072585	2.77216
15	0.039106	0.061430	2.34611
20	0.041295	0.056393	2.15376
25	0.026816	0.050480	1.92794
30	0.026786	0.046529	1.77703
35	0.017857	0.042431	1.62052
40	0.027933	0.040620	1.55135
45	0.024554	0.038834	1.48313
50	0.023464	0.037298	1.42447
55	0.020089	0.035732	1.36469
60	0.018973	0.034335	1.31133
65	0.022346	0.033414	1.27613
70	0.017857	0.032302	1.23367
75	0.010056	0.030820	1.17706
80	0.014509	0.029800	1.13811
85	0.016741	0.029031	1.10876
90	0.013408	0.028164	1.07563
95	0.010045	0.027210	1.03919
100	0.006696	0.026184	1.00000

Table 5.6

Loss Frequency: VALIDATE

Decile	Lossfrq Mean	Lossfrq Cumulative Mean	Cumulative Lift
5	0.073826	0.073826	2.84960
10	0.062500	0.068156	2.63078
15	0.051339	0.062547	2.41424
20	0.035794	0.055866	2.15638
25	0.033482	0.051385	1.98342
30	0.044643	0.050261	1.94002
35	0.022321	0.046267	1.78586
40	0.013423	0.042167	1.62761
45	0.015625	0.039216	1.51369
50	0.015625	0.036855	1.42257
55	0.017857	0.035127	1.35587
60	0.024609	0.034252	1.32209
65	0.020089	0.033162	1.28001
70	0.013393	0.031749	1.22547
75	0.008929	0.030226	1.16671
80	0.011186	0.029038	1.12085
85	0.022321	0.028643	1.10559
90	0.020089	0.028167	1.08723
95	0.008929	0.027154	1.04813
100	0.002232	0.025907	1.00000

Table 5.7

Loss Frequency: TEST

Decile	Lossfrq Mean	Lossfrq Cumulative Mean	Cumulative Lift
5	0.087248	0.087248	3.25872
10	0.040268	0.063758	2.38138
15	0.053333	0.060268	2.25100
20	0.026846	0.051926	1.93945
25	0.000000	0.041555	1.55208
30	0.013333	0.036830	1.37561
35	0.040268	0.037321	1.39392
40	0.040000	0.037657	1.40649
45	0.033557	0.037202	1.38951
50	0.033557	0.036839	1.37592
55	0.013333	0.034693	1.29577
60	0.006711	0.032366	1.20887
65	0.020000	0.031411	1.17320
70	0.026846	0.031086	1.16105
75	0.020134	0.030357	1.13384
80	0.006667	0.028870	1.07831
85	0.026846	0.028751	1.07387
90	0.013333	0.027891	1.04174
95	0.026846	0.027837	1.03969
100	0.006667	0.026774	1.00000

5.5.2 Scoring a New Data Set with the Model

The neural network model developed can be used for scoring a data set where the value of the target is not known. The **Score** node uses the model developed by the **Neural Network** node to predict the target levels and their probabilities for the Score data set.

Display 5.20 shows the process flow for scoring. The process flow consists of an **Input Data** node called LOSSDAT_SCORE which reads the data set to be scored. The **Score** node in the process flow takes the model score code generated by the **Neural Network** node and applies it to the data set to be scored.

The output generated by the **Score** node includes the predicted (*assigned*) levels of the target and their probabilities. These calculations reflect the solution of Equations 5.18 through 5.20.

Display 5.21 shows the programming statements used by the **Score** node to calculate the predicted probabilities of the target levels.

Display 5.21

```
P_LOSSFRQ2  =      1.56258835780253 * H11  +      -1.5801138416407 * H12
         +    -1.56794112123126 * H13 ;
P_LOSSFRQ1  =      1.56258835780253 * H11  +      -1.5801138416407 * H12
         +    -1.56794112123126 * H13 ;
P_LOSSFRQ0  =      1.56258835780253 * H11  +      -1.5801138416407 * H12
         +    -1.56794112123126 * H13 ;
P_LOSSFRQ2  =     -2.94734581823348 + P_LOSSFRQ2 ;
P_LOSSFRQ1  =     -0.20578811716334 + P_LOSSFRQ1 ;
DROP _EXP_BAR;
_EXP_BAR=50;
P_LOSSFRQ2  = 1.0 / (1.0 + EXP(MIN( - P_LOSSFRQ2 , _EXP_BAR)));
P_LOSSFRQ1  = 1.0 / (1.0 + EXP(MIN( - P_LOSSFRQ1 , _EXP_BAR)));
P_LOSSFRQ0  = 1. - P_LOSSFRQ1 ;
P_LOSSFRQ1  = P_LOSSFRQ1  - P_LOSSFRQ2 ;
```

Display 5.22 shows the assignment of the target levels to individual records using the profit matrix.

Display 5.22

```
*** Decision Processing;
label D_LOSSFRQ = 'Decision: LOSSFRQ' ;
label EP_LOSSFRQ = 'Expected Profit: LOSSFRQ' ;

length D_LOSSFRQ $ 9;

D_LOSSFRQ = ' ';
EP_LOSSFRQ = .;

*** Compute Expected Consequences and Choose Decision;
_decnum = 1; drop _decnum;

D_LOSSFRQ = '2' ;
EP_LOSSFRQ = P_LOSSFRQ2 * 5 + P_LOSSFRQ1 * -1 + P_LOSSFRQ0 * -2;
drop _sum;
_sum = P_LOSSFRQ2 * -1 + P_LOSSFRQ1 * 4 + P_LOSSFRQ0 * -1;
if _sum > EP_LOSSFRQ + 2.273737E-12 then do;
   EP_LOSSFRQ = _sum; _decnum = 2;
   D_LOSSFRQ = '1' ;
end;
_sum = P_LOSSFRQ2 * -2 + P_LOSSFRQ1 * -1 + P_LOSSFRQ0 * 3;
if _sum > EP_LOSSFRQ + 2.273737E-12 then do;
   EP_LOSSFRQ = _sum; _decnum = 3;
   D_LOSSFRQ = '0' ;
end;
```

The code segments shown in Display 5.23 and Display 5.24 create additional variables.

Display 5.23

```
*** **************************;
*** Writing the I_LOSSFRQ  AND U_LOSSFRQ ;
*** **************************;
_MAXP_ = P_LOSSFRQ2 ;
I_LOSSFRQ  = "2            " ;
U_LOSSFRQ  =                2;
IF( _MAXP_ LT P_LOSSFRQ1  ) THEN DO;
   _MAXP_ = P_LOSSFRQ1 ;
   I_LOSSFRQ  = "1            " ;
   U_LOSSFRQ  =                1;
END;
IF( _MAXP_ LT P_LOSSFRQ0  ) THEN DO;
   _MAXP_ = P_LOSSFRQ0 ;
   I_LOSSFRQ  = "0            " ;
   U_LOSSFRQ  =                0;
END;
```

Display 5.23 shows the target level assignment process. The variable I_LOSSFRQ is created according to the posterior probability found for each record.

Display 5.24

```
*------------------------------------------------------------*;
* TOOL: Score Node;
* TYPE: ASSESS;
* NODE: Score;
*------------------------------------------------------------*;
*------------------------------------------------------------*;
* Score: Creating Fixed Names;
*------------------------------------------------------------*;
LABEL EM_EVENTPROBABILITY = 'Probability for level 2 of LOSSFRQ';
EM_EVENTPROBABILITY = P_LOSSFRQ2;
LABEL EM_PROBABILITY = 'Probability of Classification';
EM_PROBABILITY = max(
P_LOSSFRQ2
,
P_LOSSFRQ1
,
P_LOSSFRQ0
);
LENGTH EM_CLASSIFICATION $%dmnorlen;
LABEL EM_CLASSIFICATION = "Prediction for LOSSFRQ";
EM_CLASSIFICATION = I_LOSSFRQ;
LENGTH EM_DECISION $%dmnorlen;
LABEL EM_DECISION= "Recommended Decision for LOSSFRQ";
EM_DECISION = D_LOSSFRQ;
LABEL EM_PROFIT= "Expected Profit for LOSSFRQ";
EM_PROFIT = EP_LOSSFRQ;
```

In Display 5.24, the variable EM_DECISION represents the predicted levels of the target based on the profit matrix. The variable EM_CLASSIFICATION represents predicted values based on maximum posterior probability. EM_CLASSIFICATION is the same as I_LOSSFRQ shown in Display 5.23.

Display 5.25 shows the list of variables created by the **Score** node using the **Neural Network** model.

Display 5.25

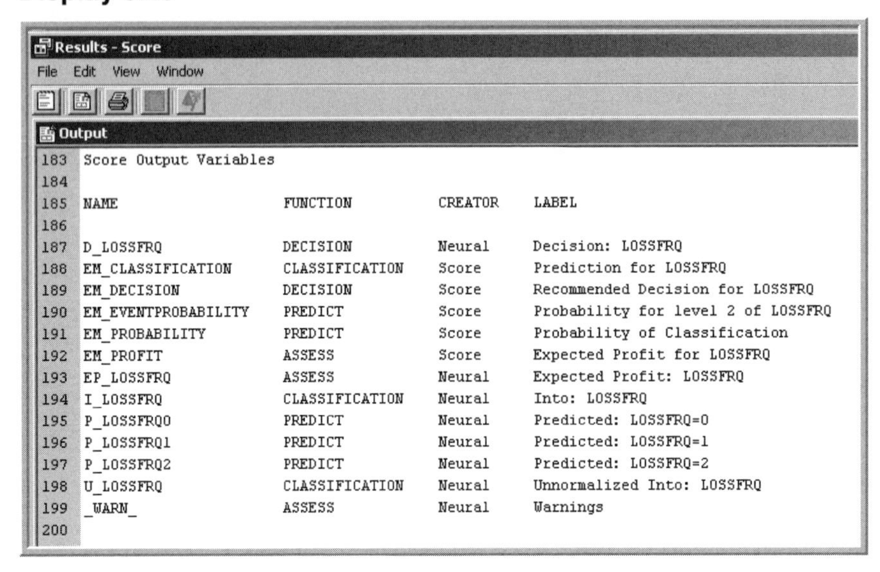

Results - Score

File Edit View Window

Output

183 Score Output Variables

NAME	FUNCTION	CREATOR	LABEL
D_LOSSFRQ	DECISION	Neural	Decision: LOSSFRQ
EM_CLASSIFICATION	CLASSIFICATION	Score	Prediction for LOSSFRQ
EM_DECISION	DECISION	Score	Recommended Decision for LOSSFRQ
EM_EVENTPROBABILITY	PREDICT	Score	Probability for level 2 of LOSSFRQ
EM_PROBABILITY	PREDICT	Score	Probability of Classification
EM_PROFIT	ASSESS	Score	Expected Profit for LOSSFRQ
EP_LOSSFRQ	ASSESS	Neural	Expected Profit: LOSSFRQ
I_LOSSFRQ	CLASSIFICATION	Neural	Into: LOSSFRQ
P_LOSSFRQ0	PREDICT	Neural	Predicted: LOSSFRQ=0
P_LOSSFRQ1	PREDICT	Neural	Predicted: LOSSFRQ=1
P_LOSSFRQ2	PREDICT	Neural	Predicted: LOSSFRQ=2
U_LOSSFRQ	CLASSIFICATION	Neural	Unnormalized Into: LOSSFRQ
WARN	ASSESS	Neural	Warnings

5.5.3 Classification of Risks for Rate Setting in Auto Insurance with Predicted Probabilities

The probabilities generated by the neural network model can be used to classify risk for two purposes:

- to select new customers
- to determine premiums to be charged according to the predicted risk

For each record in the data set, you can compute the expected *LOSSFRQ* as
$E(lossfrq \mid X_i) = Pr(lossfrq = 0 \mid X_i) * 0 + Pr(lossfrq = 1 \mid X_i) * 1 + Pr(lossfrq = 2 \mid X_i) * 2$.
The customers can be ranked by this expected frequency and assigned to different risk groups.

5.6 An Introduction to Radial Basis Functions

In Section 5.2, I described a neural network with two hidden layers and one output layer, where the hidden layers used linear combination functions (Equations 5.1 and 5.3). As mentioned earlier, by setting the **Architecture** property to **MLP** in the **Neural Network** node, you will get a neural network with one hidden layer that has a linear combination function and a sigmoid (S-shaped) activation function (*tanh*). The functions *tanh*, *logistic*, *arctan,* and *Elliot* belong to the sigmoid category. Displays 5.0B, 5.0C, 5.0D, and 5.0E depict these sigmoid functions.

I will now introduce you to another type of radial basis function that can be used as an activation function in neural networks, the radial basis function. Radial basis functions are bell-shaped, but their height and width are determined by the data in the **Neural Network** node.

Suppose I use a radial basis function for the jth hidden unit. The radial basis function is of the following form:

$$H_j = \exp(\eta_j) \tag{5.25}$$

where $$\eta_j = \log(abs(a_j)) - b_j^2 \sum_i^p (w_{ij} - x_i)^2 \tag{5.26}$$

and where x_i is the ith standardized input, the w_{ij} 's are weights representing the center (origin) of the basis function, a_j is the *height*, or altitude, b_j is a parameter sometimes referred to as the *bias*, and p is the number of inputs. Another measure that is related to *bias* is the *width*. In the SAS reference manual, width is sometimes referred to as the reciprocal of the *bias*. (See the section "Definitions of Built-in Architecture" in *Neural Network Node: Reference*, which you can access through the Enterprise Miner Help facility.) All of the charts presented here are labeled according to this definition of width. However, in other places in the SAS reference manual and in the literature in general, *width* is sometimes defined as the reciprocal of the square of the coefficient b_j. Because the graphs presented here are generated for different values of b_j, you can change the labels according to the definition that you adopt. The coefficient a_j, called the *height parameter*, measures the maximum height of the bell-shaped curve presented in Displays 5.26A through 5.26C. The subscript j is added to indicate that the radial basis function is used to calculate the output of the jth hidden unit.

When there are p inputs, there are p weights. These weights are estimated by the **Neural Network** node. The vector of inputs for any record or observation may be thought of as a point in p-dimensional space in which the vector of weights is the center. Given the height and width, the value of the radial basis function depends on the distance between the inputs and the center.

I will now plot the radial basis functions with different heights and widths for a single input that ranges between –10 and 10. Since there is only input, there will be only one weight w_{1j}, which I arbitrarily set to 0.

Display 5.26A shows a graph of the radial basis function given in Equation 5.25 with *height* = 1 and *width* = 1. That is, $a_j = 1$ and $b_j = \dfrac{1}{width} = \dfrac{1}{1} = 1$.

Display 5.26A

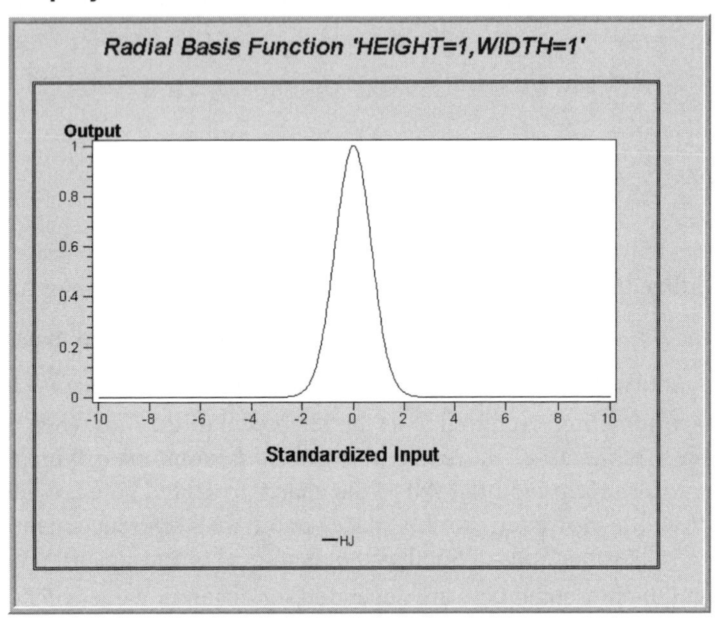

Display 5.26B shows a graph of the radial basis function given in Equation 5.25 with *height* = 1 and *width* = 4. That is, $a_j = 1$ and $b_j = \dfrac{1}{width} = \dfrac{1}{4} = 0.25.$

Display 5.26B

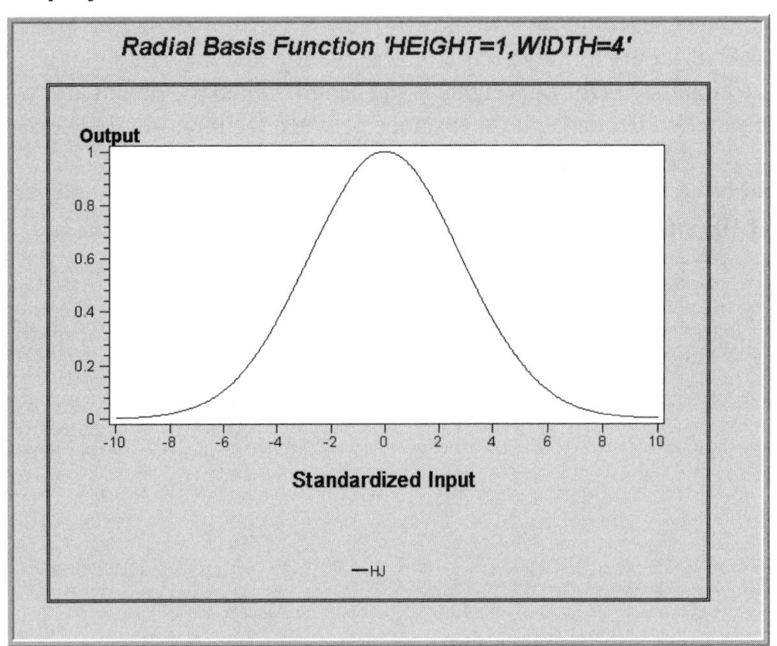

Display 5.26C shows a graph of the radial basis function given in Equation 5.25 with *height* = 5 and *width* = 4. That is, $a_j = 5$ and $b_j = \dfrac{1}{width} = \dfrac{1}{4} = 0.25.$

Display 5.26C

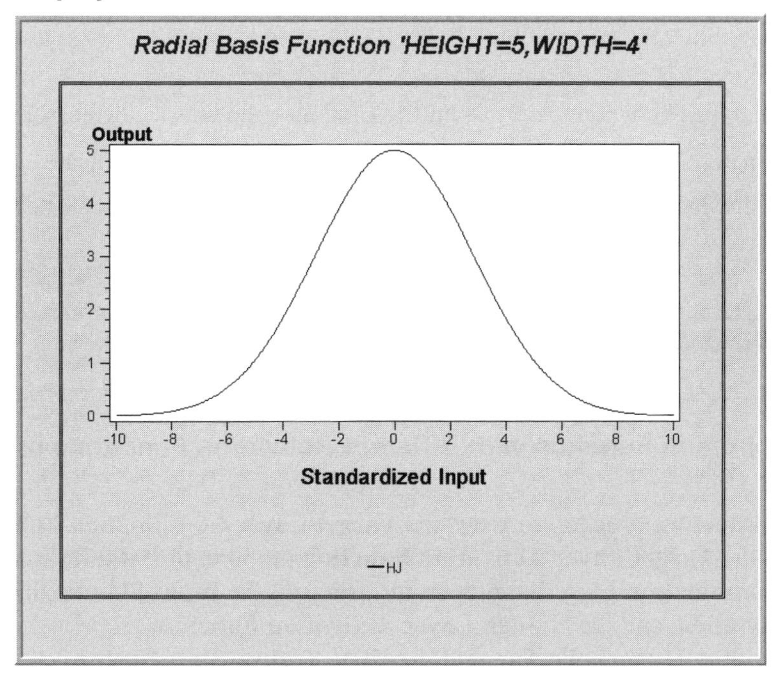

If there are three units in a hidden layer, each having a radial basis function for activation, and if the radial basis functions are constrained to sum to one, then they are called *normalized radial basis functions* in Enterprise Miner. Without this constraint, they are called *ordinary radial basis functions*.

5.7 Alternative Specifications of the Neural Network Architecture

It is possible to produce a number of neural network models[7] by specifying alternative architectures in the **Neural Network** node. You can then make a selection from among these models, based on lift or other measures, using the **Model Comparison** node.

The rest of this section presents a sample list of possible architectural specifications for a neural network in Enterprise Miner. For each specification, a single *hidden layer* with three *hidden* units is assumed. These specifications were tested for the binary target, response, i.e. (the response to a direct mail solicitation), which is described in Section 5.4. The target layer activation function was set to **Default** for all the specifications, which produced the following outputs:

$$P(y=1)=\frac{1}{1+\exp[-(w_0+w_1H_1+w_2H_2+w_3H_3)]}\text{, and}$$

[7] I have tested all the specifications listed here for the response model presented in Section 5.4, and I have determined that the default MLP specification given in Section 5.4.1 is as good as or better than the other specifications. (The full battery of results is not reported here, to avoid clutter.)

$$P(y=0) = \frac{1}{1+\exp[(w_0 + w_1 H_1 + w_2 H_2 + w_3 H_3)]},$$

where y is the target variable. The weights $w_1, w_2,$ and w_3 and also the bias w_0 differ from model to model. $H_1, H_2,$ and H_3 are the outputs of the hidden units. As these results indicate, Enterprise Miner uses a logistic activation function as the default for the target layer whenever the target is binary.

5.7.1 User-specified Architectures

5.7.1.1 Default Combination Function with Different Activation Functions for the Hidden Layer

Begin by setting the **Architecture** property to **User**, the **Target Layer Combination Function** property to **Default**, and the **Target Layer Activation Function** property to **Default**. In addition, set the **Hidden Layer Combination Function** property to **Default**, the **Input Standardization** property to **Standard deviation**, and the **Hidden Layer Activation Function** property to one of the following: **ArcTan**, **Elliot**, **Hyperbolic Tangent**, **Logistic**, **Gauss**, **Sine**, **Cosine**, **Exponential**, **Square**, **Reciprocal**, or **Softmax**.

For each of the different hidden layer activation functions listed above, I used the same hidden layer combination function by setting the **Hidden Layer Combination Function** property to **Default** in each case. The default hidden layer combination function is linear and is given by

$$\eta_j = w_{01j} + \sum_{i=1}^{p} w_{i1j} x_i \qquad (5.27)$$

for the jth hidden unit, where $w_{11j}, w_{21j}, \ldots w_{p1j}$ are the weights, $x_1, x_2, \ldots x_p$ are the standardized inputs[8], and w_{01j} is the bias coefficient. The specific numeric values that these weights and bias take will differ, as a rule, from one model to the next as I try different values from the list for the **Hidden Layer Activation Functions** property. While the hidden layer activation functions for each model take on the same functional form (Equation 5.27), the actual values of the weights and bias will differ from one model to the next.

[8] To transform an input to standardized units, the mean and standard deviation of the input across all records are computed, and then the mean is subtracted from the original input values, and the result is then divided by the standard deviation, for each individual record.

The above specifications of the hidden layer activation function produced 11 different neural network models, and these were compared to the model presented in Section 5.4 with respect to their lift charts. Based on lift, none of the specifications produced a better model than the one shown in Section 5.4. In particular, the hidden layer activation functions that I tested were

Arc Tan $H_j = (2/\pi) * \tan^{-1}(\eta_j),\ j = 1,2,3.$

Elliot $H_j = \dfrac{\eta_j}{1+|\eta_j|}$

Hyperbolic Tangent $H_j = \tanh(\eta_j)$

Logistic $H_j = \dfrac{1}{1+\exp(-\eta_j)}$

Gauss $H_j = \exp(-0.5 * \eta^2{}_j)$

Sine $H_j = \sin(\eta_j)$

Cosine $H_j = \cos(\eta_j)$

Exponential $H_j = \exp(\eta_j)$

Square $H_j = \eta_j^2$

Reciprocal $H_j = \dfrac{1}{\eta_j}$

Softmax $H_j = \dfrac{\exp(\eta_j)}{\sum_{k=1}^{3}\exp(\eta_k)},\ j = 1,2,3.$

In the Softmax expression above, the outputs of the hidden units sum to unity. If I set the value of η_3 to 0 and estimate only η_1 and η_2, the outputs will still sum to unity.

5.7.1.2 Default Activation Function with Different Combination Functions for the Hidden Layer

Following is a list of values to which the **Hidden Layer Combination Function** property can be set. For each item in the list, I show the formula for the combination function for the jth hidden unit.

ADD

$\eta_j = \sum_{i=1}^{p} x_i$, where $x_1, x_2, \ldots x_p$ are the standardized inputs.

LINEAR

$\eta_j = w_{01j} + \sum_{i=1}^{p} w_{i1j} x_i$, where w_{i1j} is the weight of ith input in the jth hidden unit and w_{01j} is the bias coefficient.

EQSLOPES

$\eta_j = w_{01j} + \sum_{i=1}^{p} w_{i1} x_i$, where the w_{i1} weights (coefficients) on the inputs do not differ from one hidden unit to the next, though the bias coefficients, given by the w_{01j}, may differ.

EQRADIAL

$\eta_j = -b^2 \sum_{i=1}^{p} (w_{ij} - x_i)^2$, where x_i is the ith standardized input, and w_{ij} and b are calculated iteratively by the **Neural Network** node. The coefficient b does not differ from one hidden unit to the next.

EHRADIAL

$\eta_j = -b_j^2 \sum_{i=1}^{p} (w_{ij} - x_i)^2$, where x_i is the ith standardized input, and w_{ij} and b_j are calculated iteratively by the **Neural Network** node.

EWRADIAL

$\eta_j = f * \log(abs(a_j)) - b^2 \sum_{i}^{p} (w_{ij} - x_i)^2$, where x_i is the ith standardized input, w_{ij}, a_j, and b are calculated iteratively by the **Neural Network** node, and f is the number of connections to the unit. In Enterprise Miner the constant f is called fan-in. The fan-in of a unit is the number of other units, from the preceding layer, feeding into that unit.

EVRADIAL

$\eta_j = f * \log(abs(b_j)) - b_j^2 \sum_{i}^{p} (w_{ij} - x_i)^2$, where w_{ij} and b_j are calculated iteratively by the **Neural Network** node and f is the number of other units feeding in to the unit.

XRADIAL

$\eta_j = f * \log(abs(a_j)) - b_j^2 \sum_{i}^{p} (w_{ij} - x_i)^2$, where w_{ij}, b_j, and a_j are calculated iteratively by the **Neural Network** node.

5.7.2 Multilayer Perceptron with Default Setting

In general, a multilayer perceptron (MLP) uses a linear combination function and a sigmoid activation function in the hidden layers. (Tanh, arctan, logistic, and Elliot functions are called sigmoid functions because they are s-shaped when plotted.) If you set the **Architecture** property to **MLP**, the **Target Layer Combination Function** property to **Default**, the **Target Layer Activation Function** property to **Default**, and the **Hidden Layer Combination Function** property to **Default**, the **Neural Network** node automatically sets the **Hidden Layer Activation Function** property to **tanh**.

5.7.3 Radial Basis Function Networks

By setting the **Architecture** property to the following values: **ORBFEQ**, **ORBFUN**, **NRBFEQ**, **NRBFEH**, **NRBFEW**, **NRBFEV**, or **NRBFUN**, and by then setting the other properties to **Default**, you can generate seven different models that all use radial basis functions as activation functions in their hidden units. Each model differs from the other with respect to the outputs of the hidden units (H_1, H_2, and H_3). The combination functions and the outputs of the hidden units are calculated as follows.

5.7.3.1 Ordinary Radial Basis Function with Equal Widths (ORBFEQ)

The combination function for the jth unit in this case is $\eta_j = -b^2 \sum_{i=1}^{p} (w_{ij} - x_i)^2$, where w_{ij} are the weights (coordinates of the basis function center), and b is a constant (the reciprocal of the width). The **Neural Network** node iteratively calculates the weights and the constant. The number of inputs is p, and there are three hidden units ($i.e., j = 1$ to 3). The output of the jth hidden unit is given by $H_j = \exp(\eta_j)$.

Display 5.26 shows the output of the three hidden units $H_j (j = 1, 2$ and $3)$ with a single standardized input AGE (p =1), and arbitrary values for w_{ij} and b.

Display 5.26

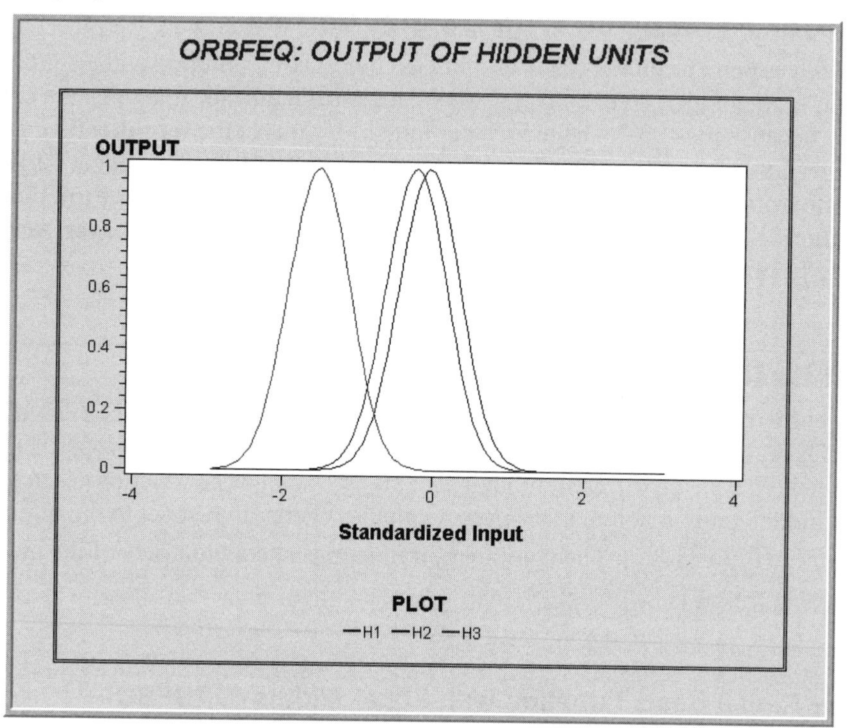

5.7.3.2 Ordinary Radial Basis Function with Unequal Widths (ORBFUN)

In this case, the combination function for the jth unit is $\eta_j = -b_j^{\,2}\sum_{i=1}^{p}(w_{ij} - x_i)^2$, where w_{ij} are the weights iteratively calculated by the **Neural Network** node, and x_i is the ith input (standardized), and b_j is the reciprocal of the width for the jth hidden unit. Unlike the ORBFEQ example in which a single value for *b* is calculated to be used with all of the hidden units, the constant b_j in this case can differ from one hidden unit to another. The output of the jth hidden unit is given by $H_j = \exp(\eta_j)$.

Display 5.27 shows the output of the three hidden units $H_j (j = 1, 2,$ and $3)$ with a single standardized input AGE ($p = 1$), and arbitrary values for w_{ij} and b_j.

Display 5.27

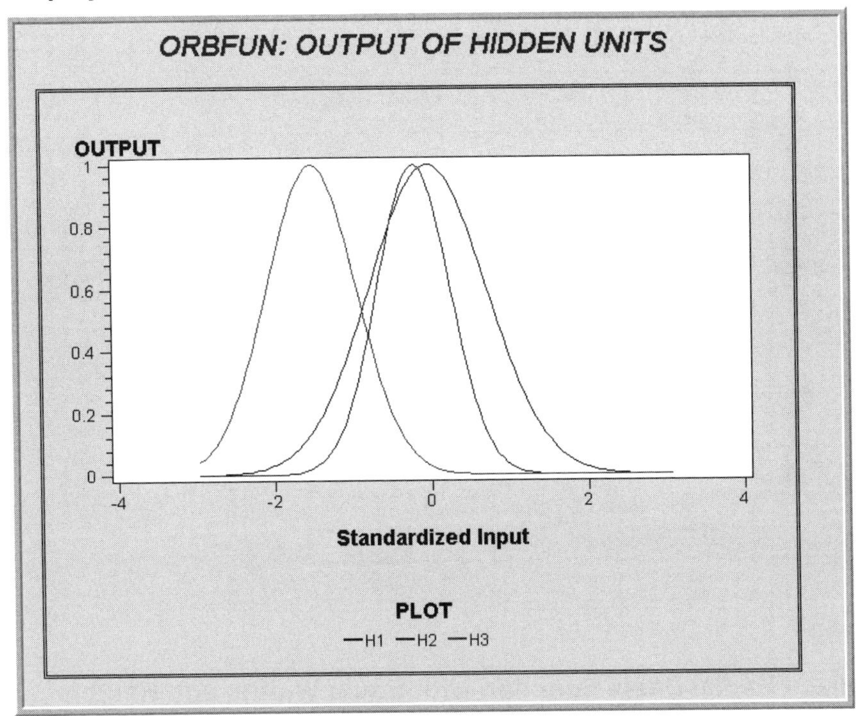

5.7.3.2.A Ordinary Radial Basis Function with Equal Widths and Unequal Heights (ORBFEW)

I have generated the outputs of the three hidden units (shown in Display 5.27A) using an ordinary radial basis function with equal widths and unequal heights in order to demonstrate how the altitude, or height, parameter affects the shape of radial basis functions in general. The list of built-in networks which are available within Enterprise Miner does not include an ordinary radial basic function network with equal widths and unequal heights (ORBFEW) as an option. However, the list does include a normalized radial basis function network with equal widths and unequal heights (NRBFEW). The outputs of the hidden units of a NRBFEW are depicted in Display 5.30 (in Section 5.7.3.5).

The formulas for ORBFEW are

$\eta_j = f * \log(abs(a_j)) - b^2 \sum_i^p (w_{ij} - x_i)^2$, where a_j is an altitude parameter representing the maximum height, f is the number of connections to the unit, and b is the reciprocal of the width. The output of the jth hidden unit is given by $H_j = \exp(\eta_j)$.

Display 5.27A

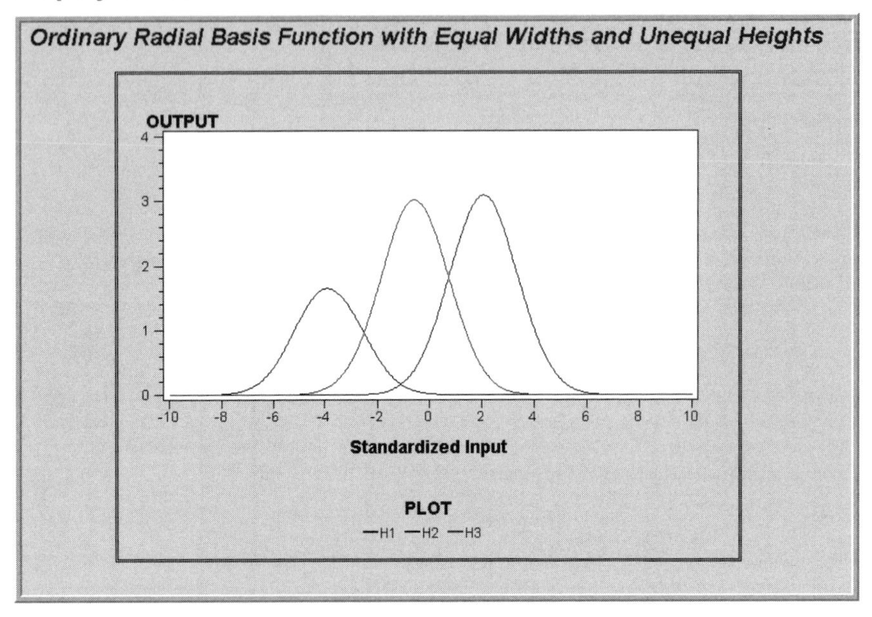

5.7.3.3 Normalized Radial Basis Function with Equal Widths and Heights (NRBFEQ)

In this case, the combination function for the jth hidden unit is $\eta_j = -b^2 \sum_{i=1}^{p} (w_{ij} - x_i)^2$, where w_{ij} are the weights iteratively calculated by the **Neural Network** node, and x_i is the ith input (standardized), and b is the reciprocal of the width.

The output of the jth hidden unit is given by $H_j = \dfrac{\exp(\eta_j)}{\sum_{k=1}^{3} \exp(\eta_k)}$. This is called the

Softmax activation function. Display 5.28 shows the output of the three hidden units $H_j (j = 1, 2,$ and $3)$ with a single standardized input AGE (p =1), and arbitrary values for w_{ij} and b.

Display 5.28

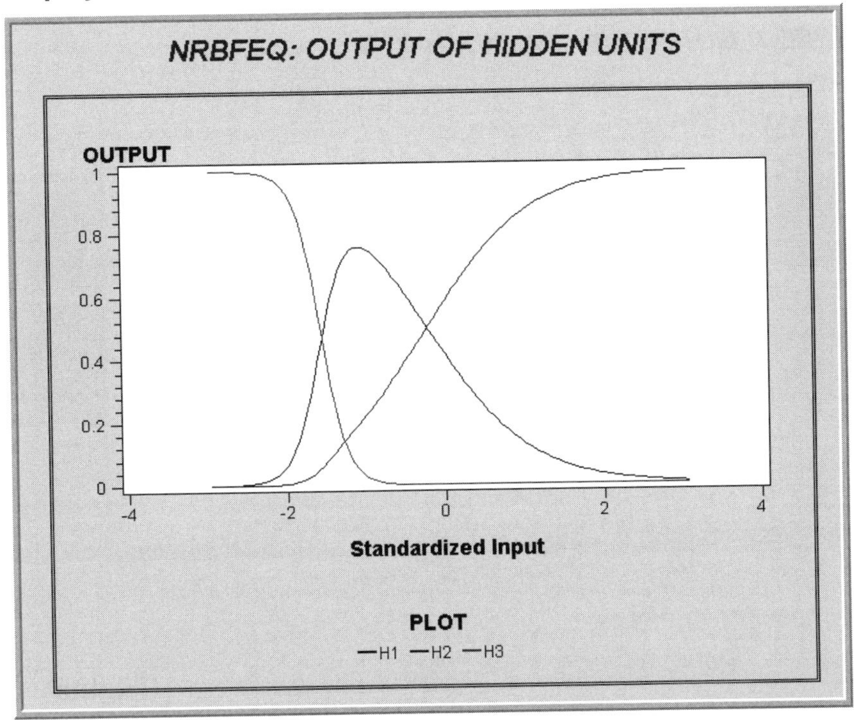

5.7.3.4 Normalized Radial Basis Function with Equal Heights and Unequal Widths (NRBFEH)

The combination function here is $\eta_j = -b_j^2 \sum_{i=1}^{p} (w_{ij} - x_i)^2$,

where w_{ij} are the weights iteratively calculated by the **Neural Network** node, and x_i is the ith input (standardized), and b_j is the reciprocal of the width for the jth hidden unit. The output of the jth hidden unit is given by $H_j = \dfrac{\exp(\eta_j)}{\sum_{k=1}^{3} \exp(\eta_k)}$.

Display 5.29 shows the output of the three hidden units $H_j (j = 1, 2, \text{ and } 3)$ with a single standardized input AGE (p =1), and arbitrary values for w_{ij} and b_j.

Display 5.29

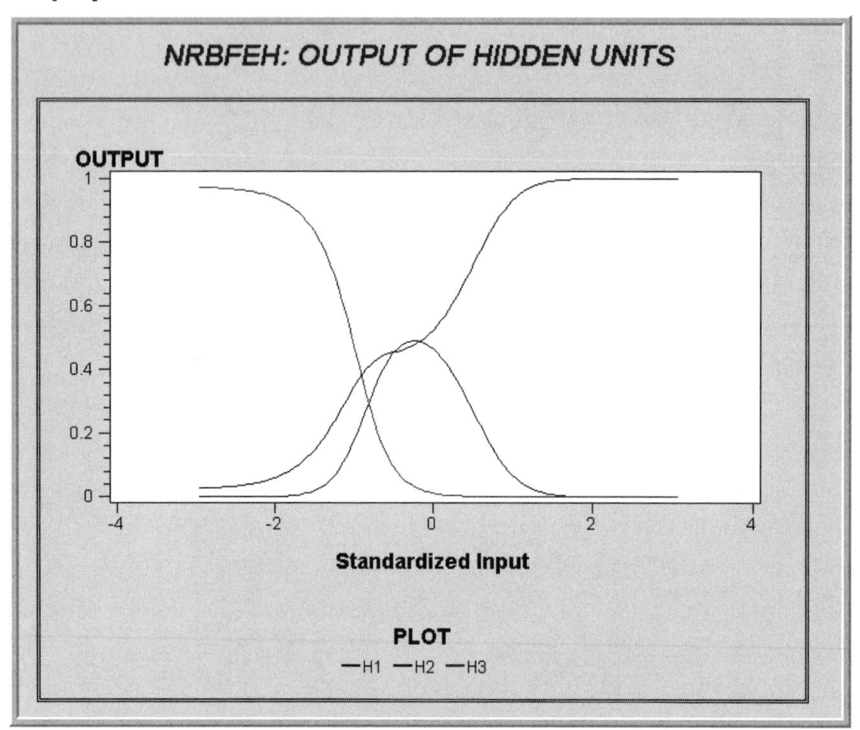

5.7.3.5 Normalized Radial Basis Function with Equal Widths and Unequal Heights (NRBFEW)

The combination function in this case is

$\eta_j = f * \log(abs(a_j)) - b^2 \sum_i^p (w_{ij} - x_i)^2$, where a_j is an altitude parameter representing the maximum height, f is the number of connections to the unit, and b is the reciprocal of the width.

The output of the jth hidden unit is given by $H_j = \dfrac{\exp(\eta_j)}{\sum_{k=1}^{3} \exp(\eta_k)}$.

Display 5.30 shows the output of the three hidden units $H_j (j = 1, 2,$ and $3)$ with a single standardized input AGE (p =1), and arbitrary values for w_{ij} , a_j , and b.

Display 5.30

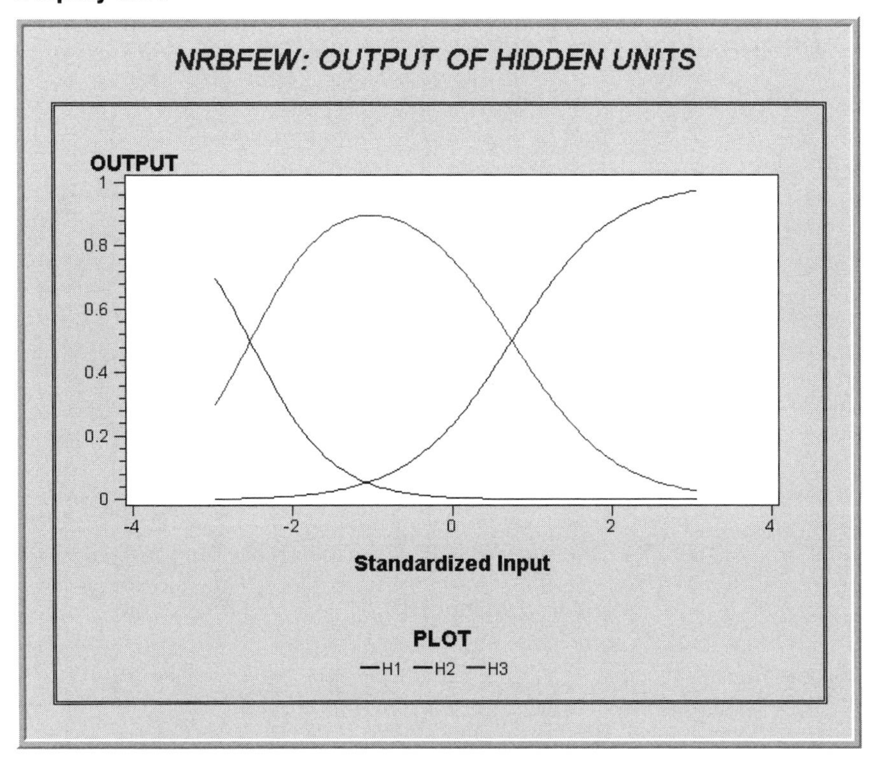

5.7.3.6 Normalized Radial Basis Function with Equal Volumes (NRBFEV)

In this case the combination function is

$$\eta_j = f * \log(abs(b_j)) - b_j^2 \sum_i^p (w_{ij} - x_i)^2$$

The output of the jth hidden unit is given by $H_j = \dfrac{\exp(\eta_j)}{\sum_{k=1}^{3} \exp(\eta_k)}$.

Display 5.31 shows the output of the three hidden units $H_j (j = 1, 2, \text{ and } 3)$ with a single standardized input AGE (p =1), and arbitrary values for w_{ij} and b_j.

Display 5.31

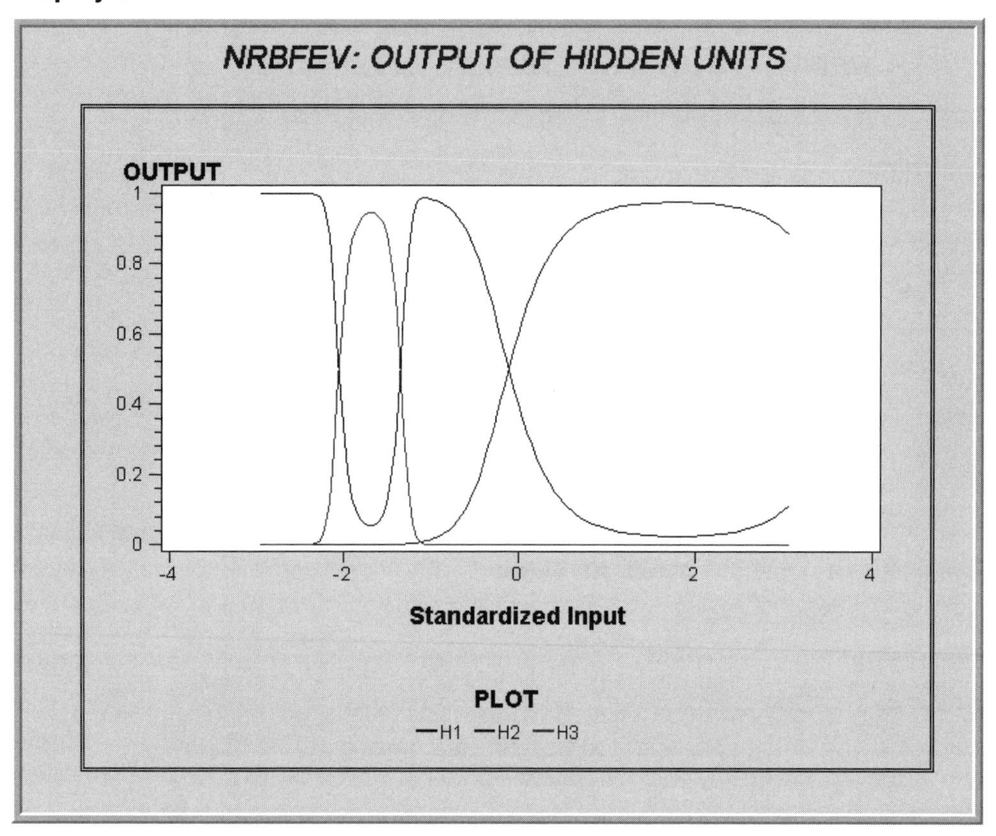

5.7.3.7 Normalized Radial Basis Function with Unequal Widths and Heights (NRBFUN)

Here, the combination function for the jth hidden unit is

$\eta_j = f * \log(abs(a_j)) - b_j^2 \sum_i^p (w_{ij} - x_i)^2$, where a_j is the altitude and b_j is the reciprocal of the width.

The output of the jth hidden unit is given by $H_j = \dfrac{\exp(\eta_j)}{\sum_{k=1}^{3} \exp(\eta_k)}$.

Display 5.32 shows the output of the three hidden units $H_j (j = 1, 2,$ and $3)$ with a single standardized input AGE (p =1), and arbitrary values for w_{ij}, a_j, and b_J.

Display 5.32

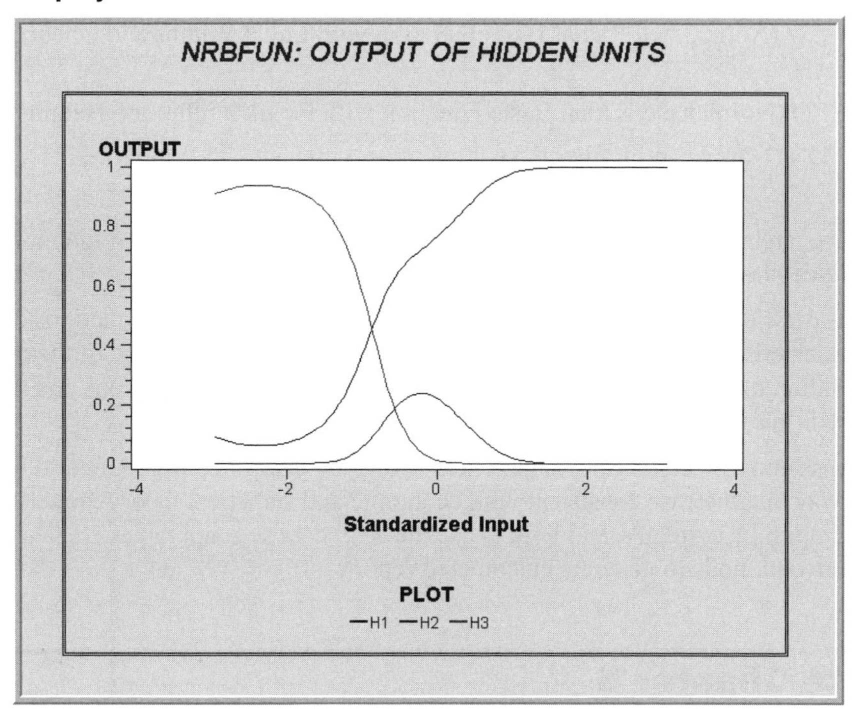

5.8 Summary

This chapter introduced you to neural networks, presenting the following concepts:

- A neural network is essentially nothing more than a complex nonlinear function of the inputs. Dividing the network into different layers and different units within each layer makes it very flexible. A very large number of nonlinear functions can be generated and fitted to the data by means of different architectural specifications.
- The combination and activation functions in the hidden layers and in the target layer are key elements of the *architecture* of a neural network.
- You specify the architecture by setting the **Architecture** property of the **Neural Network** node to **User** or to one of the built-in architecture specifications.
- When you set the **Architecture** property to **User**, you can specify different combinations of hidden layer combination functions, hidden layer activation functions, target layer combination functions, and target layer activation functions. These combinations produce a large number of potential neural network models.
- If you want to use a built-in architecture, set the **Architecture** property to one of the following values:
  - GLIM (generalized linear model, – not discussed in this book), MLP (multilayer perceptron)
  - ORBFEQ (Ordinary Radial Basis Function with Equal Widths and Heights)
  - ORBFUN (Ordinary Radial Basis Function with Unequal Widths)
  - NRBFEH (Normalized Radial Basis Function with Equal Heights and Unequal Widths)

  - NRBFEV (Normalized Radial Basis Function with Equal Volumes)
  - NRBFEW (Normalized Radial Basis Function with Equal Widths and Unequal Heights)
  - NRBFEQ (Normalized Radial Basis Function with Equal Widths and Heights)
  - NRBFUN (Normalized Radial Basis Functions with Unequal Widths and Heights)

- Each built-in architecture comes with a specific hidden layer combination function and a specific hidden layer activation function.
- While the specification of the hidden layer combination and activation functions can be based on such criteria as model fit and model generalization, the selection of the target layer activation function should also be guided by theoretical considerations and the type of output you are interested in.
- The response and risk models developed here show you how to configure neural networks in such a way that they are consistent with economic and statistical theory, how to interpret the results correctly, and how to use the SAS data sets and tables created by the **Neural Network** node to generate customized reports.

5.9 Appendix to Chapter 5

Displays 5.33, 5.34, and 5.35 show the SAS code for calculating the lift for *LOSSFRQ*.

Display 5.33

```
libname lib2 'C:\TheBook\EM_5.2\Nov2006\EM_Datasets' ;
run ;
data lib2.Ch5_risk_NN_train ;
 set &EM_import_data;
run;
data lib2.Ch5_risk_NN_validate ;
 set &EM_import_validate ;
run;

data lib2.Ch5_risk_NN_test ;
 set &EM_import_test ;
run;

 %macro lifts(ds=);
 data &ds ;
  set lib2.ch5_risk_NN_&ds ;
  keep p_lossfrq0 p_lossfrq1 p_lossfrq2 lossfrq expected_lfrq;
  expected_lfrq = 0*p_lossfrq0 + 1*p_lossfrq1 + 2*p_lossfrq2 ;
 run ;
 proc sort data=&ds;
  by descending expected_lfrq;
 run ;
proc sql noprint;
 select count(*) into : nvl from
 work.&ds ;
quit ;
```

Display 5.34

```
data &ds ;
 retain count 0 ;
 set &ds ;
 count+1 ;
 if count < (1/20)*&nvl then dec=5; else
 if count < (2/20)*&nvl then dec=10 ; else
 if count < (3/20)*&nvl then dec=15 ; else
 if count < (4/20)*&nvl then dec=20 ; else
 if count < (5/20)*&nvl then dec=25 ; else
 if count < (6/20)*&nvl then dec=30 ; else
 if count < (7/20)*&nvl then dec=35; else
 if count < (8/20)*&nvl then dec=40 ; else
 if count < (9/20)*&nvl then dec=45 ; else
 if count < (10/20)*&nvl then dec=50 ; else
 if count < (11/20)*&nvl then dec=55 ; else
 if count < (12/20)*&nvl then dec=60 ; else
 if count < (13/20)*&nvl then dec=65 ; else
 if count < (14/20)*&nvl then dec=70 ; else
 if count < (15/20)*&nvl then dec=75 ; else
 if count < (16/20)*&nvl then dec=80 ; else
 if count < (17/20)*&nvl then dec=85 ; else
 if count < (18/20)*&nvl then dec=90 ; else
 if count < (19/20)*&nvl then dec=95 ; else
 dec = 100 ;
 run ;
proc means data=&ds noprint ;
  class dec ;
  var lossfrq ;
  output out= outsum sum(lossfrq) = sum_lossfrq mean(lossfrq)=mean_lossfrq;
run ;
```

Display 5.35

```
data Total(keep=sum_lossfrq rename=(sum_lossfrq=Tot_lossfrq)) deciles ;
  set outsum ;
  if _TYPE_ = 0 then output Total;
  else output deciles ;
run ;
data tables ;
 set deciles ;
 if _N_ = 1 then set total ;
run;
data lib2.Ch5_risk_Lift_NN_&ds ;
 set tables ;
 retain cumsum 0 nobs 0;
 cumsum + sum_lossfrq ;
 capc = cumsum/tot_lossfrq ;
 gmean = tot_lossfrq/&nvl ;
 nobs+_freq_ ;
 meanc = cumsum/nobs ;
 liftc = meanc/gmean;
 label dec='Decile'
       Mean_lossfrq='Lossfrq Mean'
       Meanc='Lossfrq Cumulative Mean'
       liftc ='Cumulative Lift' ;
 RUN;
 proc print data=lib2.ch5_risk_Lift_NN_&DS label noobs ;
    var dec mean_lossfrq meanc liftc ;
 Title1 "             Loss Frequency: &DS" ;
 Title2  ;
RUN;
%mend lifts ;
 %lifts(ds=TRAIN);
 %lifts(ds=VALIDATE);
 %lifts(ds=TEST) ;
```

Display 5.36 shows the SAS code for generating Displays 5.7C, 5.7D, and 5.7E.

Display 5.36

```
ods html file='C:\TheBook\EM_5.2\Nov2006\EM_Datasets\Table5.7C.html' ;
 proc print data=emws.neural_weightds noobs ;
   where upcase(from) in ('AGE' 'CRED' 'MILEAGE') and to in ('H11' 'H12' 'H13') ;
   var _label_ from to weight ;
run ;
ods html close ;

libname T "C:\TheBook\EM_5.2\Nov2006" ;
run ;
ods html file='C:\TheBook\EM_5.2\Nov2006\EM_Datasets\Table5.7D.html' ;
proc print data=T.weights_history(obs=21) label noobs ;
  var _ITER_ H11_RESP1 H12_RESP1 H13_RESP1 BIAS_RESP1 ;
run;
ods html close ;

ods html file='C:\TheBook\EM_5.2\Nov2006\EM_Datasets\Table5.7E.html' ;
 proc print data=emws.neural_weightds noobs ;
   where upcase(from) in ('H11' 'H12' 'H13' 'BIAS') and upcase(to) in ('RESP1') ;
   var _label_ from to weight ;
run ;
ods html close ;
```

Regression Models

6.1 Introduction

This chapter explores the **Regression** node in detail using two practical business applications—one requiring a model to predict a binary target and the other requiring a model to predict a continuous target. Before developing these two models, I will present an overview of the types of models that can be developed in the **Regression** node, the theory and rationale behind each type of model, and an explanation of how to set various properties of the **Regression** node to get the desired models.

The topics covered in this chapter are arranged in the following sequence:

- The types of models you can develop using the **Regression** node. Models with
  - binary targets (e.g., response / no response)
  - ordinal targets (ordered from low to high, e.g., low, medium, and high risk)
  - nominal targets (which may indicate qualitative factors such as customer purchase of different products, customer preferences for different brands, etc.)
  - continuous targets (numerical variables, such as a customer's savings deposit dollar amounts, which can be measured to the nearest cent)

- How to set different properties of the **Regression** node and what kind of results can you expect for different values of
  - Regression Type property
  - Link Function property
  - Selection Model property
  - Selection Criterion property
- Extended example: a model with a binary target
  - regression with or without variable transformations
  - using the **Regression** node with the **Decision Tree** node to capture interactions
  - interpretation of results:
    - lift charts
    - ROC charts
    - Output window
    - custom lift tables from a Score Rankings table
- Extended example: a model with a continuous target
  - regression with or without variable transformations
  - using the **Regression** node with the **Decision Tree** node to capture interactions
  - interpretation of results:
    - lift charts
    - ROC charts
    - Output window
    - custom lift tables from a Score Rankings table

6.2 What Types of Models Can Be Developed Using the Regression Node?

6.2.1 Models with a Binary Target

When the target is binary, either numeric or character, the **Regression** node estimates a logistic regression. The **Regression** node produces SAS code to calculate the probability of the event (such as response or attrition). The computation of the probability of the event is done through a *link function.* A *link function* shows the relation between the probability of the event and a linear predictor, which is a linear combination of the inputs (explanatory variables).

The linear predictor can be written as $\beta'x$, where x is a vector of inputs and β is the vector of coefficients estimated by the **Regression** node. In the case of a response model, where the target takes on a value of 1 or 0 (1 stands for response and 0 for non-response), the *link function* is

$$\log\left(\frac{\Pr(y=1\mid x)}{1-\Pr(y=1\mid x)}\right)=\beta'x \tag{6.1}$$

where y is the target and $\Pr(y=1 \mid x)$ is the probability of the target variable taking the value 1 for a customer (or a record in the data set), given the values of the inputs or explanatory variables for that customer. The left hand side of Equation 6.1 is the logarithm of odds or log-odds of the event 'response'. Since Equation 6.1 shows the relation between the log-odds of the event and the linear predictor $\beta' x$, it is called a *logit* link.

If you solve the link function given in Equation 6.1 for $\Pr(y=1 \mid x)$, you get

$$P(y=1 \mid x) = \frac{\exp(\beta' x)}{1+\exp(\beta' x)} \tag{6.2}$$

Equation 6.2 is called the *inverse link* function. In Chapter 5, where we discussed neural networks, we called equations of this type *activation functions*.

From Equation 6.2, you can see that

$$1 - P(y=1 \mid x) = 1 - \frac{\exp(\beta' x)}{1+\exp(\beta' x)} = \frac{1}{1+\exp(\beta' x)} \tag{6.3}$$

Dividing Equation 6.3 by Equation 6.2, and taking logs I once again obtain the logit link given in Equation 6.1.

The important feature of this function for our purposes is that it produces values that lie only between 0 and 1, just as the probability of response and non-response do.

In addition to the logit link, the **Regression** node provides probit and complementary log-log link functions. These are discussed in Section 6.3.2. In addition, when the target is ordinal, you can use the cumulative logits link, and when it is nominal (without order), a generalized logit is used. All of these link functions serve the purpose of keeping the predicted probabilities between 0 and 1. In cases where it is not appropriate to constrain the predicted values to lie between 0 and 1, no link function is needed.

There are also some well-accepted theoretical rationales behind some of these link functions. For readers who are interested in pursuing this line of reasoning further, a discussion of the theory is included in Section 6.3.2.

We will learn more about modeling binary targets in Section 6.4, where a model with a binary target is developed for a marketing application.

6.2.2 Models with an Ordinal Target

Ordinal targets, also referred to as *ordered polychotomous* targets, take on more than two discrete, ordered values. An example of an ordinal target with three levels (0, 1, and 2) is the discretized version of the loss frequency variable discussed in Chapter 5.

If you want your target to be treated as an ordinal variable, you must set its measurement scale to *ordinal*. However, if the target variable has fewer than the number of levels specified in the **Advanced Advisor** metadata options, Enterprise Miner sets its measurement scale to **Nominal** (see Section 2.5, Displays 2.14A and 2.14B). If the measurement scale of your target variable is currently set to **Nominal**, but you want to model it as an ordinal target, then you must change its measurement scale setting to **Ordinal**. In the Appendix to Chapter 3, I showed how to change the measurement scale of a variable in a data source.

When the target is ordinal, then, by default, the **Regression** node produces a proportional odds (or cumulative logits) model, as discussed in Chapter 5, Section 5.5.1.1. (The example below illustrates the proportional odds model, as well.) When the target is nominal, on the other hand, it produces a generalized logits model, which is presented in Section 6.2.3. Although the default value of the **Regression Type** property is **Logistic Regression** and the default value of the **Link Function** property (link functions will be discussed in detail in Section 6.3.2) is **Logit** for both ordinal targets and nominal targets, the models produced are different for targets with different measurement scales. Hence, the setting of the measurement scale determines which type of model you get. Using some simple examples, I will demonstrate how different types of models are produced for different measurement scales even though the **Regression** node properties are all set to the same values.

Display 6.1 shows the process flow for a regression with an ordinal target.

Display 6.1

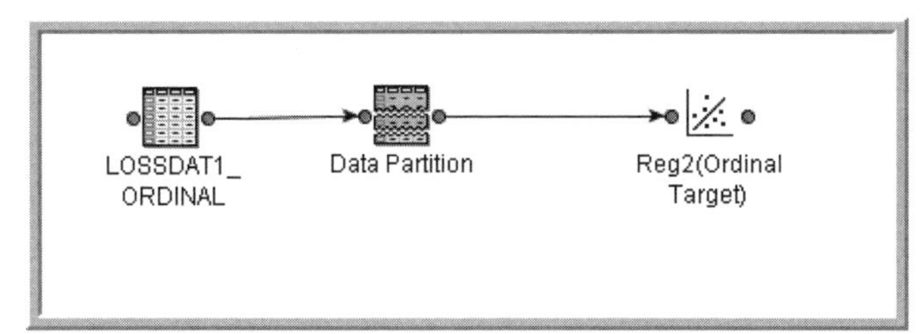

Display 6.2 shows a partial list of variables and their measurement scales, or levels. From the second and third columns in this display, it can be seen that the role of the variable *LOSSFRQ* is a *target* and its measurement scale (level) is ordinal.

Display 6.2

Variables - Ids

Name	Role	Level	Report	Order	Drop	Lower Limit	Upper Limit	Typ
GENDER	Input	Binary	No		No	.	.	C
HEQ	Input	Nominal	No		No	.	.	N
INCOME	Input	Nominal	No		No	.	.	N
LOSSFRQ	Target	Ordinal	No		No	.	.	N
MFDU	Input	Binary	No		No	.	.	N
MILEAGE	Input	Interval	No		No	.	.	N
MOB	Input	Binary	No		No	.	.	C
MRTGI	Input	Nominal	No		No	.	.	C
MS	Input	Nominal	No		No	.	.	C
NAF	Input	Nominal	No		No	.	.	N
NOSBLT	Input	Nominal	No		No	.	.	N
NPRACC	Input	Nominal	No		No	.	.	N

Explore... OK Cancel Help

Display 6.3 shows the properties panel of the **Regression** node in the process flow shown in Display 6.1.

Display 6.3

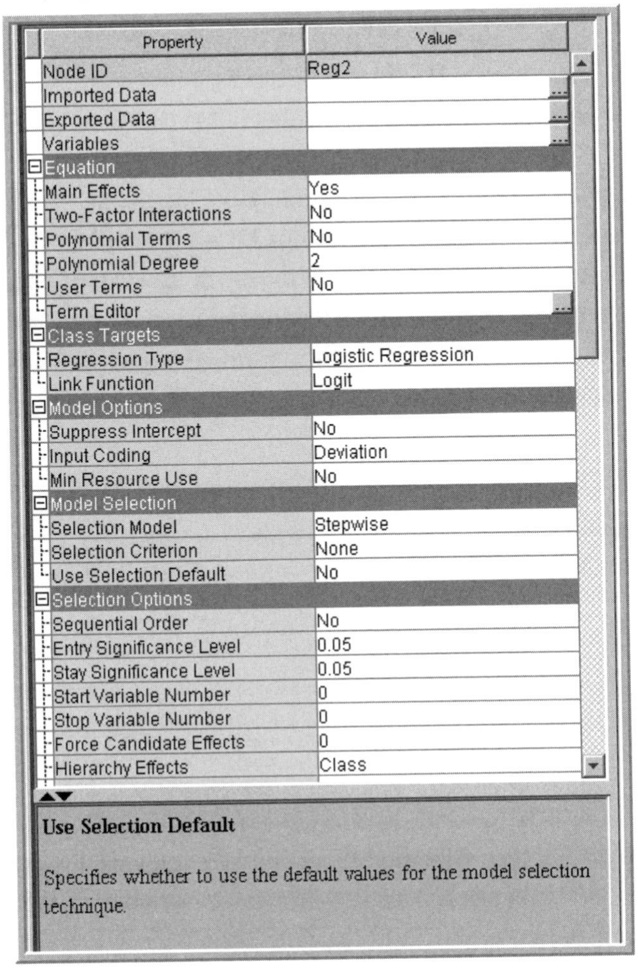

Property	Value
Node ID	Reg2
Imported Data	...
Exported Data	...
Variables	...
⊟Equation	
Main Effects	Yes
Two-Factor Interactions	No
Polynomial Terms	No
Polynomial Degree	2
User Terms	No
Term Editor	...
⊟Class Targets	
Regression Type	Logistic Regression
Link Function	Logit
⊟Model Options	
Suppress Intercept	No
Input Coding	Deviation
Min Resource Use	No
⊟Model Selection	
Selection Model	Stepwise
Selection Criterion	None
Use Selection Default	No
⊟Selection Options	
Sequential Order	No
Entry Significance Level	0.05
Stay Significance Level	0.05
Start Variable Number	0
Stop Variable Number	0
Force Candidate Effects	0
Hierarchy Effects	Class

Use Selection Default

Specifies whether to use the default values for the model selection technique.

Note that, in the **Class Targets** section of Display 6.3, the **Regression Type** property is set to **Logistic Regression** and the **Link Function** property is set to **Logit**. These are the default values for class targets (categorical targets), including binary, ordinal, and nominal targets.

Since the target variable *LOSSFRQ* is ordinal, taking on the values 0, 1, or 2 for any given record of the data set, the **Regression** node uses a cumulative logits link.

The expression $\frac{\Pr(lossfrq = 2)}{1 - \Pr(lossfrq = 2)}$ is called the *odds ratio*. It is the odds that the event *LOSSFRQ*=2 will occur. In other words, the odds ratio represents the odds of the loss frequency equaling 2. Similarly, the expression $\frac{\Pr(lossfrq = 1) + \Pr(lossfrq = 2)}{1 - \{\Pr(lossfrq = 1) + \Pr(lossfrq = 2)\}}$ represents the odds of the event *LOSSFRQ*=1 or 2. When I take the logarithms of the above expressions, they are called *log-odds* or *logits*.

Thus the logit for the event $LOSSFRQ$=2 is $\log\left(\frac{\Pr(lossfrq=2)}{1-\Pr(lossfrq=2)}\right)$, and the logit for the event $LOSSFRQ$=1 or 2 is $\log\left(\frac{\Pr(lossfrq=1)+\Pr(lossfrq=2)}{1-\{\Pr(lossfrq=1)+\Pr(lossfrq=2)\}}\right)$.

The logits shown above are called *cumulative logits*. When a cumulative logits link is used, the logits of the event $LOSSFRQ$=1 or 2 will be larger than the logits for the event $LOSSFRQ$=2 by a positive amount. In a cumulative logits model, the logits of the event $LOSSFRQ$=2 and the event $LOSSFRQ$=1 or 2 differ by a constant amount for all observations (all customers). To illustrate this phenomenon, I will run the **Regression** node with an ordinal target ($LOSSFRQ$). Display 6.4, which is taken from the **Results** window, shows the equations estimated by the **Regression** node.

Display 6.4

Analysis of Maximum Likelihood Estimates

Parameter	DF	Estimate	Standard Error	Wald Chi-Square	Pr > ChiSq	Standardized Estimate	Exp(Est)
Intercept 2	1	-2.8840	0.4145	48.42	<.0001		0.056
Intercept 1	1	0.2133	0.3236	0.43	0.5097		1.238
AGE	1	-0.0231	0.00413	31.27	<.0001	-0.1998	0.977
CRED	1	-0.00498	0.000527	89.39	<.0001	-0.2563	0.995
NPRVIO	1	0.2252	0.0665	11.47	0.0007	0.0830	1.253

From the coefficients given in Display 6.4, I can write the following two equations:

$$\log\left(\frac{\Pr(lossfrq=2)}{1-\Pr(lossfrq=2)}\right) = -2.8840 - 0.0231*AGE - 0.00498*CRED + 0.2252*NPRVIO \quad (6.4)$$

$$\log\left(\frac{\Pr(lossfrq=2)+\Pr(lossfrq=1)}{1-\{\Pr(lossfrq=2)+\Pr(lossfrq=1)\}}\right) = 0.2133 - 0.0231*AGE - 0.00498*CRED + 0.2252*NPRVIO \quad (6.5)$$

To interpret the results, I will take the example of a customer, whom I will call Customer A, who is 50 years of age (AGE=50), has a credit score of 670 (CRED=670), and has two prior traffic violations (NPRVIO=2).

From Equation 6.4, the log-odds (logit) of the event $LOSSFRQ$=2 for Customer A are

$$\log\left(\frac{\Pr(lossfrq=2)}{1-\Pr(lossfrq=2)}\right) = -2.8840 - 0.0231*50 - 0.00498*670 + 0.2252*2 \quad (6.4A)$$

Therefore, the odds of the event *LOSSFRQ*=2 for Customer A are

$$\left(\frac{\Pr(lossfrq=2)}{1-\Pr(lossfrq=2)}\right)=e^{-2.8840-0.0231*50-0.00498*670+0.2252*2} \tag{6.4B}$$

From Equation 6.5, the log-odds (logit) of the event *LOSSFRQ*=1 or 2 for Customer A are

$$\log\left(\frac{\Pr(lossfrq=2)+\Pr(lossfrq=1)}{1-\{\Pr(lossfrq=2)+\Pr(lossfrq=1)\}}\right)=0.2133-0.0231*50-0.00498*670 + 0.2252*2 \tag{6.5A}$$

The odds of the event *LOSSFRQ*=1 or 2 for Customer A are

$$\left(\frac{\Pr(lossfrq=2)+\Pr(lossfrq=1)}{1-\{\Pr(lossfrq=2)+\Pr(lossfrq=1)\}}\right)=e^{0.2133-0.0231*50-0.00498*670+0.2252*2} \tag{6.5B}$$

The difference in the logits (not the odds, but the log-odds) for the events *LOSSFRQ*=1 or 2 and *LOSSFRQ*= 2 is equal to 0.2133–(–2.8840) =3.0973, which is obtained by subtracting Equation 6.4A from Equation 6.5A. Irrespective of the values of the inputs AGE, CRED, and NPRVIO, the difference in the logits, or logit difference, for the events *LOSSFRQ*=1 or 2 and *LOSSFRQ*= 2 is the same constant 3.0973. This type of model is often referred to as a *cumulative logits* model, or a logistic regression with a *cumulative logits link.*

Now turning to the odds of the event, rather than the log-odds, if I divide Equation 6.5B by 6.4B, I get the ratio of the odds of the event *LOSSFRQ*=1 or 2 to the odds of the event *LOSSFRQ*=2. This ratio is equal to

$$\frac{e^{0.2133}}{e^{-2.8840}}=e^{0.2133+2.8840}=e^{3.0973}=22.1381.$$

By this result, I can see that the odds of the event *LOSSFRQ*=1 or 2 is proportional to the odds of the event *LOSSFRQ*= 2, and that the proportionality factor is 22.1381. This proportionality factor is the same for any values of the inputs (AGE, CRED, and NPRVIO in this example). Hence the name is *proportional odds model.*

I will now compare these results to those for Customer B, who is of the same age (50) as Customer A and has the same credit score (670) as Customer A, but who has three prior violations (NPRVIO=3) instead of two. In other words, all of the inputs except for NPRVIO are the same for Customer B and Customer A. I will show how the log-odds and the odds for the events *LOSSFRQ*=2 and *LOSSFRQ*= 1 or 2 for Customer B differ from those for Customer A.

From Equation 6.4, the log-odds (logit) of the event *LOSSFRQ*=2 for Customer B are

$$\log\left(\frac{\Pr(lossfrq=2)}{1-\Pr(lossfrq=2)}\right)=-2.8840-0.0231*50-0.00498*670 + 0.2252*3 \tag{6.4C}$$

Therefore, the odds of the event $LOSSFRQ$=2 for Customer B are

$$\left(\frac{\Pr(lossfrq=2)}{1-\Pr(lossfrq=2)}\right)=e^{-2.8840-0.0231*50-0.00498*670+0.2252*3} \tag{6.4D}$$

From Equation 6.5, the log-odds (logit) of the event $LOSSFRQ$=1 or 2 for Customer B are

$$\log\left(\frac{\Pr(lossfrq=2)+\Pr(lossfrq=1)}{1-\{\Pr(lossfrq=2)+\Pr(lossfrq=1)\}}\right)=0.2133-0.0231*50-0.00498*670+0.2252*3 \tag{6.5C}$$

The odds of the event $LOSSFRQ$=1 or 2 for Customer B are

$$\left(\frac{\Pr(lossfrq=2)+\Pr(lossfrq=1)}{1-\{\Pr(lossfrq=2)+\Pr(lossfrq=1)\}}\right)=e^{0.2133-0.0231*50-0.00498*670+0.2252*3} \tag{6.5D}$$

It is clear that the difference between log-odds of the event $LOSSFRQ$=2 for Customer B and Customer A is 0.2252, which is obtained by subtracting Equation 6.4A from 6.4C.

It is also clear that the difference between log-odds of the event $LOSSFRQ$=1 or 2 for Customer B and Customer A is also 0.2252, which is obtained by subtracting Equation 6.5A from 6.5C.

The difference in the log-odds (0.2252) is simply the coefficient of the variable NPRVIO, whose value differs by 1 for customers A and B. Generalizing this result, we can see that the coefficient of any of the inputs in the log-odds Equations 6.4 and 6.5 indicates the change in log-odds per unit of change in the input.

The odds of the event $LOSSFRQ$= 2 for Customer B relative to Customer A is $e^{0.2252}=1.252573$, which is obtained by dividing Equation 6.4D by Equation 6.4B. In this example, an increase of one prior violation indicates a 25.2% increase in the odds of the event $LOSSFRQ$=2. I leave it to you to verify that an increase of one prior violation has the same impact, a 25.2% increase in the odds of the event $LOSSFRQ$= 1 or 2 as well. (Hint: Divide Equation 6.5D by Equation 6.5B.)

Equations 6.4 and 6.5 and the identity $\Pr(lossfrq=0)+\Pr(lossfrq=1)+\Pr(lossfrq=2)=1$ are used by the **Regression** node to calculate the three probabilities $\Pr(lossfrq=0)$, $\Pr(lossfrq=1)$, and $\Pr(lossfrq=2)$. This can be seen from the SAS code generated by the **Regression** node, and shown in Displays 6.5 and 6.5A.

Display 6.5

```
*** Compute Linear Predictor;
drop _TEMP;
drop _LP0;
_LP0 = 0;

***  Effect: AGE ;
_TEMP = AGE ;
_LP0 = _LP0 + (    -0.02307201027961 * _TEMP);

***  Effect: CRED ;
_TEMP = CRED ;
_LP0 = _LP0 + (    -0.00498225050304 * _TEMP);

***  Effect: NPRVIO ;
_TEMP = NPRVIO ;
_LP0 = _LP0 + (     0.22524384228134 * _TEMP);

*** Naive Posterior Probabilities;
drop _MAXP _IY _P0 _P1 _P2;
_TEMP =     -2.88399896897927 + _LP0;
if (_TEMP < 0) then do;
   _TEMP = exp(_TEMP);
   _P0 = _TEMP / (1 + _TEMP);
end;
else _P0 = 1 / (1 + exp(-_TEMP));
_TEMP =      0.21333987443108 + _LP0;
if (_TEMP < 0) then do;
   _TEMP = exp(_TEMP);
   _P1 = _TEMP / (1 + _TEMP);
end;
else _P1 = 1 / (1 + exp(-_TEMP));
_P2 = 1.0 - _P1;
_P1 = _P1 - _P0;

REG2DR1:
```

Display 6.5A

```
*** Posterior Probabilities and Predicted Level;
label P_LOSSFRQ2 = 'Predicted: LOSSFRQ=2' ;
label P_LOSSFRQ1 = 'Predicted: LOSSFRQ=1' ;
label P_LOSSFRQ0 = 'Predicted: LOSSFRQ=0' ;
P_LOSSFRQ2 = _P0;
_MAXP = _P0;
_IY = 1;
P_LOSSFRQ1 = _P1;
if (_P1 - _MAXP > 1e-8) then do;
   _MAXP = _P1;
   _IY = 2;
end;
P_LOSSFRQ0 = _P2;
if (_P2 - _MAXP > 1e-8) then do;
   _MAXP = _P2;
   _IY = 3;
end;
```

By setting AGE=50, CRED=670, and NPRVIO=2 and using the SAS statements from Displays 6.5 and 6.5A, I generated the probabilities that *LOSSFRQ* =0, 1, and 2, as shown in Display 6.5B. The SAS code executed for generating these probabilities is given in Displays 6.41 and 6.42 in the appendix to this chapter.

Display 6.5B

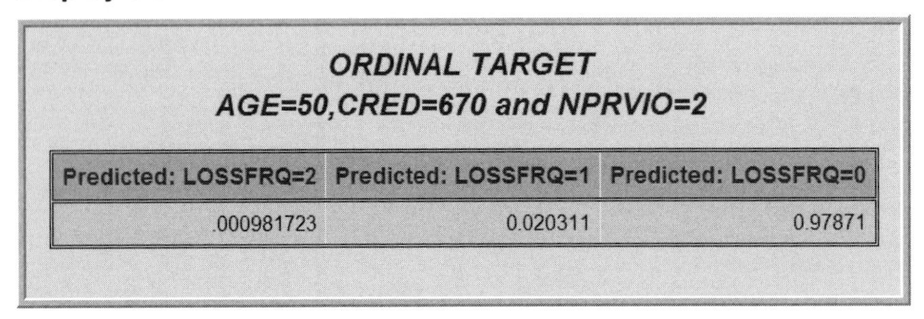

ORDINAL TARGET
AGE=50,CRED=670 and NPRVIO=2

Predicted: LOSSFRQ=2	Predicted: LOSSFRQ=1	Predicted: LOSSFRQ=0
.000981723	0.020311	0.97871

6.2.3 Models with a Nominal (Unordered) Target

Unordered polychotomous variables are nominal-scaled categorical variables where there is no particular order in which to rank the category labels. That is, you cannot assume that one category is higher or better than another. Targets of this type arise in market research where marketers are modeling customer preferences. In consumer preference studies and also in this chapter, letters such as A, B, C, and D may represent different products or different brands of the same product.

When the target is nominal categorical, the default values of the **Regression** node properties produce a model of the following type:

$$P(T = D \mid X) = \frac{\exp(\alpha_0 + \beta_0 X)}{1 + \exp(\alpha_0 + \beta_0 X) + \exp(\alpha_1 + \beta_1 X) + \exp(\alpha_2 + \beta_2 X)} \quad (6.6)$$

$$P(T = C \mid X) = \frac{\exp(\alpha_1 + \beta_1 X)}{1 + \exp(\alpha_0 + \beta_0 X) + \exp(\alpha_1 + \beta_1 X) + \exp(\alpha_2 + \beta_2 X)} \quad (6.7)$$

$$P(T = B \mid X) = \frac{\exp(\alpha_2 + \beta_2 X)}{1 + \exp(\alpha_0 + \beta_0 X) + \exp(\alpha_1 + \beta_1 X) + \exp(\alpha_2 + \beta_2 X)} \quad (6.8)$$

$$P(T = A \mid X) = \frac{1}{1 + \exp(\alpha_0 + \beta_0 X) + \exp(\alpha_1 + \beta_1 X) + \exp(\alpha_2 + \beta_2 X)} \quad (6.9)$$

where T is the target variable; A, B, C, and D represent levels of the target variable *T*; *X* is the vector of input variables; and $\alpha_0, \alpha_1, \alpha_2, \beta_0, \beta_1$, and β_2 are the coefficients estimated by the **Regression** node. Note that β_0, β_1, and β_2 are row vectors, and α_o, α_1, and α_2 are scalars. A model of the type presented in Equations 6.6, 6.7, 6.8, and 6.9 is called a *generalized logits* model or a logistic regression with a *generalized logits* link.

Using an example, I will now show that the measurement scale of the target variable determines which model is produced by the **Regression** node. I have used the same data I used in Section 6.2.2, except that I have changed the measurement scale of the target variable *LOSSFRQ* to

Nominal. Since the variable *LOSSFRQ* is a discrete version of the continuous variable Loss Frequency, it should be treated as ordinal. I have changed its measurement scale to nominal only to highlight the fact that the model produced depends on the measurement scale of the target variable.

Display 6.6 shows the flow chart for a regression with a nominal target.

Display 6.6

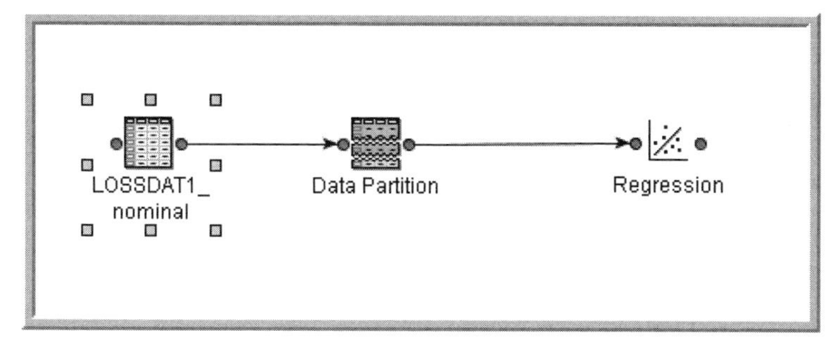

Display 6.7 shows a partial list of variables, which is identical to the list shown in Display 6.2 with the exception that the measurement scale (level) of the target variable *LOSSFRQ* is set to **Nominal**.

Display 6.7

Variables - Ids

Name	Role	Level	Report	Order	Drop	Lower Limit	Upper Limit	Typ
GENDER	Input	Binary	No		No	.	.	C
HEQ	Input	Nominal	No		No	.	.	N
INCOME	Input	Nominal	No		No	.	.	N
LOSSFRQ	Target	Nominal	No		No	.	.	N
MFDU	Input	Binary	No		No	.	.	N
MILEAGE	Input	Interval	No		No	.	.	N
MOB	Input	Binary	No		No	.	.	C
MRTGI	Input	Nominal	No		No	.	.	C
MS	Input	Nominal	No		No	.	.	C
NAF	Input	Nominal	No		No	.	.	N
NOSBLT	Input	Nominal	No		No	.	.	N
NPRACC	Input	Nominal	No		No	.	.	N

Explore... OK Cancel Help

Display 6.8 shows the properties of the **Regression** node.

Display 6.8

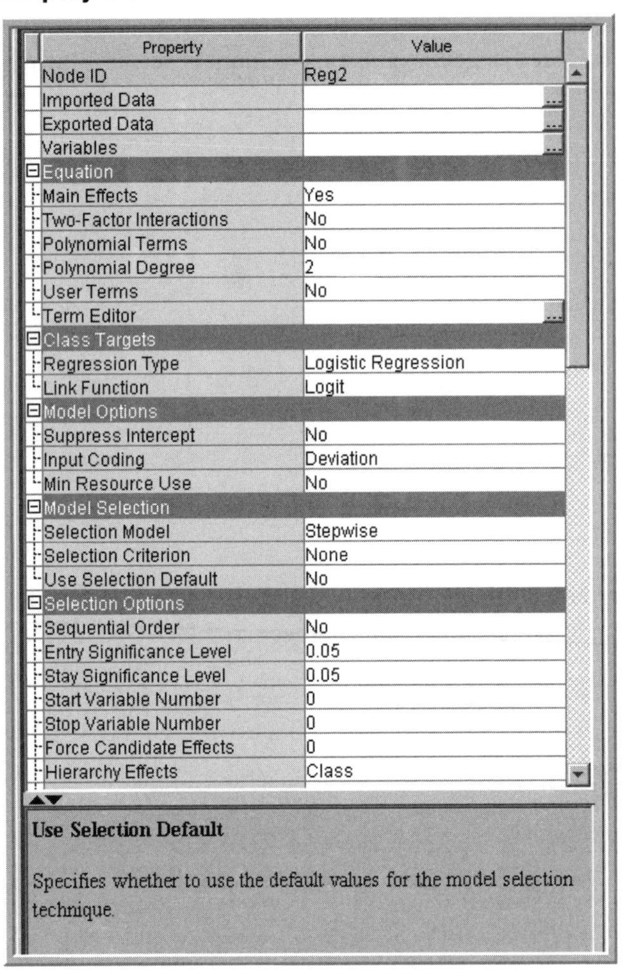

The values to which the properties are set in Display 6.8 are identical to those shown in Display 6.3.

Display 6.9 shows the coefficients of the model estimated by the **Regression** node.

Display 6.9

Analysis of Maximum Likelihood Estimates

Parameter	LOSSFRQ	DF	Estimate	Standard Error	Wald Chi-Square	Pr > ChiSq	Standardized Estimate	Exp(Est)
Intercept	2	1	1.5946	1.2965	1.51	0.2187		4.927
Intercept	1	1	-0.1186	0.3367	0.12	0.7248		0.888
AGE	2	1	-0.0324	0.0204	2.53	0.1115	-0.2805	0.968
AGE	1	1	-0.0228	0.00421	29.42	<.0001	-0.1976	0.977
CRED	2	1	-0.0134	0.00241	30.75	<.0001	-0.6871	0.987
CRED	1	1	-0.00449	0.000548	67.08	<.0001	-0.2309	0.996
NPRVIO	2	1	0.2186	0.2848	0.59	0.4426	0.0805	1.244
NPRVIO	1	1	0.2240	0.0680	10.86	0.0010	0.0825	1.251

Display 6.9 shows that the **Regression** node has produced two sets of coefficients corresponding to the two equations for calculating the probabilities of the events $LOSSFRQ=2$ and $LOSSFRQ=1$. In contrast to the coefficients shown in Display 6.4, the two equations shown in

Display 6.9 have different intercepts *and* different slopes (i.e., the coefficients of the variables AGE, $CRED$, and $NPRVIO$ are also different).

The **Regression** node gives the following formulas for calculating the probabilities of the events $LOSSFRQ$=2, $LOSSFRQ$=1, and $LOSSFRQ$=0.

$$\Pr(LOSSFRQ = 2) =$$

$$\frac{\exp(1.5946 - 0.0324 * AGE - 0.0134 * CRED + 0.2186 * NPRVIO)}{1 + \exp(1.5946 - 0.0324 * AGE - 0.0134 * CRED + 0.2186 * NPRVIO) + \exp(-0.1186 - 0.0228 * AGE - 0.00449 * CRED + 0.2240 * NPRVIO)} \quad (6.10)$$

$$\Pr(LOSSFRQ = 1) =$$

$$\frac{\exp(-0.1186 - 0.0228 * AGE - 0.00449 * CRED + 0.2240 * NPRVIO)}{1 + \exp(1.5946 - 0.0324 * AGE - 0.0134 * CRED + 0.2186 * NPRVIO) + \exp(-0.1186 - 0.0228 * AGE - 0.00449 * CRED + 0.2240 * NPRVIO)} \quad (6.11)$$

$$\Pr(LOSSFRQ = 0) =$$

$$\frac{1}{1 + \exp(1.5946 - 0.0324 * AGE - 0.0134 * CRED + 0.2186 * NPRVIO) + \exp(-0.1186 - 0.0228 * AGE - 0.00449 * CRED + 0.2240 * NPRVIO)} \quad (6.12)$$

Equation 6.12 implies that

$$\Pr(LOSSFRQ = 0) = 1 - \Pr(LOSSFRQ = 1) - \Pr(LOSSFRQ = 2)\,. \quad (6.13)$$

It is possible to check the SAS code produced by the **Regression** node to verify that the model produced by the **Regression** node is the same as Equations 6.10, 6.11, and 6.12.

Displays 6.10 and 6.10A show the SAS code for the model, which is taken from the output of the **Results** window of the **Regression** node.

Display 6.10

```
*** Compute Linear Predictor;
drop _TEMP;
drop _LP0  _LP1;
_LP0 = 0;
_LP1 = 0;

***  Effect: AGE ;
_TEMP = AGE ;
_LP0 = _LP0 + (    -0.0323912354634 * _TEMP);
_LP1 = _LP1 + (    -0.0228181055783 * _TEMP);

***  Effect: CRED ;
_TEMP = CRED ;
_LP0 = _LP0 + (   -0.01335715577514 * _TEMP);
_LP1 = _LP1 + (   -0.00448783417963 * _TEMP);

***  Effect: NPRVIO ;
_TEMP = NPRVIO ;
_LP0 = _LP0 + (    0.21864487608312 * _TEMP);
_LP1 = _LP1 + (    0.22396412290576 * _TEMP);

*** Naive Posterior Probabilities;
drop _MAXP _IY _P0 _P1 _P2;
drop _LPMAX;
_LPMAX= 0;
_LP0 =     1.59463711057944 + _LP0;
if _LPMAX < _LP0 then _LPMAX = _LP0;
_LP1 =    -0.11855123381699 + _LP1;
if _LPMAX < _LP1 then _LPMAX = _LP1;
_LP0 = exp(_LP0 - _LPMAX);
_LP1 = exp(_LP1 - _LPMAX);
_LPMAX = exp(-_LPMAX);
_P2 = 1 / (_LPMAX + _LP0 + _LP1);
_P0 = _LP0 * _P2;
_P1 = _LP1 * _P2;
_P2 = _LPMAX * _P2;

REGDR1:
```

Display 6.10A

```
*** Posterior Probabilities and Predicted Level;
label P_LOSSFRQ2 = 'Predicted: LOSSFRQ=2' ;
label P_LOSSFRQ1 = 'Predicted: LOSSFRQ=1' ;
label P_LOSSFRQ0 = 'Predicted: LOSSFRQ=0' ;
P_LOSSFRQ2 = _P0;
_MAXP = _P0;
_IY = 1;
P_LOSSFRQ1 = _P1;
if (_P1 - _MAXP > 1e-8) then do;
   _MAXP = _P1;
   _IY = 2;
end;
P_LOSSFRQ0 = _P2;
if (_P2 - _MAXP > 1e-8) then do;
   _MAXP = _P2;
   _IY = 3;
end;
```

By setting AGE=50, CRED=670, and NPRVIO=2, and by using the statements shown in Displays 6.10 and 6.10A, I generated the probabilities that *LOSSFRQ* =0, 1, and 2, as shown in Display 6.10B. The SAS code executed for generating these probabilities is given in Displays 6.43 and 6.44 in the appendix to this chapter.

Display 6.10B

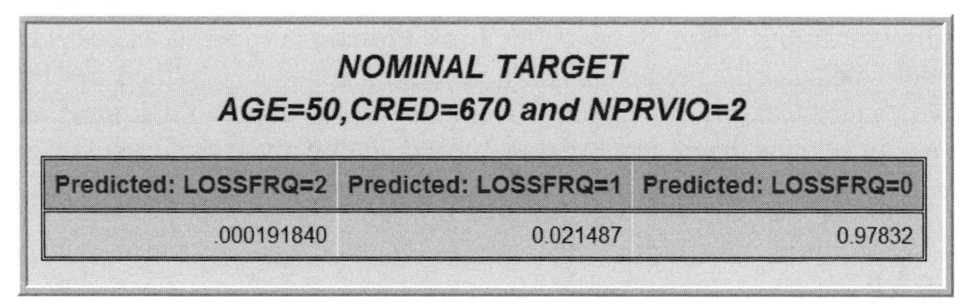

NOMINAL TARGET
AGE=50,CRED=670 and NPRVIO=2

Predicted: LOSSFRQ=2	Predicted: LOSSFRQ=1	Predicted: LOSSFRQ=0
.000191840	0.021487	0.97832

6.2.4 A Word about Models with Continuous Targets

When the target is continuous, the **Regression** node estimates a linear equation (linear in parameters). The score function is:

$E(Y/X)=\beta'X$, where Y is the target variable (such as expenditure, income, savings, etc.), where X is a vector of inputs (explanatory variables), and β is the vector of coefficients estimated by the **Regression** node.

6.3 An Overview of Some Properties of the Regression Node

A clear understanding of **Regression Type**, **Link Function**, **Selection Model**, and **Selection Criterion** properties is essential for developing predictive models.

In this section, I present a description of these properties in detail.

6.3.1 Regression Type Property

The choices available for this property are Logistic Regression and Linear Regression.

6.3.1.1 Logistic Regression

If your target is categorical (binary, ordinal, or nominal), Logistic Regression is the default regression type. If the target is binary (either numeric or character), the **Regression** node gives you a Logistic Regression with logit link by default. If the target is categorical with more than two categories, and if its measurement scale (referred to as *level* in the Variables table of the **Input Data Source** node) is declared as ordinal, the **Regression** node by default gives you a model based on a cumulative logits link, as shown in Section 6.2.2. If the target has more than two categories, and if its measurement scale is set to nominal, then the **Regression** node gives, by default, a model with a generalized logits link, as shown in Section 6.2.3.

6.3.1.2 Linear Regression

If your target is continuous or if its measurement scale is interval, then the **Regression** node gives you an ordinary least squares regression by default. A unit link function is used.

6.3.2 Link Function Property

In this section, I discuss various values to which the **Link Function** property can be set and how the **Regression** node calculates the predictions of the target for each value of the **Link Function** property. I start with an explanation of the theory behind the logit and probit link functions for a binary target. You may skip the theoretical explanation and go directly to the formula that the **Regression** node uses for calculating the target for each specified value of the **Link Function** property.

I will start with the example of a response model, where the target variable takes one of the two values, response and non-response (1 and 0, respectively).

Assume that there is a latent variable y^* for each record or customer in the data set. Now assume that if the value of $y^*>0$, then the customer responds. The value of y^* depends on the customer's characteristics as represented by the inputs or explanatory variables, which I can denote by the vector x. Assume that

$$y^* = \beta' x + U \tag{6.14}$$

where β is the vector of coefficients and U is a random variable. Different assumptions about the distribution of the random variable U give rise to different link functions. I will show how this is so.

The probability of response is

$$\Pr(y=1 \mid x) = \Pr(y^* > 0 \mid x) = \Pr(\beta' x + U > 0) = \Pr(U > -\beta' x) \tag{6.15}$$

where y is the target variable.

Therefore,

$$\Pr(y=1 \mid x) = \Pr(U > -\beta' x) = 1 - F(-\beta' x) \tag{6.16}$$

where $F(.)$ is the Cumulative Distribution Function (CDF) of the random variable U.

6.3.2.1 Logit Link

If I choose the Logit value for the **Link Function** property, then the CDF above takes on the following value:

$$F(-\beta' x) = \frac{1}{1+e^{\beta' x}} \text{ and } 1 - F(-\beta' x) = 1 - \frac{1}{1+e^{\beta' x}} = \frac{1}{1+e^{-\beta' x}} \tag{6.17}$$

Hence, the probability of response is calculated as

$$\Pr(y=1\,|\,x)=\frac{1}{1+e^{-\beta' x}}=\frac{e^{\beta' x}}{1+e^{\beta' x}} \tag{6.18}$$

From Equation 6.18 you can see that the link function is

$$\log\left(\frac{\Pr(y=1\,|\,x)}{1-\Pr(y=1\,|\,x)}\right)=\beta' x. \tag{6.19}$$

$\beta' x$ is called a *linear predictor* since it is a linear combination of the inputs.

The **Regression** node estimates the linear predictor and then applies Equation 6.18 to get the probability of response.

6.3.2.2 Probit Link

If I choose the Probit value for the **Link Function** property, then it is assumed that the random variable U in Equation 6.14 has a normal distribution with mean=0 and standard deviation=1. In this case I have

$$\Pr(y=1\,|\,x)=1-F(-\beta' x)=F(\beta' x) \tag{6.20}$$

$$\text{where } F(\beta' x)=\int_{-\infty}^{\beta' x}\frac{1}{\sqrt{2\pi}}e^{-u^2}\,du\,. \tag{6.21}$$

To estimate the probability of response, the **Regression** node uses the *probnorm* function to calculate the probability above in Equation 6.20. Thus, it calculates the probability that y = 1 as

$\Pr(y=1\,|\,x)=probnorm(\beta' x)$, where the *probnorm* function gives $F(\beta' x)$.

6.3.2.3 Complementary Log-Log Link (Cloglog)

In this case the **Regression** node calculates the probability of response as

$$\Pr(y=1\,|\,x)=1-\exp(-\exp(\beta' x))\,. \tag{6.22}$$

which is the cumulative extreme value distribution.

6.3.2.4 Identity Link

When the target is interval-scaled, the **Regression** node uses an identity link. This means that the predicted value of the target variable is equal to the linear predictor $\beta' x$. As a result, the predicted value of the target variable is not restricted to the range 0 to 1.

6.3.3 Selection Model Property

The value of this property determines the model selection method used for selecting the variables for inclusion in a regression, whether it is a logistic or linear regression. The value of this property can be set to one of the following: **None**, **Backward**, **Forward**, or **Stepwise**.

If you set the value to **None**, all inputs (sometimes referred to as **effects**) are included in the model, and there is only one step in the model selection process.

6.3.3.1 Backward Elimination Method

To use this method, set the **Selection Model** property to **Backward**, set the **Stay Significance Level** property to a desired threshold level of significance (such as 0.025, 0.05, 0.10, etc.) and then set the **Selection Default** property to **No**. In this method, the process of model fitting starts with Step 0, where all inputs are included in the model. For each parameter estimated, a Wald Chi-Square test statistic is computed. In Step 1, the input whose coefficient is least significant, and which also does not meet the threshold set by the **Stay Significance Level** property (i.e., the *p*-value of its Chi-Square test statistic is greater than the value to which the **Stay Significance Level** property is set), is removed. Also, from the remaining variables, a new model is estimated and a Wald Chi-Square test statistic is computed for the coefficient of each input in the estimated model. In Step 2, the input whose coefficient is least significant (and which does not meet the **Stay Significance Level** threshold) is removed, and a new model is built. This process is repeated until no more variables fail to meet the threshold specified. Once a variable is removed at any step, it does not enter the equation again.

If this selection process takes ten steps (including Step 0) before it comes to a stop, then ten models are available when the process stops. From these ten models, the **Regression** node selects one model. The criterion used for the selection of a final model is determined by the value to which the **Selection Criterion** property is set. If the **Selection Criterion** property is set to **None**, then the final model selected is the one from the last step of the backward elimination process. If the **Selection Criterion** property is set to any value other than **None**, then the **Regression** node uses the criterion specified by the value of the **Selection Criterion** property to select the final model. Alternative values of the **Selection Criterion** property and a description of the procedures the **Regression** node follows to select the final model according to each of these criteria will be discussed in Section 6.3.4.

6.3.3.1A Backward Elimination Method When the Target Is Binary

Displays 6.11A, 6.11B, and 6.11C illustrate the backward elimination method using a data set with only ten inputs and a binary target. I set the **Stay Significance Level** property to 0.05 and the **Selection Criterion** property to **None**.

Display 6.11A shows that at Step 0 all of the ten inputs are included in the model.

Display 6.11A

```
Backward Elimination Procedure


Step 0: The following effects were entered.

Intercept  CVAR001  NVAR103  NVAR179  NVAR2  NVAR253  NVAR27  NVAR305  NVAR5  NVAR7  NVAR8
```

Display 6.11B shows that in Step 1 the variable NVAR5 was removed because its Wald Chi-Square statistic had the highest *p*-value (0.8540) which is above the threshold significance level of 0.05 specified in the **Stay Significance Level** property. Next, a model with only nine inputs (all except NVAR5) was estimated, and a Wald Chi-Square statistic was computed for the coefficient of each input. Because, of the remaining variables, NVAR8 had the highest *p*-value for the Wald Test statistic, this input was removed from Step 2, and a logistic regression was estimated with the remaining eight inputs. This process of elimination continued until two more variables NVAR2 and NVAR179 were removed at Steps 3 and 4, respectively, as shown in Display 6.11B.

Display 6.11B

NOTE: No (additional) effects met the 0.05 significance level for removal from the model.

Summary of Backward Elimination

Step	Effect Removed	DF	Number In	Wald Chi-Square	Pr > ChiSq
1	NVAR5	1	9	0.0338	0.8540
2	NVAR8	1	8	0.0750	0.7842
3	NVAR2	1	7	0.1001	0.7517
4	NVAR179	1	6	2.5626	0.1094

The Wald Test statistic was then computed for the coefficient of each input included in the model, and at this stage none of the *p*-values were above the threshold value of 0.05. Hence, the final model is the model developed at the last step, namely Step 4. Display 6.11C shows the variables included in the final model and shows that the *p*-values for all of the included inputs are below the **Stay Significance Level** property value of 0.05.

Display 6.11C

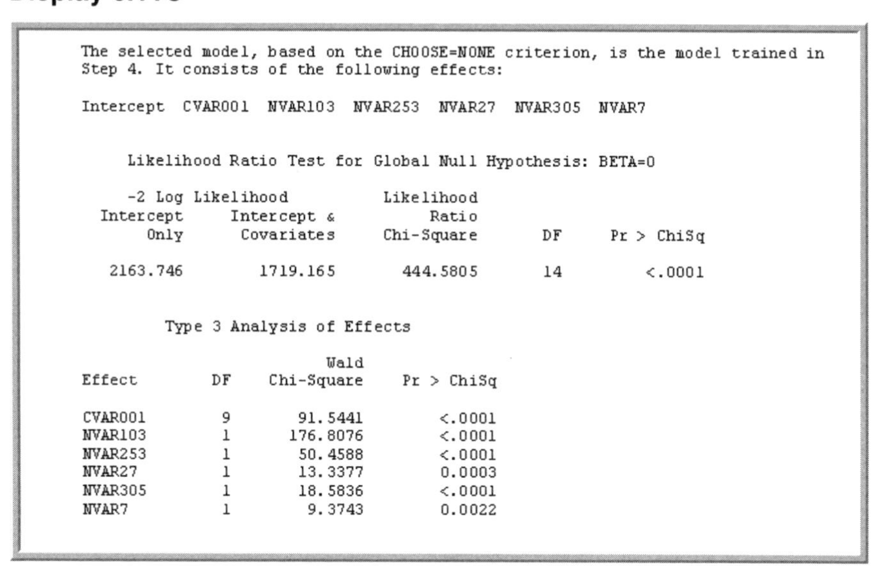

The selected model, based on the CHOOSE=NONE criterion, is the model trained in Step 4. It consists of the following effects:

Intercept CVAR001 NVAR103 NVAR253 NVAR27 NVAR305 NVAR7

Likelihood Ratio Test for Global Null Hypothesis: BETA=0

-2 Log Likelihood Intercept Only	-2 Log Likelihood Intercept & Covariates	Likelihood Ratio Chi-Square	DF	Pr > ChiSq
2163.746	1719.165	444.5805	14	<.0001

Type 3 Analysis of Effects

Effect	DF	Wald Chi-Square	Pr > ChiSq
CVAR001	9	91.5441	<.0001
NVAR103	1	176.8076	<.0001
NVAR253	1	50.4588	<.0001
NVAR27	1	13.3377	0.0003
NVAR305	1	18.5836	<.0001
NVAR7	1	9.3743	0.0022

6.3.3.1B Backward Elimination Method When the Target Is Continuous

The backward elimination method for a continuous target is similar to that for a binary target. However, when the target is continuous, an *F* statistic is used instead of Wald Chi-Square statistic in order to calculate the *p*-values.

To demonstrate the backward elimination property for a continuous target, I used a data set with 16 inputs. I set the **Stay Significance Level** property to 0.025 and the **Selection Criterion** property to **None**.

In Step 0 (not shown), all 16 inputs were included in the model. In the backward elimination process, four inputs were removed, one at each step. These results are shown in Display 6.11D.

Display 6.11D

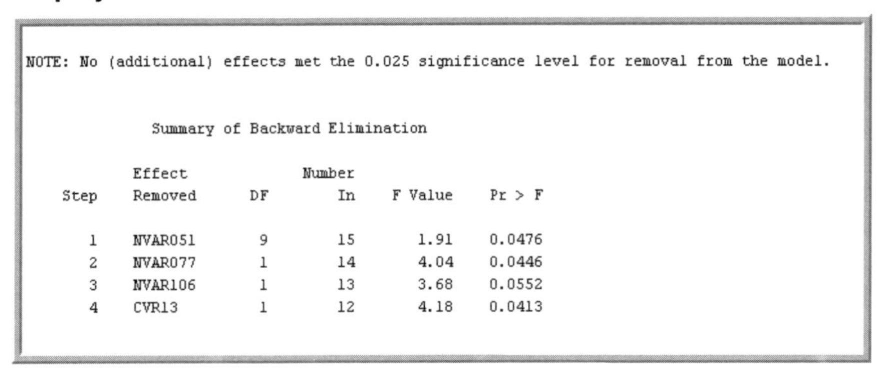

NOTE: No (additional) effects met the 0.025 significance level for removal from the model.

Summary of Backward Elimination

Step	Effect Removed	DF	Number In	F Value	Pr > F
1	NVAR051	9	15	1.91	0.0476
2	NVAR077	1	14	4.04	0.0446
3	NVAR106	1	13	3.68	0.0552
4	CVR13	1	12	4.18	0.0413

Display 6.11E shows inputs included in the final model.

Display 6.11E

The selected model, based on the CHOOSE=NONE criterion, is the model trained in Step 4. It consists of the following effects:

Intercept CVAR05 CVAR14 NVAR009 NVAR027 NVAR048 NVAR054 NVAR084 NVAR085 NVAR112 NVAR148 NVAR174 NVAR190

Analysis of Variance

Source	DF	Sum of Squares	Mean Square	F Value	Pr > F
Model	29	819.067387	28.243703	14.64	<.0001
Error	926	1785.906019	1.928624		
Corrected Total	955	2604.973407			

Model Fit Statistics

R-Square	0.3144	Adj R-Sq	0.2930
AIC	657.4266	BIC	660.2396
SBC	803.3093	C(p)	47.2135

Type 3 Analysis of Effects

Effect	DF	Sum of Squares	F Value	Pr > F
CVAR05	4	23.6150	3.06	0.0161
CVAR14	1	75.8157	39.31	<.0001
NVAR009	1	54.8664	28.45	<.0001
NVAR027	1	16.3370	8.47	0.0037
NVAR048	1	35.4411	18.38	<.0001
NVAR054	1	16.2707	8.44	0.0038
NVAR084	3	22.3877	3.87	0.0091
NVAR085	1	10.8962	5.65	0.0177
NVAR112	11	45.7402	2.16	0.0149
NVAR148	1	10.6582	5.53	0.0189
NVAR174	3	20.1009	3.47	0.0157
NVAR190	1	29.5385	15.32	<.0001

Display 6.11E shows the variables included in the final model, along with the *F* statistics and *p*-values for the variable coefficients. The *F* statistics are partial *F*'s since they are adjusted for the other effects already included in the model. In other words, the *F* statistic shown for each variable is based on a partial sum of squares, which measures the increase in the model sum of squares due to the addition of the variable to a model that already contains all the other main effects (11 in this case) plus any interaction terms (0 in this case). This type of analysis, where the contribution of each variable is adjusted for all other effects, including the interaction effects, is called a Type 3 analysis.

At this point, it may be worth mentioning Type 2 sum of squares. According to Littell, Freund, and Spector, "Type 2 sum of squares for a particular variable is the increase in Model sum of square due to adding the variable to a model that already contains all other variables. If the model contains only main effects, then type 3 and type 2 analyses are the same."[1] For a discussion of partial sums of squares and different types of analysis, see *SAS System for Linear Models*.

6.3.3.2 Forward Selection Method

To use this method, set the **Selection Model** property to **Forward**, set the **Entry Significance Level** property to a desired threshold level of significance, such as 0.025 0.05, etc. Then set the **Selection Default** property to **No**.

6.3.3.2A Forward Selection Method When the Target Is Binary

At Step 0 of the forward selection process, the model consists of only the intercept and no other inputs. At the beginning of Step 1, the **Regression** node calculates the score Chi-Square statistic[2] for each variable not included in the model and selects the variable with the largest score statistic, provided that it also meets the Entry Significance Level threshold value. At the end of Step 1, the model consists of the intercept plus the first variable selected. In Step 2, the **Regression** node again calculates the score Chi-Square statistic for each variable not included in the model and similarly selects the best variable. This process is repeated until none of the remaining variables meets the Entry Significance Level threshold. Once a variable is entered into the model at any step, it is never removed.

To demonstrate the Forward Selection method, I used a data set with a binary target. I set the **Selection Model** property to **Forward**, the **Entry Significance Level** property to 0.025, and the **Selection Criterion** property to **None**. Displays 6.12A, 6.12B, and 6.12C show how the Forward Selection method works when the target is binary.

Display 6.12A

```
Forward Selection Procedure


Step 0: Intercept entered.
```

[1] Ramon C. Littell, Walter W. Stroupe, and Rudolf J. Freund. *SAS System for Linear Models,* 4th ed. (Cary, NC: SAS Institute Inc., 1991), 21, 22, 156.

[2] For the computation of the score Chi-Square statistic, see "The Logistic Procedure" in the *SAS/STAT User's Guide*, pp. 1948-1949.

Display 6.12B

NOTE: No (additional) effects met the 0.025 significance level for entry into the model.

Summary of Forward Selection

Step	Effect Entered	DF	Number In	Score Chi-Square	Pr > ChiSq
1	NVAR103	1	1	148.1507	<.0001
2	IMP_NVAR253	1	2	112.7779	<.0001
3	CVAR001	9	3	130.2618	<.0001
4	IMP_NVAR305	1	4	40.2100	<.0001
5	IMP_NVAR179	1	5	11.2827	0.0008
6	NVAR27	1	6	11.5922	0.0007
7	IMP_NVAR299	1	7	9.0667	0.0026
8	IMP_NVAR306	1	8	7.8575	0.0051
9	IMP_NVAR6	1	9	7.3933	0.0065
10	IMP_CVAR7	1	10	6.9745	0.0083
11	IMP_NVAR16	6	11	16.9559	0.0094
12	IMP_NVAR238	1	12	6.7879	0.0092
13	IMP_NVAR177	1	13	5.7585	0.0164

Display 6.12C

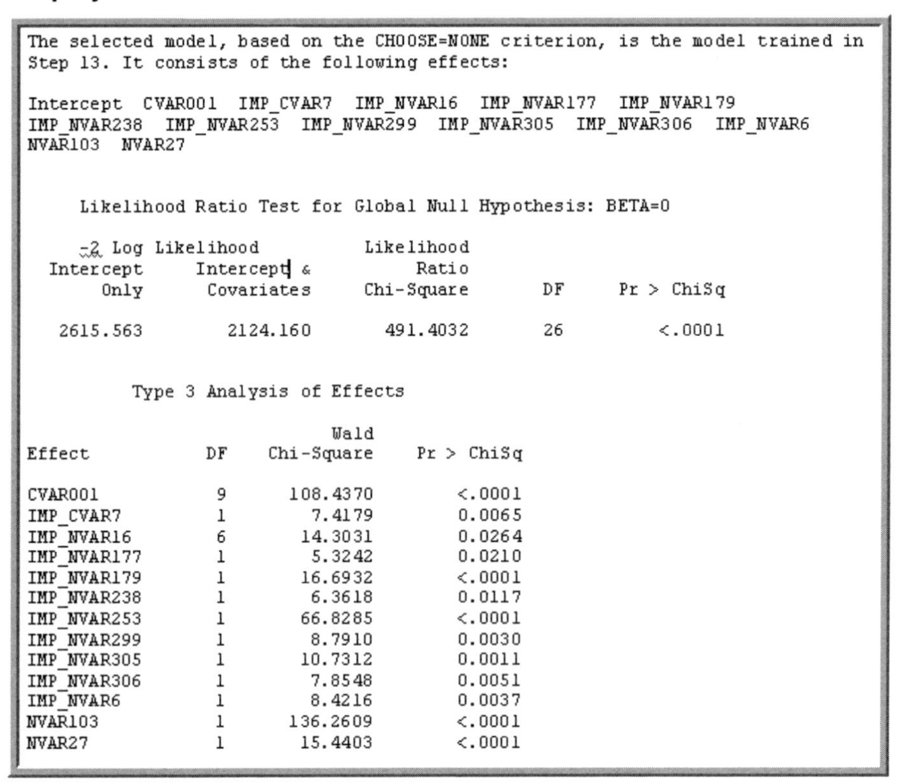

The selected model, based on the CHOOSE=NONE criterion, is the model trained in Step 13. It consists of the following effects:

Intercept CVAR001 IMP_CVAR7 IMP_NVAR16 IMP_NVAR177 IMP_NVAR179 IMP_NVAR238 IMP_NVAR253 IMP_NVAR299 IMP_NVAR305 IMP_NVAR306 IMP_NVAR6 NVAR103 NVAR27

Likelihood Ratio Test for Global Null Hypothesis: BETA=0

-2 Log Likelihood Intercept Only	-2 Log Likelihood Intercept & Covariates	Likelihood Ratio Chi-Square	DF	Pr > ChiSq
2615.563	2124.160	491.4032	26	<.0001

Type 3 Analysis of Effects

Effect	DF	Wald Chi-Square	Pr > ChiSq
CVAR001	9	108.4370	<.0001
IMP_CVAR7	1	7.4179	0.0065
IMP_NVAR16	6	14.3031	0.0264
IMP_NVAR177	1	5.3242	0.0210
IMP_NVAR179	1	16.6932	<.0001
IMP_NVAR238	1	6.3618	0.0117
IMP_NVAR253	1	66.8285	<.0001
IMP_NVAR299	1	8.7910	0.0030
IMP_NVAR305	1	10.7312	0.0011
IMP_NVAR306	1	7.8548	0.0051
IMP_NVAR6	1	8.4216	0.0037
NVAR103	1	136.2609	<.0001
NVAR27	1	15.4403	<.0001

It can be seen from Display 6.12C that all the variables in the final model, except IMP_NVAR16, have *p*-values below the threshold level of 0.025. These variables were significant when they entered the equation, and remained significant even after other variables entered the equation at the subsequent steps. Even IMP_NVAR16 has a *p*-value (only slightly) above the threshold level. This desirable result may be attributable to the fact that, prior to running the forward selection process, I cleaned the data set and eliminated variables which are likely to be collinear with other variables or have a highly skewed distribution or both.

6.3.3.2B Forward Selection Method When the Target Is Continuous

Once again, at Step 0 of the forward selection process, the model consists only of the intercept. At Step 1, the **Regression** node calculates, for each variable not included in the model, the reduction in the sum of squared residuals that would result if that variable were included in the model. It then selects the variable which, when added to the model, results in the largest reduction in the sum of squared residuals, provided that the p-value of the F statistic for the variable is less than or equal to the Entry Significance Level threshold. At Step 2 the variable selected in Step 1 is added to the model, and the **Regression** node again calculates, for each variable not included in the model, the reduction in the sum of squared residuals that would result if that variable were included in the model. Again, the variable whose inclusion results in the largest reduction in the sum of squared residuals is added to the model, and so on, until none of the remaining variables meets the Entry Significance Level threshold value. As with the binary target, once a variable is entered into the model at any step, it is never removed.

In order to demonstrate the above process, I used a data set with a continuous target. I set the **Selection Model** property to **Forward**, the **Entry Significance Level** property to 0.05, and **Selection Criterion** property to **None**. Displays 6.12D and 6.12E show how the forward selection method works when the target is continuous.

Display 6.12D

NOTE: No (additional) effects met the 0.05 significance level for entry into the model.

Summary of Forward Selection

Step	Effect Entered	DF	Number In	F Value	Pr > F
1	CVAR14	1	1	208.63	<.0001
2	NVAR048	1	2	38.11	<.0001
3	NVAR009	1	3	29.61	<.0001
4	NVAR174	3	4	7.27	<.0001
5	NVAR190	1	5	8.00	0.0048
6	NVAR054	1	6	7.81	0.0053
7	NVAR084	3	7	4.25	0.0054
8	NVAR027	1	8	7.02	0.0082
9	NVAR112	11	9	2.14	0.0159
10	CVAR05	4	10	3.03	0.0170
11	NVAR085	1	11	5.47	0.0196
12	NVAR148	1	12	5.53	0.0189
13	CVR13	1	13	4.18	0.0413

Display 6.12E

```
The selected model, based on the CHOOSE=NONE criterion, is the model trained in
Step 13. It consists of the following effects:

Intercept  CVAR05  CVAR14  CVR13  NVAR009  NVAR027  NVAR048  NVAR054  NVAR084
NVAR085  NVAR112  NVAR148  NVAR174  NVAR190


                              Analysis of Variance

                                      Sum of
Source                  DF           Squares      Mean Square    F Value    Pr > F

Model                   30        827.092669        27.569756      14.34    <.0001
Error                  925       1777.880738         1.922033
Corrected Total        955       2604.973407


              Model Fit Statistics

R-Square          0.3175      Adj R-Sq         0.2954
AIC             655.1210      BIC            658.2445
SBC             805.8665      C(p)            44.9751


             Type 3 Analysis of Effects

                               Sum of
Effect          DF            Squares    F Value    Pr > F

CVAR05           4            23.4906       3.06    0.0162
CVAR14           1            72.9383      37.95    <.0001
CVR13            1             8.0253       4.18    0.0413
NVAR009          1            52.9248      27.54    <.0001
NVAR027          1            15.8249       8.23    0.0042
NVAR048          1            30.1796      15.70    <.0001
NVAR054          1            16.6272       8.65    0.0034
NVAR084          3            22.8133       3.96    0.0081
NVAR085          1            11.1219       5.79    0.0163
NVAR112         11            45.7621       2.16    0.0144
NVAR148          1            10.7603       5.60    0.0182
NVAR174          3            21.1381       3.67    0.0121
NVAR190          1            29.3891      15.29    <.0001
```

6.3.3.3 Stepwise Selection Method

To use this method, set the **Selection Model** property to **Stepwise**, the **Entry Significance Level** property to a desired threshold level of significance (such as 0.025, 0.05, etc.), the **Stay Significance Level** property to a desired value (such as 0.025, 0.05, etc.), and finally, set the **Selection Default** property to **No**.

The stepwise selection procedure consists of both forward and backward steps, although the backward steps may or may not yield any change in the forward path. The variable selection is similar to the forward selection method except that the variables already in the model do not necessarily remain in the model. When a new variable is entered at any step, some or all variables which are already included in the model may become insignificant. Such variables may be removed by the same procedures used in the backward method. Thus, variables are entered by forward steps and removed by backward steps. A forward step may be followed by more than one backward step.

6.3.3.3A Stepwise Selection Method When the Target Is Binary

I used a data set with a binary target, and I set the **Selection Model** property to **Stepwise**, the **Entry Significance Level** property to 0.025, the **Stay Significance Level** property to 0.025, and the **Selection Criterion** property to **None**. The results are shown in Displays 6.13A, 6.13B, and 6.13C.

Display 6.13A

```
Stepwise Selection Procedure

Step 0: Intercept entered.
```

Display 6.13B

NOTE: Model building terminates because the last effect entered is removed by the Wald test criterion.

Summary of Stepwise Selection

Step	Effect Entered	Effect Removed	DF	Number In	Score Chi-Square	Wald Chi-Square	Pr > ChiSq
1	NVAR103		1	1	148.1507		<.0001
2	IMP_NVAR253		1	2	112.7779		<.0001
3	CVAR001		9	3	130.2618		<.0001
4	IMP_NVAR305		1	4	40.2100		<.0001
5	IMP_NVAR179		1	5	11.2827		0.0008
6	NVAR27		1	6	11.5922		0.0007
7	IMP_NVAR299		1	7	9.0667		0.0026
8	IMP_NVAR306		1	8	7.8575		0.0051
9	IMP_NVAR6		1	9	7.3933		0.0065
10	IMP_CVAR7		1	10	6.9745		0.0083
11	IMP_NVAR16		6	11	16.9559		0.0094
12		IMP_NVAR16	6	10		14.1052	0.0285

Display 6.13C

The selected model, based on the CHOOSE=NONE criterion, is the model trained in Step 12. It consists of the following effects:

Intercept CVAR001 IMP_CVAR7 IMP_NVAR179 IMP_NVAR253 IMP_NVAR299 IMP_NVAR305 IMP_NVAR306 IMP_NVAR6 NVAR103 NVAR27

Likelihood Ratio Test for Global Null Hypothesis: BETA=0

-2 Log Likelihood Intercept Only	-2 Log Likelihood Intercept & Covariates	Likelihood Ratio Chi-Square	DF	Pr > ChiSq
2615.563	2156.583	458.9797	18	<.0001

Type 3 Analysis of Effects

Effect	DF	Wald Chi-Square	Pr > ChiSq
CVAR001	9	108.4109	<.0001
IMP_CVAR7	1	6.8509	0.0089
IMP_NVAR179	1	17.4004	<.0001
IMP_NVAR253	1	70.1089	<.0001
IMP_NVAR299	1	7.8834	0.0050
IMP_NVAR305	1	31.0677	<.0001
IMP_NVAR306	1	8.5660	0.0034
IMP_NVAR6	1	8.5996	0.0034
NVAR103	1	137.8948	<.0001
NVAR27	1	13.6509	0.0002

Comparing the variables shown in Display 6.13C with those shown in Display 6.12C, it can be seen that the stepwise selection method produced a slightly more parsimonious (and therefore better) solution than the forward selection method. The forward selection method selected 13 variables (Display 6.12C), while the stepwise selection method selected 10 variables.

6.3.3.3B Stepwise Selection Method When the Target Is Continuous

The stepwise procedure in the case of a continuous target is similar to the stepwise procedure used when the target is binary. In the forward step, the R-square, along with the *F* statistic, is used to evaluate the input variables. In the backward step, partial *F* statistics are used.

I used a data set with a continuous target and set the **Selection Model** property to **Stepwise**, the **Entry Significance Level** property to 0.05, the **Stay Significance Level** property to 0.025, and the **Selection Criterion** property to **None**. The results are shown in Displays 6.13D and 6.13E.

Display 6.13D

Summary of Stepwise Selection

Step	Effect Entered	Effect Removed	DF	Number In	F Value	Pr > F
1	CVAR14		1	1	208.63	<.0001
2	NVAR048		1	2	38.11	<.0001
3	NVAR009		1	3	29.61	<.0001
4	NVAR174		3	4	7.27	<.0001
5	NVAR190		1	5	8.00	0.0048
6	NVAR054		1	6	7.81	0.0053
7	NVAR084		3	7	4.25	0.0054
8	NVAR027		1	8	7.02	0.0082
9	NVAR112		11	9	2.14	0.0159
10		NVAR174	3	8	3.09	0.0262
11	NVAR148		1	9	5.61	0.0180
12	CVAR05		4	10	2.86	0.0225
13	NVAR174		3	11	3.23	0.0219
14	NVAR085		1	12	5.65	0.0177
15	CVR13		1	13	4.18	0.0413
16		CVR13	1	12	4.18	0.0413

Display 6.13E

```
The selected model, based on the CHOOSE=NONE criterion, is the model trained in
Step 16. It consists of the following effects:

Intercept  CVAR05  CVAR14  NVAR009  NVAR027  NVAR048  NVAR054  NVAR084  NVAR085
NVAR112  NVAR148  NVAR174  NVAR190


                          Analysis of Variance

                                    Sum of
Source                  DF         Squares     Mean Square    F Value    Pr > F

Model                   29      819.067387       28.243703      14.64    <.0001
Error                  926     1785.906019        1.928624
Corrected Total        955     2604.973407


                 Model Fit Statistics

R-Square          0.3144     Adj R-Sq          0.2930
AIC             657.4266     BIC             660.2396
SBC             803.3093     C(p)             47.2135


                Type 3 Analysis of Effects

                               Sum of
Effect          DF            Squares    F Value    Pr > F

CVAR05           4            23.6150       3.06    0.0161
CVAR14           1            75.8157      39.31    <.0001
NVAR009          1            54.8664      28.45    <.0001
NVAR027          1            16.3370       8.47    0.0037
NVAR048          1            35.4411      18.38    <.0001
NVAR054          1            16.2707       8.44    0.0038
NVAR084          3            22.3877       3.87    0.0091
NVAR085          1            10.8962       5.65    0.0177
NVAR112         11            45.7402       2.16    0.0149
NVAR148          1            10.6582       5.53    0.0189
NVAR174          3            20.1009       3.47    0.0157
NVAR190          1            29.5385      15.32    <.0001
```

6.3.4 Selection Criterion Property[3]

In Section 6.3.3, I pointed out that during the backward, forward, and stepwise selection methods, the **Regression** node creates one model at each step. Thus, when using any of these selection methods, the **Regression** node selects the final model based on the value to which the **Selection Criterion** property is set. If the **Selection Criterion** property is set to **None**, then the final model is simply the one created in the last step of the selection process. But any value other than **None** requires the **Regression** node to use the criterion specified by the value of the **Selection Criterion** property to select the final model.

[3] My explanations of the procedures described here are not written from the actual computational algorithms used inside Enterprise Miner. I have no access to those. There may be some variance between my descriptions and the actual computations in some places. Nonetheless, these explanations should provide a clear view of how the various criteria of model selection are applied in general, and can most certainly be applied using Enterprise Miner.

You can set the value of the **Selection Criterion** property to any one of the following options:

- Akaike Information Criterion (AIC)
- Schwarz Bayesian Criterion (SBC)
- Validation Error
- Validation Misclassification
- Cross Validation Error
- Cross Validation Misclassification
- Validation Profit/Loss
- Profit/Loss
- Cross Validation Profit/Loss

I demonstrate how each of these criteria work below using a small data set. For all of these examples, I set the **Regression Type** property to **Logistic Regression**, the **Link Function** property to **Logit**, the **Selection Model** property to **Backward**, the **Stay Significance Level** property to 0.05, and the **Selection Criterion** property to one of the values listed above. During a preliminary examination of the data set inputs, using the Enterprise Miner **StatExplore** and **MultiPlot** nodes, I found the variable NVAR179 to have a skewed distribution, and to be insignificant when included with other variables. However, I tried the *equalize* and *optimal* methods of transformation using the **Transform Variables** node. The significance level improved considerably when the *optimal* method was used. Also, the distributions of the target variable within each bin were much more even than they were in the original quantiles of NVAR179.

To understand the Akaike Information Criterion and the Schwarz Bayesian Criterion, it is helpful to first review the computation of the errors in models with binary targets (such as response to a direct mail marketing campaign).

Let $\hat{\pi}(x_i)$ represent the predicted probability of response for the ith record, where x_i is the vector of inputs for the ith customer record. Let y_i represent the actual outcome for the ith record, where y_i=1 if the customer responded and y_i=0 otherwise. For a responder, the error is measured as the difference between the log of the actual value for y_i and the log of the predicted probability of response($\hat{\pi}(x_i)$). Since y_i =1 ,

$$Error_i = \log\left(\frac{y_i}{\hat{\pi}(x_i)}\right) = -\log\left(\frac{\hat{\pi}(x_i)}{y_i}\right) = -y_i \log\left(\frac{\hat{\pi}(x_i)}{y_i}\right).$$

For a non-responder the error is represented as

$$Error_i = -(1-y_i)\log\left(\frac{1-\hat{\pi}(x_i)}{1-y_i}\right).$$

Combining both the response and non-response possible outcomes, the error for the ith customer record can be represented by

$$Error_i = -\left[y_i \log\left(\frac{\hat{\pi}_i(x_i)}{y_i} \right) + (1 - y_i) \log\left(\frac{1 - \hat{\pi}_i(x_i)}{1 - y_i} \right) \right].$$

Notice that the second term in the brackets will go to zero in the case of a responder. Hence, the first term will give the value of the error in that case. And in the case of a non-responder, the first term will go to zero and the error will be given by the second term alone.

Summing the errors for all the records in the data set, I get

$$ErrorSum = -\sum_{i=1}^{n} \left[y_i \log\left(\frac{\hat{\pi}_i(x_i)}{y_i} \right) + (1 - y_i) \log\left(\frac{1 - \hat{\pi}_i(x_i)}{1 - y_i} \right) \right], \tag{6.23}$$

where n is the number of records in the data set.

The expression given in Equation 6.23 is multiplied by two to obtain a quantity with a known distribution, giving the expression

$$2ErrorSum = -2\sum_{i=1}^{n} \left[y_i \log\left(\frac{\hat{\pi}_i(x_i)}{y_i} \right) + (1 - y_i) \log\left(\frac{1 - \hat{\pi}_i(x_i)}{1 - y_i} \right) \right]. \tag{6.24}$$

In general, Equation 6.24 is used as the measure of model error for models with a binary target. By taking the anti-logs of Equation 6.24, I get

$$-2\prod_{i=1}^{n} \frac{(\hat{\pi}_i(x_i))^{y_i}}{y_i^{y_i}} \frac{(1 - \hat{\pi}_i(x_i))^{1-y_i}}{(1 - y_i)^{1-y_i}}, \tag{6.25}$$

which can be rewritten as

$$-2\frac{\prod_{i=1}^{n} (\hat{\pi}_i(x_i))^{y_i} (1 - \hat{\pi}_i(x_i))^{1-y_i}}{\prod_{i=1}^{n} y_i^{y_i} (1 - y_i)^{1-y_i}}. \tag{6.26}$$

The denominator of Equation 6.26 is 1, and the numerator is simply the likelihood of the model, often represented by the symbol L, so I can write

$$-2\prod_{i=1}^{n}\left(\hat{\pi}_i(x_i)\right)^{y_i}\left(1-\hat{\pi}_i(x_i)\right)^{1-y_i}=2L \tag{6.27}$$

Taking logs of the terms on both sides in Equation 6.27, I get

$$-2LogL=-2\sum_{i=1}^{n}\left[y_i\log\left(\hat{\pi}_i(x_i)\right)+(1-y_i)\log\left(1-\hat{\pi}_i(x_i)\right)\right]. \tag{6.28}$$

The quantity shown in Equation 6.28 is used as a measure of model error for comparing models when the target is binary.

In the case of a continuous target the model error is the sum of the squares of the errors of all the individual records of the data set, where the errors are the simple differences between the actual and predicted values.

6.3.4.1 The Akaike Information Criterion

When used for comparing different models, the Akaike Information Criterion (AIC) adjusts the quantity shown in Equation 6.28 as described below.

The Akaike Information Criterion (AIC) statistic is

$$AIC=-2LogL+2(k+s)\ , \tag{6.29}$$

where k is the number of response levels minus one, and s is the number of explanatory variables in the model.

When I performed a model selection on my example data set (using backward elimination) and set the **Selection Criterion** property to Akaike's Information Criterion, there were five steps in the model selection process, and the model from Step 4 was selected. The results are shown in Display 6.14A.

Display 6.14A

NOTE: No (additional) effects met the 0.05 significance level for removal from the model.

Summary of Backward Elimination

Step	Effect Removed	DF	Number In	Wald Chi-Square	Pr > ChiSq
1	NVAR5	1	10	0.0290	0.8649
2	NVAR2	1	9	0.0677	0.7947
3	NVAR8	1	8	0.1017	0.7498
4	NVAR179	1	7	0.2000	0.6548
5	OPT_NVAR179	1	6	3.8337	0.0502

The selected model, based on the CHOOSE=AIC criterion, is the model trained in Step 4. It consists of the following effects:

Intercept CVAR001 NVAR103 NVAR253 NVAR27 NVAR305 NVAR7 OPT_NVAR179

Likelihood Ratio Test for Global Null Hypothesis: BETA=0

-2 Log Likelihood Intercept Only	-2 Log Likelihood Intercept & Covariates	Likelihood Ratio Chi-Square	DF	Pr > ChiSq
2163.746	1715.876	447.8696	15	<.0001

Type 3 Analysis of Effects

Effect	DF	Wald Chi-Square	Pr > ChiSq
CVAR001	9	92.4868	<.0001
NVAR103	1	177.9873	<.0001
NVAR253	1	51.2445	<.0001
NVAR27	1	13.9899	0.0002
NVAR305	1	16.8519	<.0001
NVAR7	1	9.1248	0.0025
OPT_NVAR179	1	3.8337	0.0502

Models with smaller values of AIC are better. When this criterion is used the **Regression** node selects the model with the smallest value of the Akaike Information Criterion statistic.

6.3.4.2 The Schwarz Bayesian Criterion

The Schwarz Bayesian Criterion (SBC) adjusts $-2LogL$ in the following way:

$$SBC = -2LogL + (k + s)n, \quad (6.30)$$

where k and s are same as in Equation 6.29 and n is the number of observations.

Models with smaller values of SBC are better. When this criterion is used, the **Regression** node selects the model with the smallest value for the Schwarz Bayesian Criterion statistic.

As Display 6.14B shows, the model selection process again took only five steps, but this time the model from Step 5 is selected.

Display 6.14B

NOTE: No (additional) effects met the 0.05 significance level for removal from the model.

Summary of Backward Elimination

Step	Effect Removed	DF	Number In	Wald Chi-Square	Pr > ChiSq
1	NVAR5	1	10	0.0290	0.8649
2	NVAR2	1	9	0.0677	0.7947
3	NVAR8	1	8	0.1017	0.7498
4	NVAR179	1	7	0.2000	0.6548
5	OPT_NVAR179	1	6	3.8337	0.0502

The selected model, based on the CHOOSE=SBC criterion, is the model trained in Step 5. It consists of the following effects:

Intercept CVAR001 NVAR103 NVAR253 NVAR27 NVAR305 NVAR7

Likelihood Ratio Test for Global Null Hypothesis: BETA=0

-2 Log Likelihood Intercept Only	-2 Log Likelihood Intercept & Covariates	Likelihood Ratio Chi-Square	DF	Pr > ChiSq
2163.746	1719.165	444.5805	14	<.0001

Type 3 Analysis of Effects

Effect	DF	Wald Chi-Square	Pr > ChiSq
CVAR001	9	91.5441	<.0001
NVAR103	1	176.8076	<.0001
NVAR253	1	50.4588	<.0001
NVAR27	1	13.3377	0.0003
NVAR305	1	18.5836	<.0001
NVAR7	1	9.3743	0.0022

6.3.4.3 The Validation Error Criterion

Validation error, in the case of a binary target, is the error calculated by the negative log likelihood expression shown in Equation 6.28. The error is calculated from the validation data set for each of the models generated at the various steps of the selection process. The parameters of each model are used to calculate the predicted probability of response for each record in the data set. In Equation 6.28 this is shown as $\hat{\pi}(x_i)$, where x_i is the vector of inputs for the ith record in the validation data set. (In the case of a continuous target the validation error is the sum of the squares of the errors calculated from the validation data set, where the errors are the simple differences between the actual and predicted values for the continuous target.) The model with the smallest validation error is selected, as shown in Display 6.14C.

Display 6.14C

```
NOTE: No (additional) effects met the 0.05 significance level for removal from
the model.


                     Summary of Backward Elimination

              Effect                   Number          Wald
    Step      Removed         DF          In      Chi-Square     Pr > ChiSq

       1      NVAR5            1          10          0.0290         0.8649
       2      NVAR2            1           9          0.0677         0.7947
       3      NVAR8            1           8          0.1017         0.7498
       4      NVAR179          1           7          0.2000         0.6548
       5      OPT_NVAR179      1           6          3.8337         0.0502

The selected model, based on the CHOOSE=VERROR criterion, is the model trained
in Step 5. It consists of the following effects:

Intercept  CVAR001  NVAR103  NVAR253  NVAR27  NVAR305  NVAR7


     Likelihood Ratio Test for Global Null Hypothesis: BETA=0

     -2 Log Likelihood             Likelihood
  Intercept      Intercept &           Ratio
       Only       Covariates      Chi-Square        DF      Pr > ChiSq

   2163.746         1719.165        444.5805        14          <.0001


          Type 3 Analysis of Effects

                                 Wald
Effect            DF      Chi-Square      Pr > ChiSq

CVAR001            9         91.5441          <.0001
NVAR103            1        176.8076          <.0001
NVAR253            1         50.4588          <.0001
NVAR27             1         13.3377          0.0003
NVAR305            1         18.5836          <.0001
NVAR7              1          9.3743          0.0022
```

6.3.4.4 The Validation Misclassification Criterion

For analyses with a binary target, such as response, a record in the validation data set is classified as a responder if the posterior probability of response is greater than the posterior probability of non-response. Similarly, if the posterior probability of non-response is greater than the posterior probability of response, the record is classified as a non-responder. These posterior probabilities are calculated for each customer in the validation data set using the parameters estimated for each model generated at each step of the model selection process. Since the validation data includes the actual customer response for each record, you can check whether the classification is correct or not. If a responder is correctly classified as a responder, or a non-responder is correctly classified as a non-responder, then there is no misclassification; otherwise there is a misclassification. The error rate for each model can be calculated by dividing the total number of misclassifications in the validation data set by the total number of records. The validation misclassification rate is calculated for each model generated at the various steps of the selection process, and the model with the smallest validation misclassification rate is selected.

Display 6.14D shows the model selected by the validation misclassification criterion.

Display 6.14D

NOTE: No (additional) effects met the 0.05 significance level for removal from the model.

Summary of Backward Elimination

Step	Effect Removed	DF	Number In	Wald Chi-Square	Pr > ChiSq
1	NVAR5	1	10	0.0290	0.8649
2	NVAR2	1	9	0.0677	0.7947
3	NVAR8	1	8	0.1017	0.7498
4	NVAR179	1	7	0.2000	0.6548
5	OPT_NVAR179	1	6	3.8337	0.0502

The selected model, based on the CHOOSE=VMISC criterion, is the model trained in Step 5. It consists of the following effects:

Intercept CVAR001 NVAR103 NVAR253 NVAR27 NVAR305 NVAR7

Likelihood Ratio Test for Global Null Hypothesis: BETA=0

-2 Log Likelihood Intercept Only	-2 Log Likelihood Intercept & Covariates	Likelihood Ratio Chi-Square	DF	Pr > ChiSq
2163.746	1719.165	444.5805	14	<.0001

Type 3 Analysis of Effects

Effect	DF	Wald Chi-Square	Pr > ChiSq
CVAR001	9	91.5441	<.0001
NVAR103	1	176.8076	<.0001
NVAR253	1	50.4588	<.0001
NVAR27	1	13.3377	0.0003
NVAR305	1	18.5836	<.0001
NVAR7	1	9.3743	0.0022

6.3.4.5 The Cross Validation Error Criterion

In Enterprise Miner, cross validation errors are calculated from the training data set. One observation is omitted at a time, and the model is re-estimated with the remaining observations. The re-estimated model is then used to predict the value of the target for the omitted observation, and then the error of the prediction, in the case of a binary target, is calculated using the formula given in Equation 6.28 with $n=1$. (In the case of a continuous target, the error is the squared difference between the actual and predicted values.) Next, another observation is omitted, the model is re-estimated and used to predict the omitted observation, and this new prediction error is calculated. This process is repeated for each observation in the training data set, and the aggregate error is found by summing the individual errors.

If you do not have enough observations for training and validation, then you can select the cross validation error criterion for model selection. The data sets I used have several thousand observations. Therefore, in my example, it may take a long time to calculate the cross validation errors, especially because there are several models. Hence this criterion is not tested here.

6.3.4.6 The Cross Validation Misclassification Rate Criterion

The cross validation misclassification rate is also calculated from the training data set. As in the case described in Section 6.3.4.5, one observation is omitted at a time, and the model is re-estimated every time. The re-estimated model is used to calculate the posterior probability of response for the omitted observation. The observation is then assigned a target level (such as response or non-response) based on maximum posterior probability. Enterprise Miner then checks the actual response of the omitted customer to see if this re-estimated model misclassifies the omitted customer. This process is repeated for each of the observations, after which the misclassifications are counted, and the misclassification rate is calculated. For the same reason cited in Section 6.3.4.5, I have not tested this criterion.

6.3.4.7 The Validation Profit/Loss Criterion

In order to use the validation profit/loss criterion, you must provide Enterprise Miner with a profit matrix. To demonstrate this method, I have provided the profit matrix shown in Display 6.14E.

Display 6.14E

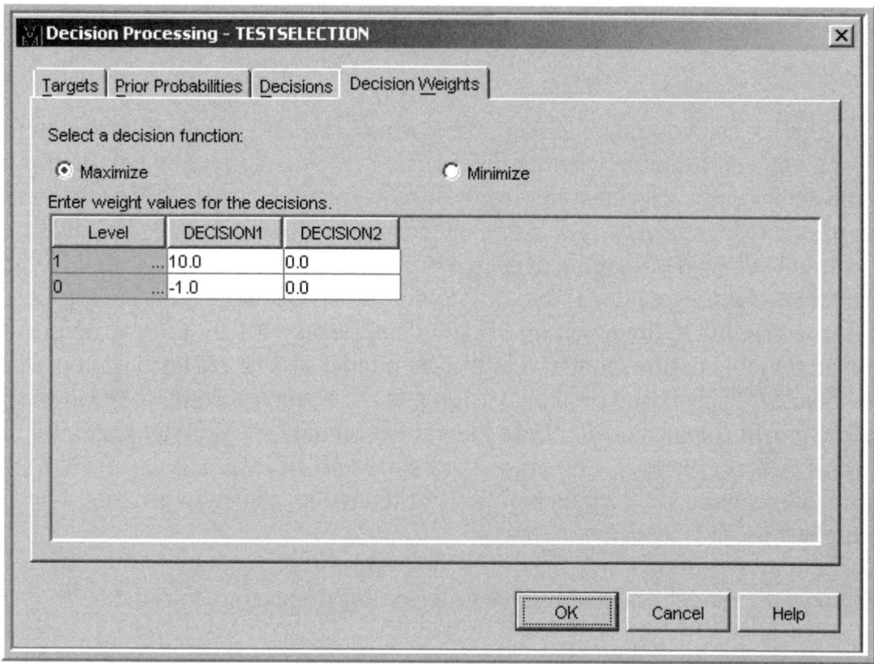

When calculating the profit/loss associated with a given model, the validation data set is used. From each model produced by the backward elimination process, in this example, a formula for calculating posterior probabilities of response and non-response is generated. Using these posterior probabilities, the expected profits under decision 1 (assigning the record to the responder category) and decision 2 (assigning the record to the non-responder category) are calculated. If the expected profits under decision 1 is greater than the expected profit under decision 2, then the record is assigned the target level responder; otherwise it is set to non-responder. Thus, each record in the validation data set is assigned to a target level.

The following equations show how the expected profit is calculated given the posterior probabilities calculated from a model and the profit matrix shown in Display 6.14E.

The expected profit under decision 1 is

$$E_{1i} = \$10 * \hat{\pi}(x_i) + (-\$1)(1 - \hat{\pi}(x_i)), \quad (6.31)$$

where E_{1i} is the expected profit under decision 1 for the ith record , $\hat{\pi}(x_i)$ is the posterior probability of response of the ith record, and x_i is the vector of inputs for the ith record in the validation data set.

The expected profit under decision 2 (assigning the record to the target level non-responder) is

$$E_{2i} = \$0 * \hat{\pi}(x_i) + \$0(1 - \hat{\pi}(x_i)) \ , \quad (6.32)$$

which is always zero in this case, given the profit matrix I have chosen.

The rule for classifying customers, then, is

If $E_{1i} > E_{2i}$, then assign the record to the target level responder. Otherwise, assign the record to the target level non-responder.

Having assigned a target category such as responder or non-responder to each record in the validation data set, you can examine whether the decision taken is correct by comparing the assignment to the actual value. If a customer record is assigned the target class of responder, and the customer is truly a responder, then the company has made a profit of $10. This is the actual profit which I would have made—not the expected profit calculated before. If, on the other hand, we assigned the target class responder and the customer turned out to be a non-responder, then I incur a loss of $1. According to the profit matrix given in Display 6.14E, there is no profit or loss if I assigned the target level non-responder to a true responder or a target level of non-responder to a true non-responder. Thus, based on the decision and the actual outcome for each record, I can calculate the actual profit for each customer in the validation data set. And by summing the profits across all of the observations, I obtain the validation profit/loss. These calculations are made for each model generated at various iterations of the model selection process. The model that yields the highest profit is selected.

Display 6.14F shows the model selected by this criterion for my example problem.

Display 6.14F

```
NOTE: No (additional) effects met the 0.05 significance level for removal from
the model.


                   Summary of Backward Elimination

            Effect                   Number         Wald
    Step    Removed          DF          In   Chi-Square     Pr > ChiSq

       1    NVAR5             1          10       0.0290         0.8649
       2    NVAR2             1           9       0.0677         0.7947
       3    NVAR8             1           8       0.1017         0.7498
       4    NVAR179           1           7       0.2000         0.6548
       5    OPT_NVAR179       1           6       3.8337         0.0502

The selected model, based on the CHOOSE=VDECDATA criterion, is the model
trained in Step 5. It consists of the following effects:

Intercept  CVAR001  NVAR103  NVAR253  NVAR27  NVAR305  NVAR7


     Likelihood Ratio Test for Global Null Hypothesis: BETA=0

     -2 Log Likelihood           Likelihood
  Intercept     Intercept &           Ratio
       Only      Covariates      Chi-Square       DF      Pr > ChiSq

   2163.746        1719.165        444.5805       14          <.0001


         Type 3 Analysis of Effects

                                 Wald
Effect            DF       Chi-Square     Pr > ChiSq

CVAR001            9          91.5441         <.0001
NVAR103            1         176.8076         <.0001
NVAR253            1          50.4588         <.0001
NVAR27             1          13.3377         0.0003
NVAR305            1          18.5836         <.0001
NVAR7              1           9.3743         0.0022
```

6.3.4.8 The Profit/Loss Criterion

When this criterion is selected, the profit/loss is calculated in the same way as the validation profit/loss above, but it is done using the training data set instead of the validation data. Again, it is possible to resort to this option only when there is insufficient data to create separate training and validation data sets. Display 6.14G shows the results of applying this option to my example data.

Display 6.14G

```
NOTE: No (additional) effects met the 0.05 significance level for removal from
the model.

                    Summary of Backward Elimination

            Effect                      Number            Wald
    Step    Removed            DF         In        Chi-Square      Pr > ChiSq

      1     NVAR5               1         10          0.0290          0.8649
      2     NVAR2               1          9          0.0677          0.7947
      3     NVAR8               1          8          0.1017          0.7498
      4     NVAR179             1          7          0.2000          0.6548
      5     OPT_NVAR179         1          6          3.8337          0.0502

The selected model, based on the CHOOSE=TDECDATA criterion, is the model
trained in Step 3. It consists of the following effects:

Intercept  CVAR001  NVAR103  NVAR179  NVAR253  NVAR27  NVAR305  NVAR7
OPT_NVAR179


        Likelihood Ratio Test for Global Null Hypothesis: BETA=0

       -2 Log Likelihood          Likelihood
    Intercept     Intercept &        Ratio
       Only       Covariates      Chi-Square        DF      Pr > ChiSq

     2163.746        1715.679       448.0667        16         <.0001


            Type 3 Analysis of Effects

                                Wald
Effect              DF    Chi-Square    Pr > ChiSq

CVAR001              9       92.3626        <.0001
NVAR103              1      178.1561        <.0001
NVAR179              1        0.2000        0.6548
NVAR253              1       51.4491        <.0001
NVAR27               1       14.0130        0.0002
NVAR305              1       16.9380        <.0001
NVAR7                1        8.3406        0.0039
OPT_NVAR179          1        1.1980        0.2737
```

6.3.4.9 The Cross Validation Profit/Loss Criterion

The process by which the cross validation profit/loss criterion is calculated is similar to the cross validation misclassification rate method described in Section 6.3.4.6. One observation is omitted at a time, the model is re-estimated every time, and the re-estimated model is used to calculate posterior probabilities of the target levels (such as response or no-response) for the omitted observation. Based on the posterior probability and the profit matrix that is specified, the omitted observation is assigned a target level in such a way that the expected profit is maximized for that observation, as described in Section 6.3.4.7. Actual profit for the omitted record is calculated by comparing the actual target level and the assigned target level. This process is repeated for all the observations in the data set. By summing the profits across all of the observations, I obtain the cross validation profit/loss. These calculations are made for each model generated at various iterations of the model selection process. The model that yields the highest profit is selected. The selected model is shown in Display 6.14H.

Display 6.14H

```
NOTE: No (additional) effects met the 0.05 significance level for removal from
the model.


                   Summary of Backward Elimination

          Effect                   Number         Wald
   Step   Removed           DF         In    Chi-Square      Pr > ChiSq

      1   NVAR5              1         10        0.0290          0.8649
      2   NVAR2              1          9        0.0677          0.7947
      3   NVAR8              1          8        0.1017          0.7498
      4   NVAR179            1          7        0.2000          0.6548
      5   OPT_NVAR179        1          6        3.8337          0.0502


The selected model, based on the CHOOSE=XDECDATA criterion, is the model
trained in Step 3. It consists of the following effects:

Intercept  CVAR001  NVAR103  NVAR179  NVAR253  NVAR27  NVAR305  NVAR7
OPT_NVAR179


     Likelihood Ratio Test for Global Null Hypothesis: BETA=0

     -2 Log Likelihood            Likelihood
  Intercept     Intercept &          Ratio
       Only      Covariates     Chi-Square        DF     Pr > ChiSq

   2163.746        1715.679       448.0667        16         <.0001


          Type 3 Analysis of Effects

                              Wald
Effect            DF    Chi-Square    Pr > ChiSq

CVAR001            9       92.3626        <.0001
NVAR103            1      178.1561        <.0001
NVAR179            1        0.2000        0.6548
NVAR253            1       51.4491        <.0001
NVAR27             1       14.0130        0.0002
NVAR305            1       16.9380        <.0001
NVAR7              1        8.3406        0.0039
OPT_NVAR179        1        1.1980        0.2737
```

6.4 Business Applications

The purpose of this section is to illustrate the use of the **Regression** node with two business applications, one requiring a binary target, and the other a continuous target.

The first application is concerned with identifying the current customers of a hypothetical bank who are most likely to respond to an invitation to sign up for Internet banking that carries a monetary reward with it. A pilot study was already conducted in which a sample of customers were sent the invitation. The task of the modeler is to use the results of the pilot study by developing a predictive model which can be used to score all customers and send mail to those customers who are most likely to respond (a *binary* target).

The second application is concerned with identifying customers of a hypothetical bank who are likely to increase their savings deposits by the largest dollar amounts if the bank increases the interest paid to them by a preset number of basis points. A sample of customers was tested first. Based on the results of the test, the task of the modeler is to develop a model to use for predicting the amount each customer would increase his savings deposits (a *continuous* target), given the increase in the interest paid. (Had the test also included several points representing decrease, increase, and no change from the current rate, then the estimated model could be used to test the price sensitivity in any direction. However, in the current study the bank is interested only in the consequences of a rate increase to customer's savings deposits.)

Before using the **Regression** node for modeling, you need to clean the data. This very important step includes purging some variables, imputing of missing values, transforming of the inputs, and checking for spurious correlation between the inputs and target. Spurious correlations arise when input variables are derived in part from the target variables, or generated from target variable behavior or closely connected to the target variable. Post-event inputs (inputs recorded after the event being modeled had occurred) might also cause spurious correlations.

I performed a prior examination of the variables using the **StatExplore** node and **MultiPlot** node, and set the roles of some variables to **rejected**, such as the one shown in Displays 6.14I.

Display 6.14I

Table of CVAR12 by resp			
	resp		
CVAR12	0	1	Total
N	1212	0	1212
Y	8278	1079	9357
Total	9490	1079	10569

In Table 6.14I the variable CVAR12 is a categorical input, taking the values & and N. A look at the distribution of the variable shown in the 2 x 2 table suggests that it might be derived from the target variable, or connected to it somehow, or it might be an input with an uneven spread across the target levels. Variables such as these can produce undesirable results, and they should be excluded from the model if their correlation with the target is truly spurious. However, there is also the possibility that this variable has valuable information for predicting the target. Clearly, only a modeler who knows where his data came from and how the data were constructed and what each variable represents in terms of the characteristics and behaviors of customers will be able to detect spurious correlations. As a precaution before applying the modeling tool, every variable included in the final model (at a minimum) should be examined for this type of problem.

In order to demonstrate alternative types of transformations, I present three models for each type of target that is discussed. The first model is based on untransformed inputs. This type of model can be used to make a preliminary identification of important inputs. For the second model, the transformations are done by the **Transform Variables** node, and for the third model transformations are done by the **Decision Tree** node.

The **Transform Variables** node has a variety of transformations (discussed in detail in Section 2.4.7 of Chapter 2 and in Section 3.2 of Chapter 3). I use some of these transformation methods here for demonstration purposes.

The transformations produced by the **Decision Tree** node yield a new categorical variable, and this happens in the following way. First, a decision tree model is built based on the inputs in order to predict the target (or classify the records). Next, a categorical variable is created, where the value (category) of the variable for a given record is simply the leaf to which the record belongs. The **Regression** node then uses this categorical variable as a class input. The **Regression** node creates a dummy variable for each category and uses it in the regression equation. These dummy variables capture the interactions between different inputs included in the tree, which are potentially very important.

If there are a large number of variables, you can use the **Variable Selection** node to screen important variables prior to using the **Regression** node. However, since the **Regression** node itself has several methods of variable selection, as shown above in Sections 6.3.3 and 6.3.4, and since a moderate number of inputs is contained in each of my data sets, I did not use the **Variable Selection** node prior to using the **Regression** node in these examples.

Two data sets are used in this chapter: one for modeling the binary target and the other for modeling the continuous target. Both of these are simulated. The results may therefore seem "too good to be true" at times, and not so good at other times.

6.4.1 Logistic Regression for Predicting Response to a Mail Campaign

For predicting response, I explore three different ways of developing a logistic regression model. The models developed are Model_B1, Model_B2, and Model_B3. The process flow for each model is shown in Display 6.15.

Display 6.15

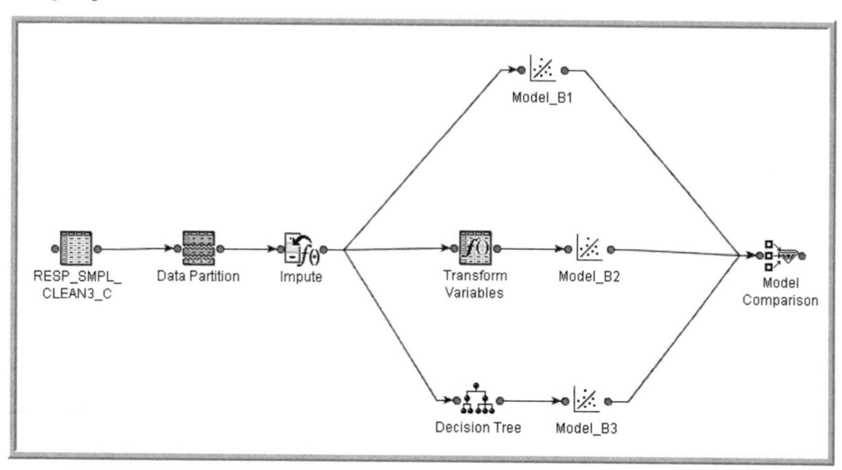

In the process flow, the first node is the **Input Data Source** node that reads the modeling data. From this modeling data set, the **Data Partition** node creates three data sets **Part_Train**, **Part_Validate**, and **Part_Test** according to the **Training**, **Validation,** and **Test** properties, which I set at 40%, 30%, and 30%, respectively. These data sets, with the roles of **Train**, **Validate**, and **Test**, are passed to the **Impute** node. In the **Impute** node, I set the **Default Input Method** property to **Median** for the interval variables, and I set it to **Count** for the class variables. The **Impute** node creates three data sets **Impt_Train**, **Impt_Validate**, and **Impt_Test** including the imputed variables, and passes them to the next node. The last node in the process flow shown in Display 6.15 is the **Model Comparison** node. This node generates various measures such as lift, capture rate, etc., from the training, validation, and test data sets.

Model_B1

The process flow for Model_B1 is at the top segment of Display 6.15. The **Regression** node uses the data set **Impt_Train** for effect selection. Each step of the effects selection method generates a model. In this example, I chose to use the stepwise selection method by setting the **Selection**

Model property to **Stepwise**. I set the **Entry Significance Level** property to 0.1, the **Stay Significance Level** property to 0.05, the **Use Selection Default** property to **No**, and **Selection Criteria** property to **Validation Error**. After trying alternative values for the **Selection Model** property, it was found that the stepwise selection method produces a more parsimonious model than other methods.

The validation data set is used by the **Regression** node to select the best model from those generated at several steps of the selection process. Finally, the data set **Impt_Test** is used to get an independent assessment of the model. This is done in the **Model Comparison** node at the end of the process flow.

Display 6.16 shows the settings of the **Regression** node properties for Model_B1.

Display 6.16

Property	Value
Node ID	Reg2
Imported Data	
Exported Data	
Variables	
Equation	
Main Effects	Yes
Two-Factor Interactions	No
Polynomial Terms	No
Polynomial Degree	2
User Terms	No
Term Editor	
Class Targets	
Regression Type	Logistic Regression
Link Function	Logit
Model Options	
Suppress Intercept	No
Input Coding	Deviation
Min Resource Use	No
Model Selection	
Selection Model	Stepwise
Selection Criterion	Validation Error
Use Selection Default	No
Selection Options	
Sequential Order	No
Entry Significance Level	0.1
Stay Significance Level	0.05
Start Variable Number	0
Stop Variable Number	0
Force Candidate Effects	0
Hierarchy Effects	Class
Moving Effect Rule	None
Maximum Number of Steps	0
Optimization Options	
Technique	Default
Default Optimization	Yes

Results

By opening the **Results** window and scrolling through the output, you can see the models created at different steps of the effect selection process and the final model selected from these steps. In this example, the **Regression** node took 19 steps to complete the effect selection. The model created at Step 13 is selected as the final model since it minimized the validation error.

Display 6.17 shows the variables included in the final model.

Display 6.17

```
The selected model, based on the CHOOSE=VERROR criterion, is the model trained
in Step 13. It consists of the following effects:

Intercept  CVAR001  IMP_NVAR100  IMP_NVAR179  IMP_NVAR253  IMP_NVAR278
IMP_NVAR286  IMP_NVAR306  IMP_NVAR6  NVAR103  NVAR27  NVAR304


     Likelihood Ratio Test for Global Null Hypothesis: BETA=0

     -2 Log Likelihood           Likelihood
  Intercept     Intercept &          Ratio
       Only      Covariates      Chi-Square        DF      Pr > ChiSq

   2789.110        2293.734        495.3756        24          <.0001


          Type 3 Analysis of Effects

                                     Wald
Effect               DF      Chi-Square      Pr > ChiSq

CVAR001               9        126.4748          <.0001
IMP_NVAR100           1          7.9498          0.0048
IMP_NVAR179           1         12.0276          0.0005
IMP_NVAR253           1         66.9916          <.0001
IMP_NVAR278           1          6.1975          0.0128
IMP_NVAR286           1          6.9065          0.0086
IMP_NVAR306           1          8.7280          0.0031
IMP_NVAR6             1          9.6545          0.0019
NVAR103               1        135.6620          <.0001
NVAR27                1         14.5192          0.0001
NVAR304               6         42.5853          <.0001
```

Display 6.17 shows a partial view of the output in the **Results** window. Display 6.18 shows the lift charts for the final Model_B1. These charts are displayed under **Score Rankings** in the **Results** window.

Display 6.18

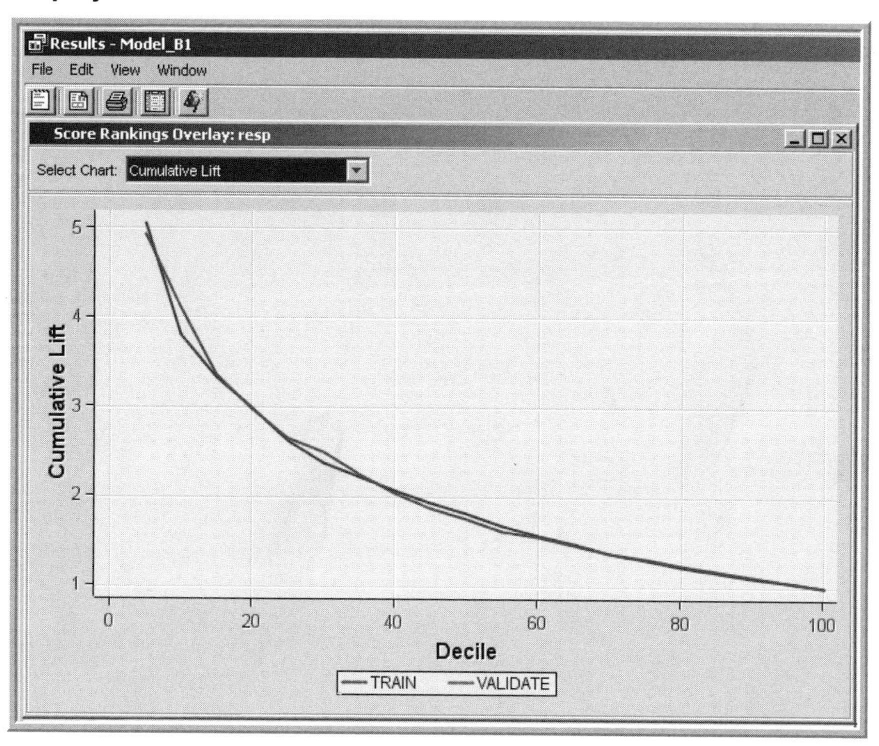

If you click anywhere under **Score Rankings Overlay** in the **Results** window and select the **Table** icon, you will see the table shown in Display 6.19.

Display 6.19

Score Rankings Overlay: resp - EMWS5.Reg2_EMRANK

Bin	Observation...	Number of E...	Gain	Lift	Cumulative ...	% Response	Cumulative ...	% Captured...
.	.	.	.	.	.	.	.	.
1	211.4	32.05528	405.4443	5.054443	5.054443	15.16333	15.16333	25.27222
2	211.4	16.09694	279.6296	2.538149	3.796296	7.614448	11.38889	12.69075
3	211.4	15.26778	233.3333	2.407407	3.333333	7.222222	10	12.03704
4	211.4	12.62528	199.7685	1.990741	2.997685	5.972222	8.993056	9.953704
5	211.4	7.046667	162.037	1.111111	2.62037	3.333333	7.861111	5.555556
6	211.4	7.046667	136.8827	1.111111	2.368827	3.333333	7.106481	5.555556
7	211.4	8.221111	121.5608	1.296296	2.215608	3.888889	6.646825	6.481481
8	211.4	6.753056	107.1759	1.064815	2.071759	3.194444	6.215278	5.324074
9	211.4	4.991389	92.90123	0.787037	1.929012	2.361111	5.787037	3.935185
10	211.4	4.991389	81.48148	0.787037	1.814815	2.361111	5.444444	3.935185
11	211.4	2.055278	67.92929	0.324074	1.679293	0.972222	5.037879	1.62037
12	211.4	2.6425	57.40741	0.416667	1.574074	1.25	4.722222	2.083333
13	211.4	1.468056	47.07977	0.231481	1.470798	0.694444	4.412393	1.157407
14	211.4	0.880833	37.56614	0.138889	1.375661	0.416667	4.126984	0.694444
15	211.4	1.174444	29.62963	0.185185	1.296296	0.555556	3.888889	0.925926
16	211.4	0.880833	22.39583	0.138889	1.223958	0.416667	3.671875	0.694444
17	211.4	0.587222	15.74074	0.092593	1.157407	0.277778	3.472222	0.462963
18	211.4	0.880833	10.0823	0.138889	1.100823	0.416667	3.302469	0.694444
19	211.4	1.174444	5.263158	0.185185	1.052632	0.555556	3.157895	0.925926

The table shown in Display 6.19 contains several measures such as lift, cumulative lift, capture rate, etc., by bin. In Table 6.19 a bin is the same as a demi-decile. This table is saved is a SAS data set, and it can be used for making custom tables such as Table 6.1.

Table 6.1, which was created from the saved SAS data set, shows the cumulative lift and cumulative capture rate by demi-decile for the training and validation data sets. The SAS program used for generating Table 6.1 is given in the appendix to this chapter.

Table 6.1

Model_B1

Bin	Demi-Decile	Cumulative Lift Training	Cumulative Lift Validation	Cumulative % Captured Response Training	Cumulative % Captured Response Validation
1	5	5.05444	4.93090	25.272	24.654
2	10	3.79630	4.10494	37.963	41.049
3	15	3.33333	3.35391	50.000	50.309
4	20	2.99769	2.97840	59.954	59.568
5	25	2.62037	2.64555	65.509	66.139
6	30	2.36883	2.50000	71.065	75.000
7	35	2.21561	2.24868	77.546	78.704
8	40	2.07176	2.04475	82.870	81.790
9	45	1.92901	1.87929	86.806	84.568
10	50	1.81481	1.75309	90.741	87.654
11	55	1.67929	1.62177	92.361	89.198
12	60	1.57407	1.54835	94.444	92.901
13	65	1.47080	1.46724	95.602	95.370
14	70	1.37566	1.37566	96.296	96.296
15	75	1.29630	1.31276	97.222	98.457
16	80	1.22396	1.23843	97.917	99.074
17	85	1.15741	1.16921	98.380	99.383
18	90	1.10082	1.10768	99.074	99.691
19	95	1.05263	1.05263	100.000	100.000
20	100	1.00000	1.00000	100.000	100.000

The lift for any demi-decile is the ratio of the observed response rate in the demi-decile to the overall observed response rate in the validation sample. The capture rate for any demi-decile shows the ratio of the number of observed responders within that demi-decile to the total number of observed responders in the sample. The cumulative lift and the lift are same for the first demi-decile. However, the cumulative lift for the second demi-decile is the ratio of the observed response rate in the first two demi-deciles combined with the overall observed response rate. The observed response rate for the first two demi-deciles is calculated as the ratio of the total number of observed responders in the first two demi-deciles to the total number of records in the first two demi-deciles, and so on. The cumulative capture rate for the second demi-decile is the ratio of the number of observed responders in the first two demi-deciles to the total number of observed responders in the sample, and so on for the remaining demi-deciles.

For example, the last column for demi-decile 40 shows that 81.79% of all responders are in the top 40% of the ranked data set. The ranking is done by the scores (predicted probability).

Model_B2

The middle section of the Display 6.15 shows the process flow for Model_B2. This model is built by using transformed variables. All interval variables are transformed by the **Maximum Normal** transformation. To use this transformation the value of the **Interval Inputs** property is set to **Maximum Normal**. In addition, the value of the **Class Inputs** property is set to **Group rare levels**.

To include the original as well as transformed variables for consideration in the model, you can reset the **Hide** and **Reject** properties to **No**. However, in this example, I left these properties at the default value of **Yes**, so that Model_B2 is built solely on the transformed variables. The reason for this is that, earlier, when I did pass both the transformed and the original variables to the

Regression node, the result did not yield a better model. Display 6.20 shows the settings of the **Transform Variables** node properties.

Display 6.20

Property	Value
Node ID	Trans
Imported Data	...
Exported Data	...
Variables	...
Formula Builder	...
Interactions Editor	...
⊟Default Methods	
Interval Inputs	Maximum Normal
Interval Targets	None
Class Inputs	Group rare levels
Class Targets	None
⊟Sample Properties	
Method	Top
Size	Default
Random Seed	12345
⊟Grouping Method	
Cutoff Value	0.5
Group Missing	No
⊟Original Variables	
Hide	Yes
Reject	Yes
Missing Value	Use in Search
Add Minimum Value to Offset Value	Yes
Offset Value	1.0
Summary Variables	Transformed and New Variables
⊟Status	
Time of Creation	2/10/07 4:31 PM
Run Id	7d3444f0-f8fe-4ab7-8352-47cf543
Last Error	
Last Status	Complete
Needs Updating	No
Needs to Run	No
Time of Last Run	2/15/07 3:24 PM
Run Duration	0 Hr. 3 Min. 53.00 Sec.
Grid Host	

The value of the **Interval Inputs** property is set to **Maximum Normal**. This means that for each input X, the transformations $\log(X)$, $X^{1/4}$, $sqrt(X)$, X^2, X^4, and e^X are evaluated. The transformation that yields sample quantiles that are closest to the theoretical quantiles of a normal distribution is selected. Suppose, by using one of these transformations, you get the transformed variable Y. This means that, the 0.75_quantile for Y from the sample is that value of Y such that 75% of the observations are at or below that value of Y. The 0.75_quantile for a standard normal distribution is 0.6745 given by $P(Z \leq 0.6745) = 0.75$, where Z is a normal random variable with mean 0 and standard deviation 1. The 0.75_quantile for Y is compared with 0.6745; likewise, the other quantiles are compared with the corresponding quantiles of the standard normal distribution. To compare the sample quantiles with the quantiles of a standard normal distribution, you must first standardize the variables. Display 6.21 shows the standardization of the variable, and the transformation that yielded maximum normality for each variable. In the code given in this display the variable _SCALEVAR_ is the standardized variable.

Display 6.21

SAS Code

```
25  *------------------------------------------------------------*;
26  label LOG_IMP_NVAR108 = 'Transformed: Imputed NVAR108';
27  _SCALEVAR_ = (IMP_NVAR108 - 0)/(653 - 0);
28  LOG_IMP_NVAR108 = log(_SCALEVAR_ + 1);
29  *------------------------------------------------------------*;
30  * TRANSFORM: IMP_NVAR11 , log(_SCALEVAR_ + 1);
31  *------------------------------------------------------------*;
32  label LOG_IMP_NVAR11 = 'Transformed: Imputed NVAR11';
33  _SCALEVAR_ = (IMP_NVAR11 - 0)/(1685 - 0);
34  LOG_IMP_NVAR11 = log(_SCALEVAR_ + 1);
35  *------------------------------------------------------------*;
36  * TRANSFORM: IMP_NVAR110 , Sqrt(_SCALEVAR_);
37  *------------------------------------------------------------*;
38  label SQRT_IMP_NVAR110 = 'Transformed: Imputed NVAR110';
39  _SCALEVAR_ = (IMP_NVAR110 - 0)/(1500 - 0);
40  SQRT_IMP_NVAR110 = Sqrt(_SCALEVAR_);
41  *------------------------------------------------------------*;
42  * TRANSFORM: IMP_NVAR111 , Sqrt(_SCALEVAR_);
43  *------------------------------------------------------------*;
44  label SQRT_IMP_NVAR111 = 'Transformed: Imputed NVAR111';
45  _SCALEVAR_ = (IMP_NVAR111 - 0)/(1500 - 0);
46  SQRT_IMP_NVAR111 = Sqrt(_SCALEVAR_);
47  *------------------------------------------------------------*;
48  * TRANSFORM: IMP_NVAR112 , Sqrt(_SCALEVAR_);
49  *------------------------------------------------------------*;
50  label SQRT_IMP_NVAR112 = 'Transformed: Imputed NVAR112';
51  _SCALEVAR_ = (IMP_NVAR112 - 0)/(6000 - 0);
52  SQRT_IMP_NVAR112 = Sqrt(_SCALEVAR_);
53  *------------------------------------------------------------*;
```

From Display 6.21 it can be seen that the variables are first scaled, then transformed, and the transformation that yielded quantiles closest to the theoretical normal quantiles is selected.

Results

The **Regression** node properties settings are the same as those for Model_B1.

Display 6.22 shows the variables selected in the final model.

Display 6.22

```
Diagram 6.22

The selected model, based on the CHOOSE=VERROR criterion, is the model trained
in Step 8. It consists of the following effects:

Intercept  CVAR001  LOG_IMP_NVAR253  LOG_IMP_NVAR305  LOG_IMP_NVAR306
PWR_IMP_NVAR205  PWR_IMP_NVAR23  SQRT_IMP_NVAR286  SQRT_NVAR103


        Likelihood Ratio Test for Global Null Hypothesis: BETA=0

        -2 Log Likelihood          Likelihood
     Intercept     Intercept &         Ratio
          Only      Covariates     Chi-Square       DF     Pr > ChiSq

      2789.110        2337.347       451.7624       16         <.0001


              Type 3 Analysis of Effects

                                         Wald
Effect                    DF       Chi-Square     Pr > ChiSq

CVAR001                    9         123.8626         <.0001
LOG_IMP_NVAR253            1          65.1270         <.0001
LOG_IMP_NVAR305            1          39.0889         <.0001
LOG_IMP_NVAR306            1           7.1625         0.0074
PWR_IMP_NVAR205            1          10.7527         0.0010
PWR_IMP_NVAR23             1           9.9010         0.0017
SQRT_IMP_NVAR286           1          19.5572         <.0001
SQRT_NVAR103               1         122.2522         <.0001
```

Display 6.23 shows the cumulative lift for the training and validation data sets.

Display 6.23

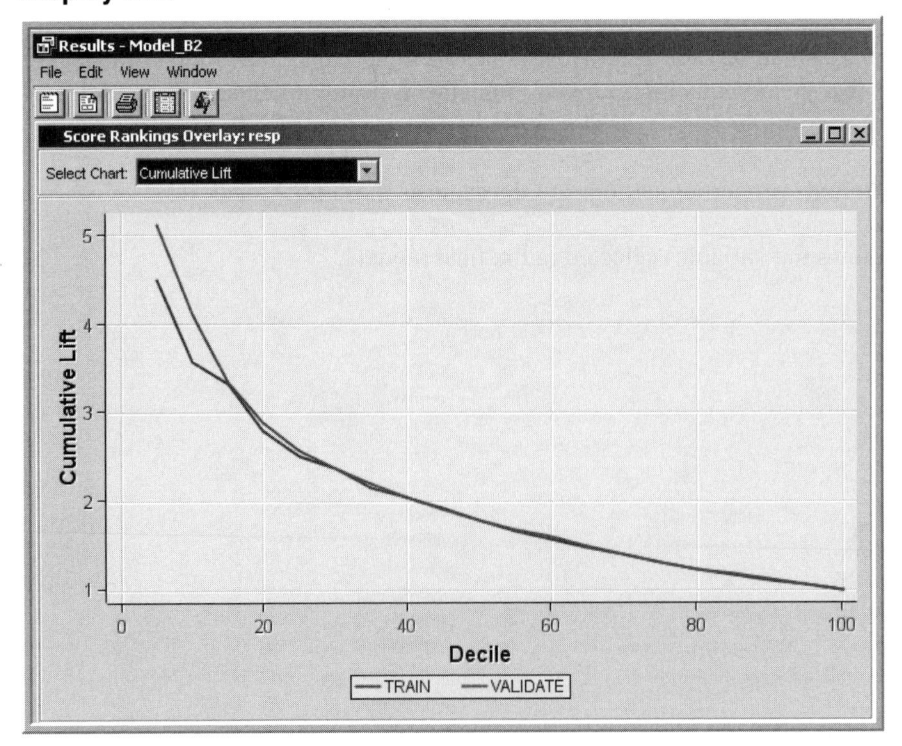

From Display 6.23 it can be seen that the model seems to make a better prediction for the validation sample than it does for the training data set. This does happen occasionally, and is no cause for alarm. It may be caused by the fact that the final model selection is based on the minimization of the validation error, which is calculated from the validation data set, since I set the **Selection Criterion** property to **Validation Error**.

When I allocated the data between the training, validation, and test data sets, I did use the default method, which is stratified partitioning with the target variable serving as the stratification variable. Partitioning done this way usually avoids the situations where the model performs better on the validation data set than on the training data set because of the differences in outcome concentrations.

The lift table corresponding to Display 6.23 and the cumulative capture rate are shown in Table 6.2.

Table 6.2

Model_B2

Bin	Demi-Decile	Cumulative Lift Training	Cumulative Lift Validation	Cumulative % Captured Response Training	Cumulative % Captured Response Validation
1	5	4.49074	5.12346	22.454	25.617
2	10	3.57371	4.10494	35.737	41.049
3	15	3.31790	3.33333	49.769	50.000
4	20	2.80093	2.88580	56.019	57.716
5	25	2.50926	2.56790	62.731	64.198
6	30	2.36111	2.36626	70.833	70.988
7	35	2.15608	2.19577	75.463	76.852
8	40	2.04861	2.04475	81.944	81.790
9	45	1.91872	1.89986	86.343	85.494
10	50	1.77315	1.78395	88.657	89.198
11	55	1.64983	1.66667	90.741	91.667
12	60	1.56250	1.58951	93.750	95.370
13	65	1.46724	1.48148	95.370	96.296
14	70	1.38889	1.38889	97.222	97.222
15	75	1.30247	1.30864	97.685	98.148
16	80	1.22106	1.23071	97.685	98.457
17	85	1.15741	1.17284	98.380	99.691
18	90	1.10082	1.11111	99.074	100.000
19	95	1.05263	1.05263	100.000	100.000
20	100	1.00000	1.00000	100.000	100.000

Model_B3

The bottom segment of Display 6.15 shows the process flow for Model_B3. In this case, the **Decision Tree** node is used for selecting the inputs. The inputs that give significant splits in growing the tree are selected. In addition, inputs selected for creating the surrogate rules are also saved and passed to the **Regression** node with the **Role** input. In addition, the **Decision Tree** node creates a categorical variable _NODE_. The categories of this variable identify the leaves of the tree. The value of the _NODE_ variable, for a given record, indicates the leaf of the tree to which the record belongs. The **Regression** node uses the _NODE_ as a class variable.

Display 6.24 shows the property settings of the **Decision Tree** node.

Display 6.24

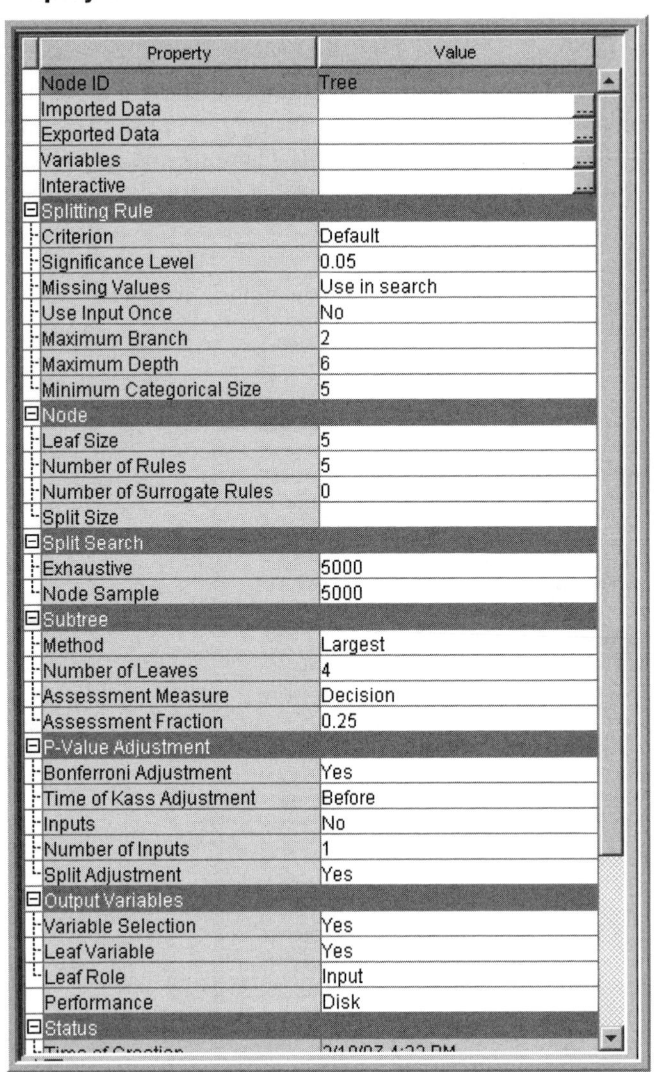

Property	Value
Node ID	Tree
Imported Data	
Exported Data	
Variables	
Interactive	
Splitting Rule	
Criterion	Default
Significance Level	0.05
Missing Values	Use in search
Use Input Once	No
Maximum Branch	2
Maximum Depth	6
Minimum Categorical Size	5
Node	
Leaf Size	5
Number of Rules	5
Number of Surrogate Rules	0
Split Size	
Split Search	
Exhaustive	5000
Node Sample	5000
Subtree	
Method	Largest
Number of Leaves	4
Assessment Measure	Decision
Assessment Fraction	0.25
P-Value Adjustment	
Bonferroni Adjustment	Yes
Time of Kass Adjustment	Before
Inputs	No
Number of Inputs	1
Split Adjustment	Yes
Output Variables	
Variable Selection	Yes
Leaf Variable	Yes
Leaf Role	Input
Performance	Disk
Status	

As Display 6.24 shows, the **Splitting Criterion** property of the **Decision Tree** node is set to **Default**, which is **ProbChisq** for binary targets. The **Significance Level** property is set at 0.05. The **Significance Level**, **Leaf Size,** and the **Maximum Depth** properties control the growth of the tree, as discussed in as in Chapter 4, "Building Decision Tree Models to Predict Response and Risk." No pruning is done, since the **Subtree Method** property is set to **Largest**. Therefore, the validation data set is not utilized in the tree development. In the **Output Variables** group, the **Leaf Variable** and **Variable Selection** properties are set to **Yes**, and the **Leaf Role** property is set to **Input.** As a result, the **Decision Tree** node creates a class variable _NODE_, which indicates the leaf node to which a record is assigned. _NODE_ will be passed to the next node, the **Regression** node, as a categorical input, along with the other selected variables.

In the **Model Selection** group of the **Regression** node, the **Selection Model** property is set to **Stepwise**, and the **Selection Criterion** property to **Validation Error**. These are the same settings used earlier for Model_B1 and Model_B2.

Results

The variables selected are given in Display 6.25.

Display 6.25

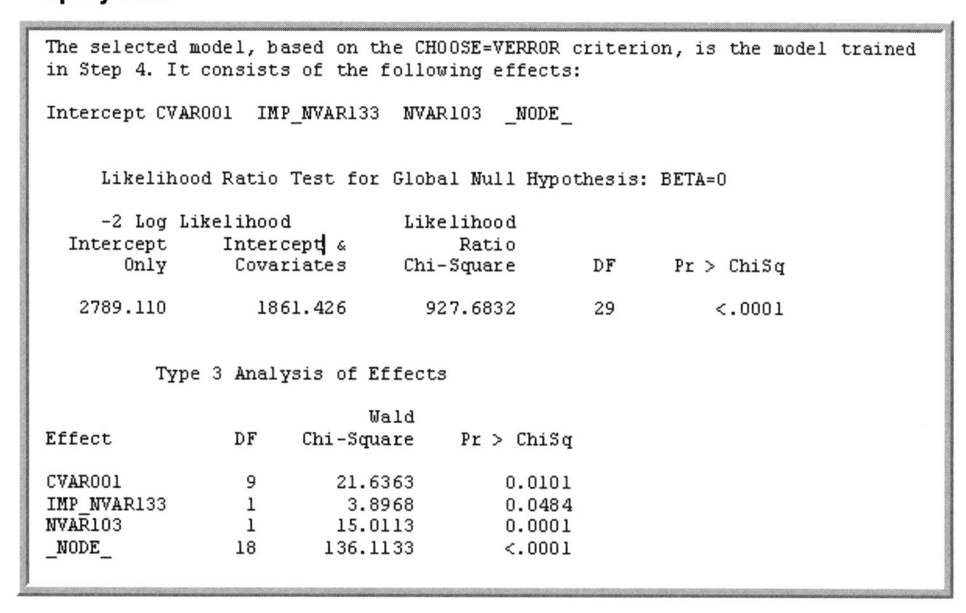

The selected model, based on the CHOOSE=VERROR criterion, is the model trained in Step 4. It consists of the following effects:

Intercept CVAR001 IMP_NVAR133 NVAR103 _NODE_

Likelihood Ratio Test for Global Null Hypothesis: BETA=0

-2 Log Likelihood Intercept Only	-2 Log Likelihood Intercept & Covariates	Likelihood Ratio Chi-Square	DF	Pr > ChiSq
2789.110	1861.426	927.6832	29	<.0001

Type 3 Analysis of Effects

Effect	DF	Wald Chi-Square	Pr > ChiSq
CVAR001	9	21.6363	0.0101
IMP_NVAR133	1	3.8968	0.0484
NVAR103	1	15.0113	0.0001
NODE	18	136.1133	<.0001

Display 6.26 shows the lift chart for the training and validation data sets. Enterprise Miner saved the data behind these charts in the SAS data set EMWS5.Reg5_EMRANK. Table 6.3 is created from this SAS data set.

Display 6.26

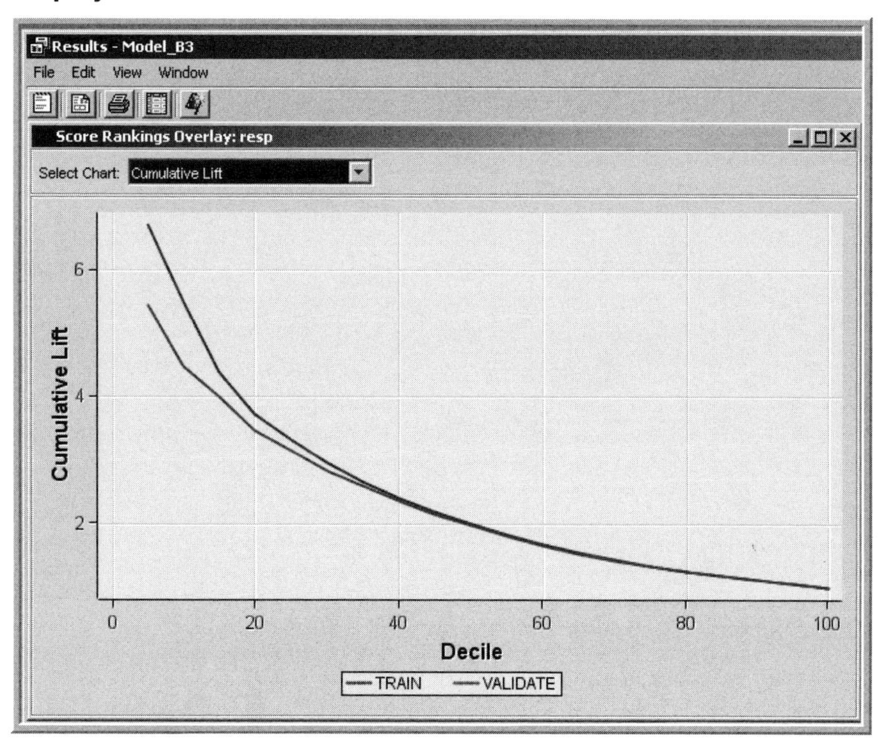

Table 6.3

Model_B3

Bin	Demi-Decile	Cumulative Lift Training	Cumulative Lift Validation	Cumulative % Captured Response Training	Cumulative % Captured Response Validation
1	5	6.71296	5.43210	33.565	27.160
2	10	5.50926	4.47646	55.093	44.765
3	15	4.39815	4.01235	65.972	60.185
4	20	3.72746	3.47222	74.549	69.444
5	25	3.31481	3.14815	82.870	78.704
6	30	2.95525	2.82922	88.657	84.877
7	35	2.65356	2.58377	92.875	90.432
8	40	2.40162	2.35340	96.065	94.136
9	45	2.18107	2.15364	98.148	96.914
10	50	1.98611	1.97531	99.306	98.765
11	55	1.81818	1.79574	100.000	98.765
12	60	1.66667	1.65123	100.000	99.074
13	65	1.53846	1.52993	100.000	99.446
14	70	1.42857	1.42178	100.000	99.525
15	75	1.33333	1.32805	100.000	99.604
16	80	1.25000	1.24604	100.000	99.683
17	85	1.17647	1.17368	100.000	99.762
18	90	1.11111	1.10935	100.000	99.842
19	95	1.05263	1.05180	100.000	99.921
20	100	1.00000	1.00000	100.000	100.000

The three models are compared in the **Model Comparison** node.

Display 6.27 shows the lift charts for all three models together for the training, validation, and test data sets.

Display 6.27

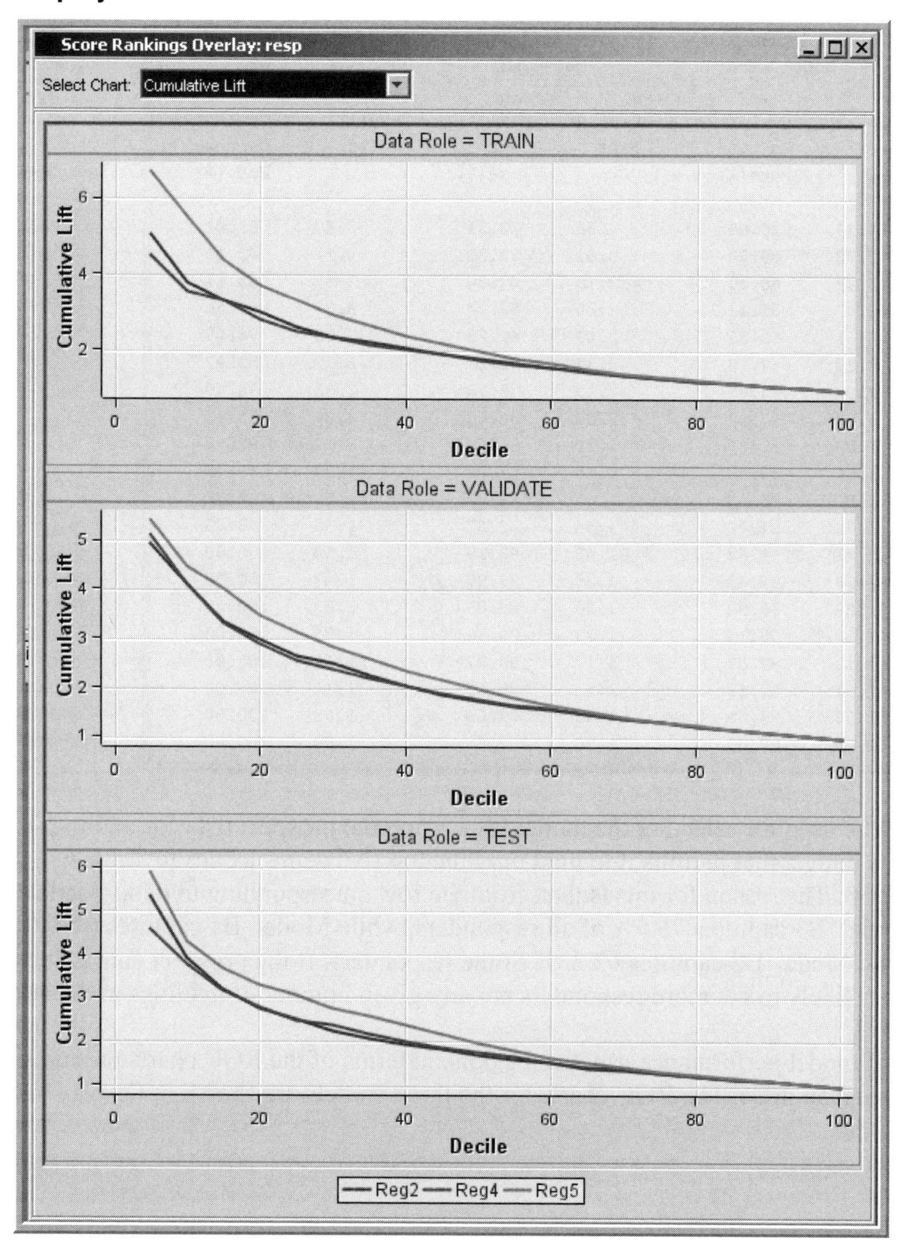

In Display 6.27 Reg2 is Model_B1, Reg4 is Model_B2, and Reg5 is Model_B3. By clicking on the **Table** icon that appears below the menu bar in the **Results** window, you can see the data behind the charts. Enterprise Miner saves this data as a SAS data set. In this example, the data set is named EMWS5.MdlComp_EMRANK. Table 6.4 is created from this data set. It shows a comparison of lift charts and capture rates for the test data set.

Table 6.4

	Model_B1		Model_B2		Model_B3	
Demi-Decile	Cumulative Lift	Cumulative Capture Rate(%)	Cumulative Lift	Cumulative Capture Rate(%)	Cumulative Lift	Cumulative Capture Rate(%)
5	5.33	26.63	4.58	22.91	5.88	29.41
10	3.93	39.32	3.81	38.08	4.31	43.13
15	3.22	48.30	3.18	47.68	3.67	55.11
20	2.77	55.42	2.80	56.04	3.33	66.56
25	2.51	62.85	2.49	62.23	3.06	76.47
30	2.24	67.18	2.37	71.21	2.77	82.97
35	2.09	73.07	2.19	76.78	2.58	90.40
40	1.93	77.09	2.02	80.80	2.39	95.67
45	1.83	82.35	1.86	83.59	2.17	97.52
50	1.69	84.52	1.76	87.93	1.97	98.45
55	1.60	88.24	1.65	91.02	1.80	98.76
60	1.52	91.02	1.54	92.57	1.65	98.76
65	1.41	91.64	1.43	93.19	1.54	100.00
70	1.35	94.43	1.38	96.90	1.43	100.00
75	1.29	96.90	1.30	97.83	1.33	100.00
80	1.23	98.14	1.23	98.76	1.25	100.00
85	1.17	99.07	1.17	99.07	1.18	100.00
90	1.11	99.69	1.11	99.69	1.11	100.00
95	1.05	99.69	1.05	99.69	1.05	100.00
100	1.00	100.00	1.00	100.00	1.00	100.00

This table can be used for selecting the model for scoring the prospect data for mailing. If the company has a budget that permits it to mail only the top five demi-deciles in Table 6.4, it should select Model_B3. The reason for this is that, from the row corresponding to demi-decile 25 in Table 6.4, Model_B3 captures 76.5% of all responders, while Model_B1 captures 62.85% of the responders, and Model_B2 captures 62.23% of the responders. If the prospect data is scored using Model_B3, it is likely to get more responders per any given number of mailings at decile 25.

Comparison of model performance can also be done in terms of the ROC charts produced by the **Model Comparison** node. The ROC charts for the three models are shown in Display 6.28.

Display 6.28

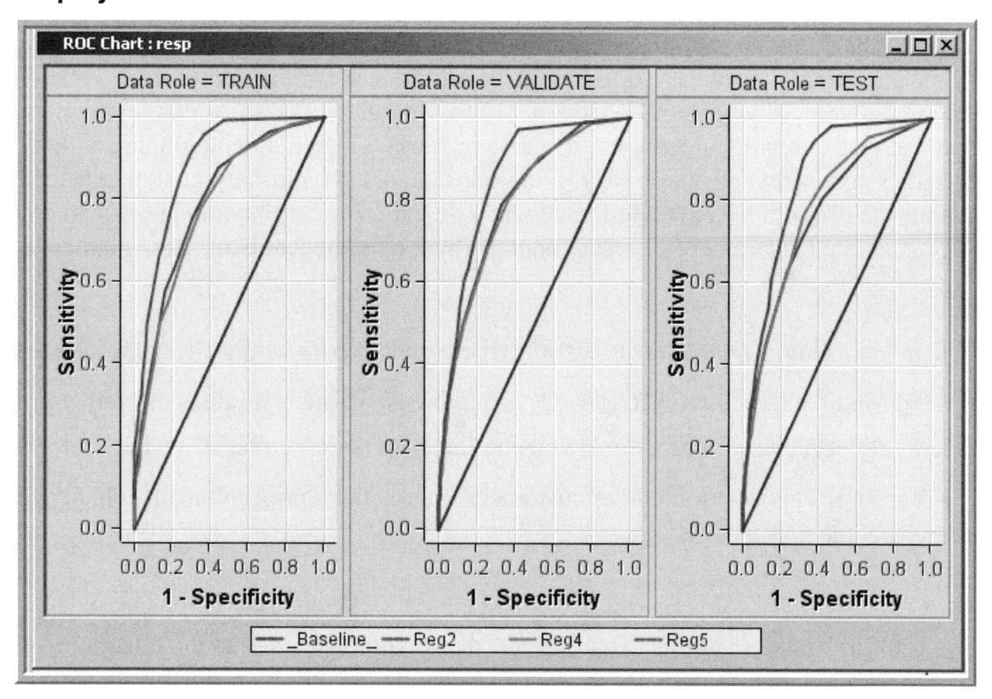

Display 6.29 gives a partial view of the data behind the ROC charts.

Display 6.29

ROC Chart : resp - EMWS5.MdlComp_EMROC

Predecesso...	Model Node	Target Vari...	Data Role	Event	1 - Specificity	Sensitivity
Reg2	Reg2	resp	TRAIN	1	0.000263	0.05787
Reg2	Reg2	resp	TRAIN	1	0.000527	0.060185
Reg2	Reg2	resp	TRAIN	1	0.00079	0.064815
Reg2	Reg2	resp	TRAIN	1	0.001581	0.071759
Reg2	Reg2	resp	TRAIN	1	0.002107	0.074074
Reg2	Reg2	resp	TRAIN	1	0.002898	0.087963
Reg2	Reg2	resp	TRAIN	1	0.003688	0.094907
Reg2	Reg2	resp	TRAIN	1	0.004215	0.104167
Reg2	Reg2	resp	TRAIN	1	0.005269	0.111111
Reg2	Reg2	resp	TRAIN	1	0.00764	0.12963
Reg2	Reg2	resp	TRAIN	1	0.010801	0.134259
Reg2	Reg2	resp	TRAIN	1	0.013435	0.148148
Reg2	Reg2	resp	TRAIN	1	0.01765	0.152778
Reg2	Reg2	resp	TRAIN	1	0.019758	0.164352
Reg2	Reg2	resp	TRAIN	1	0.024499	0.189815
Reg2	Reg2	resp	TRAIN	1	0.031085	0.210648
Reg2	Reg2	resp	TRAIN	1	0.038462	0.24537
Reg2	Reg2	resp	TRAIN	1	0.046628	0.266204
Reg2	Reg2	resp	TRAIN	1	0.06138	0.30787
Reg2	Reg2	resp	TRAIN	1	0.074289	0.347222
Reg2	Reg2	resp	TRAIN	1	0.100369	0.405093

The name of the SAS data set that contains the details of the ROC charts is EMWS5.MdlComp_EMROC, as shown in Display 6.29.

6.4.2 Regression for a Continuous Target

Now I turn to the second business application discussed earlier which involves developing a model with a continuous target. As I did in the case of binary targets, I built three versions of the model described above: Model_C1, Model_C2, and Model_C3. All three predict the increase in customers' savings deposits in response to a rate increase, given her demographic and lifestyle characteristics, assets owned, and past banking transactions. My primary goal here is to demonstrate the use of the **Regression** node when the target is continuous. In order to do this, I have used simulated data of a price test conducted by a hypothetical bank, as explained at the beginning of Section 6.4.

The basic model is $\log(\Delta Deposits_{it}) = \beta' X_i$, where $\Delta Deposits_{it}$ is the change in deposits in the ith account during the time interval $(t, t + \Delta t)$. An increase in interest rate is offered at time t. β is the vector of coefficients to be estimated using the **Regression** node. X_i is the vector of inputs for the ith account. This vector includes customers' transactions prior to the rate change, and other customer characteristics.

The target variable LDELBAL is $\log(\Delta Deposits_{it})$ for the models developed here. Here, my objective is to find answers to this question: "Of those who responded to the rate increase, who are likely to increase their savings the most?" The change in deposits is positive by definition, but it is very skewed in the sample created. Hence, I made a log transformation of the dependent variable. When the model is created, it is possible to score the prospects using either the estimated value of $\log(\Delta Deposits_{it})$ or $\Delta Deposits_{it}$.

As I did in the case of binary targets, I built three versions of the model described above. I referred to these three versions as Model_C1, Model_C2, and Model_C3.

Display 6.30 shows the process flow for the three models to be tested.

Display 6.30

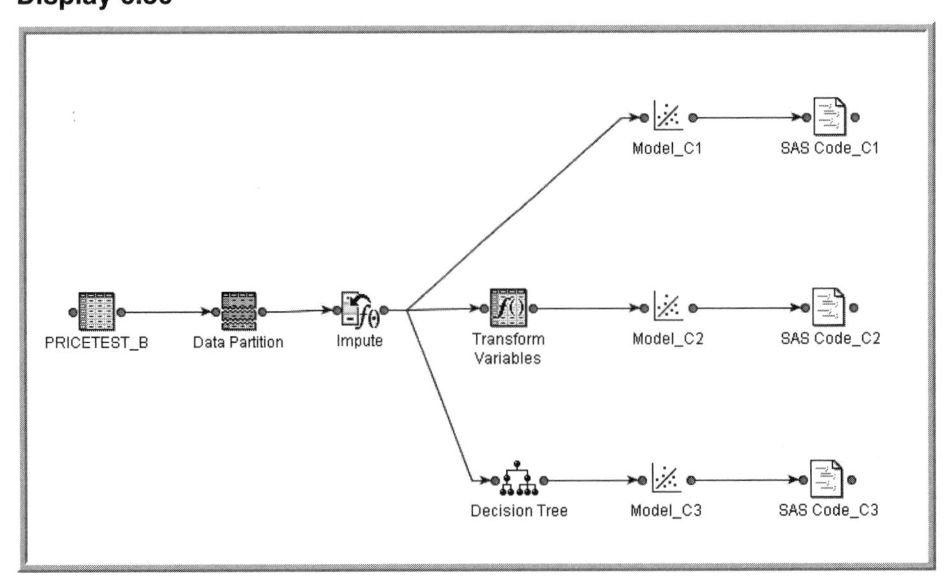

The SAS data set containing the price test results is read by the **Input Data Source** node. From this data set, the **Data Partition** node creates the three data sets **Part_Train**, **Part_Validate**, and **Part_Test** with the roles **Train**, **Validate**, and **Test**, respectively. 40% of the records are randomly allocated for training, 30% for validation and 30% for test.

Display 6.31 shows the property settings for the **Regression** node. It is the same for all three models.

Display 6.31

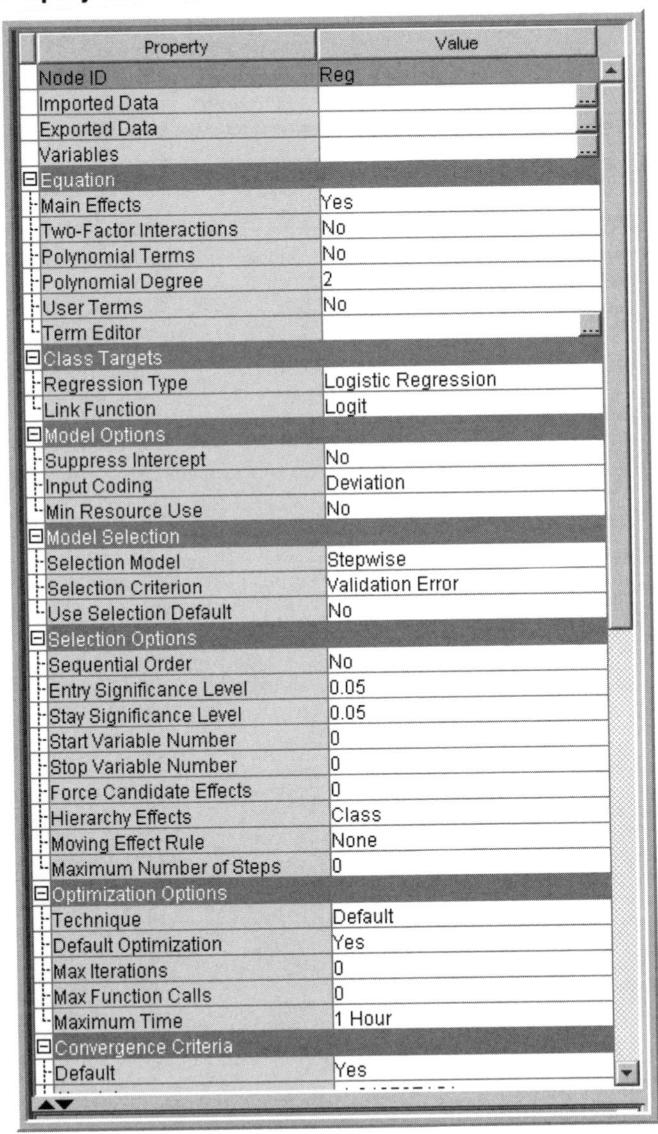

Property	Value
Node ID	Reg
Imported Data	
Exported Data	
Variables	
Equation	
Main Effects	Yes
Two-Factor Interactions	No
Polynomial Terms	No
Polynomial Degree	2
User Terms	No
Term Editor	
Class Targets	
Regression Type	Logistic Regression
Link Function	Logit
Model Options	
Suppress Intercept	No
Input Coding	Deviation
Min Resource Use	No
Model Selection	
Selection Model	Stepwise
Selection Criterion	Validation Error
Use Selection Default	No
Selection Options	
Sequential Order	No
Entry Significance Level	0.05
Stay Significance Level	0.05
Start Variable Number	0
Stop Variable Number	0
Force Candidate Effects	0
Hierarchy Effects	Class
Moving Effect Rule	None
Maximum Number of Steps	0
Optimization Options	
Technique	Default
Default Optimization	Yes
Max Iterations	0
Max Function Calls	0
Maximum Time	1 Hour
Convergence Criteria	
Default	Yes

The property settings for the **Model Selection** options are the same for all three models. The **Selection Model** property is set to **Stepwise**, the **Selection Criterion** property is set to **Validation Error**, and the **Use Selection Default** property is set **No**.

Model_C1

The top section of Display 6.16 shows the process flow for Model_C1. In this model, the inputs are used without any transformations. The target is continuous; therefore, the coefficients of the regression are obtained by minimizing the error sums of squares.

Results

The stepwise selection process took 16 steps to complete the effect selection. After Step 16, no additional inputs met the 0.05 significance level for entry into the model. Since the **Selection Criterion** property is set to **Validation Error**, I get the following message under the output in the **Results** window as shown in Display 6.32.

Display 6.32

```
The selected model, based on the CHOOSE=VERROR criterion, is the model trained
in Step 5. It consists of the following effects:

Intercept  CVAR14  IMP_NVAR048  IMP_NVAR054  NVAR009  NVAR174
```

Display 6.33 shows the estimated equation.

Display 6.33

Analysis of Variance

Source	DF	Sum of Squares	Mean Square	F Value	Pr > F
Model	7	791.282729	113.040390	55.78	<.0001
Error	1160	2350.923422	2.026658		
Corrected Total	1167	3142.206151			

Model Fit Statistics

R-Square	0.2518	Adj R-Sq	0.2473
AIC	833.0339	BIC	834.5024
SBC	873.5383	C(p)	55.9571

Type 3 Analysis of Effects

Effect	DF	Sum of Squares	F Value	Pr > F
CVAR14	1	140.2208	69.19	<.0001
IMP_NVAR048	1	37.6777	18.59	<.0001
IMP_NVAR054	1	22.6154	11.16	0.0009
NVAR009	1	41.5166	20.49	<.0001
NVAR174	3	92.1909	15.16	<.0001

Display 6.34 shows the plot of the **Mean for Predicted**.

Display 6.34

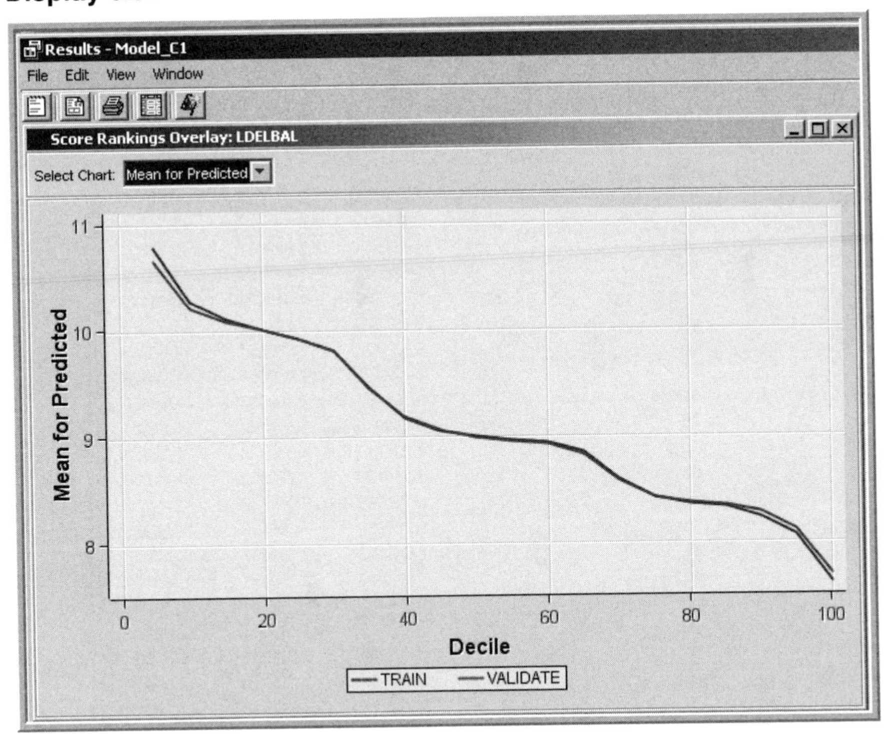

The data behind these charts can be seen by clicking on the **Table** icon in the task bar shown in Display 6.34. A partial view of this data is shown in Display 6.35.

Display 6.35

Score Rankings Overlay: LDELBAL - EMWS8.Reg_EMRANK

Target Vari...	Data Role	Decile	...	...	...	Bin	...	Target Mean	Target Min	Target Max
LDELBAL	TRAIN	5	.	.	.	1	...	10.29856	7.515165	13.44599
LDELBAL	TRAIN	10	.	.	.	2	...	10.32273	7.662275	12.37709
LDELBAL	TRAIN	15	.	.	.	3	...	10.07987	5.181615	12.09409
LDELBAL	TRAIN	20	.	.	.	4	...	9.942014	7.042505	12.78754
LDELBAL	TRAIN	25	.	.	.	5	...	10.06478	6.445831	12.53496
LDELBAL	TRAIN	30	.	.	.	6	...	9.760318	6.147806	13.11604
LDELBAL	TRAIN	35	.	.	.	7	...	9.759561	2.639057	12.34831
LDELBAL	TRAIN	40	.	.	.	8	...	9.688324	6.492134	13.79367
LDELBAL	TRAIN	45	.	.	.	9	...	9.193627	4.618284	13.99689
LDELBAL	TRAIN	50	.	.	.	10	...	9.407306	4.127618	11.97375
LDELBAL	TRAIN	55	.	.	.	11	...	8.633178	2.639057	11.60213
LDELBAL	TRAIN	60	.	.	.	12	...	8.528457	2.639057	11.08376
LDELBAL	TRAIN	65	.	.	.	13	...	8.659537	2.639057	11.08458
LDELBAL	TRAIN	70	.	.	.	14	...	9.149968	3.614156	11.32305
LDELBAL	TRAIN	75	.	.	.	15	...	8.799713	4.970438	11.27362
LDELBAL	TRAIN	80	.	.	.	16	...	8.336135	3.823192	11.41666
LDELBAL	TRAIN	85	.	.	.	17	...	8.121309	3.089678	11.112
LDELBAL	TRAIN	90	.	.	.	18	...	8.380664	4.65129	11.18365
LDELBAL	TRAIN	95	.	.	.	19	...	7.694182	2.639057	10.37007
LDELBAL	TRAIN	100	.	.	.	20	...	7.174026	2.639057	9.882122
LDELBAL	VALIDATE	0	...	...	...	.	.	.	.	.
LDELBAL	VALIDATE	5	.	.	.	1	...	10.3622	4.735057	12.54918
LDELBAL	VALIDATE	10	.	.	.	2	...	10.18869	6.165061	12.8686
LDELBAL	VALIDATE	15	.	.	.	3	...	10.42165	7.173061	13.07805
LDELBAL	VALIDATE	20	.	.	.	4	...	9.937464	7.929702	11.30019
LDELBAL	VALIDATE	25	.	.	.	5	...	10.32017	4.703204	12.84832
LDELBAL	VALIDATE	30	.	.	.	6	...	9.457114	5.830327	12.88795
LDELBAL	VALIDATE	35	.	.	.	7	...	10.18994	6.983095	14.18289
LDELBAL	VALIDATE	40	.	.	.	8	...	9.872962	5.736572	13.74705
LDELBAL	VALIDATE	45	.	.	.	9	...	8.82142	5.139556	11.12774
LDELBAL	VALIDATE	50	.	.	.	10	...	8.85985	4.039184	11.12774
LDELBAL	VALIDATE	55	.	.	.	11	...	9.209667	6.505156	11.5239
LDELBAL	VALIDATE	60	.	.	.	12	...	8.942394	5.201366	11.82811
LDELBAL	VALIDATE	65	.	.	.	13	...	8.792561	4.219361	12.26487
LDELBAL	VALIDATE	70	.	.	.	14	...	9.246386	5.747417	11.95087
LDELBAL	VALIDATE	75	.	.	.	15	...	8.887136	7.058879	11.07956
LDELBAL	VALIDATE	80	.	.	.	16	...	8.410534	2.639057	10.59756
LDELBAL	VALIDATE	85	.	.	.	17	...	8.248051	3.109061	11.59288
LDELBAL	VALIDATE	90	.	.	.	18	...	8.144461	2.639057	10.6191
LDELBAL	VALIDATE	95	.	.	.	19	...	7.723096	2.639057	10.82486
LDELBAL	VALIDATE	100	.	.	.	20	...	7.570988	2.639057	10.91606

The **Regression** node appends the predicted and actual values of the target variable for each record in the training, validation, and test data sets. These data sets are passed to the SAS code node.

Table 6.5 is constructed from the validation data set. When you run the **Regression** node, it calculates the predicted values of the target variable for each record in the training, validation, and test data sets. The predicted value of the target is P_LDELBAL in this case. This new variable is added to the three data sets.

In the SAS Code node, I use the validation data set's predicted values in P_LDELBAL to create the decile, lift, and capture rates for the deciles. The program to do these tasks is shown in Displays 6.37A and 6.37B. In the **SAS code** node I include a %INCLUDE statement to read this program as shown in Display 6.36.

Table 6.5 includes special lift and capture rates that are suitable for continuous variables in general, and for banking applications in particular.

Table 6.5

Model_C1

INCREASE IN DEPOSITS

Demi-Decile	Sum	Mean	Cumulative Sum	Cumulative Mean	Cumulative Lift	Cumulative Capture Rate(%)
5	$933,344	$21,706	$933,34	$21,706	1.90	9.3
10	$916,241	$20,824	$1,849,58	$21,260	1.86	18.5
15	$1,042,720	$23,698	$2,892,30	$22,079	1.93	28.9
20	$483,247	$10,983	$3,375,55	$19,289	1.69	33.8
25	$1,084,964	$25,232	$4,460,51	$20,461	1.79	44.6
30	$558,924	$12,703	$5,019,44	$19,158	1.68	50.2
35	$1,252,687	$28,470	$6,272,12	$20,497	1.80	62.7
40	$889,228	$20,210	$7,161,35	$20,461	1.79	71.6
45	$203,852	$4,633	$7,365,20	$18,693	1.64	73.7
50	$261,945	$6,092	$7,627,15	$17,453	1.53	76.3
55	$313,386	$7,122	$7,940,53	$16,508	1.45	79.4
60	$364,935	$8,294	$8,305,47	$15,820	1.39	83.1
65	$352,314	$8,007	$8,657,78	$15,216	1.33	86.6
70	$398,061	$9,047	$9,055,85	$14,773	1.29	90.6
75	$191,015	$4,442	$9,246,86	$14,096	1.23	92.5
80	$147,506	$3,352	$9,394,37	$13,421	1.18	93.9
85	$210,867	$4,792	$9,605,23	$12,910	1.13	96.1
90	$150,189	$3,413	$9,755,42	$12,380	1.08	97.6
95	$138,201	$3,141	$9,893,62	$11,891	1.04	98.9
100	$106,373	$2,418	$10,000,00	$11,416	1.00	100.0

The capture rate shown in the last column of Table 6.5 indicates that if the rate incentive generated an increase of $10 million in deposits, 9.3% of this increase would be coming from the accounts in the top demi-decile, 18.5% would be coming from the top two demi-deciles, and so on. Therefore, by targeting only the top 50%, the bank can get 76.3% of the potential increase in deposits.

By looking at the last row of Cumulative Mean column, you can see that the average increase in deposits for the entire data set is $11,416. However, the average increase for the top demi-decile is $21,076—almost double the overall mean. Hence, the lift for the top demi-decile is 1.90. The cumulative lift for the top two demi-deciles is 1.86, and so on.

The data included in Table 6.5 is generated by the SAS code shown in Display 6.36.

Display 6.36

Macro Variables | Macros

```
%let model = Model_C1 ;
%include 'c:/TheBook/EM_5.2/Programs/deciles_lift_Ch6_ptest.sas';
```

The SAS program included in the **SAS Code** node is shown in Displays 6.37A and 6.37B.

Display 6.37A

```
libname Lib2 'C:\TheBook\EM_5.2\EM_datasets' ;
proc sort data=&em_import_validate out=validate;
 by descending p_ldelbal ;
run ;
proc sql noprint;
 select count(*) into : nv1 from
 work.validate ;
quit ;
%let scale = 10000000;
data validate ;
 retain count 0 ;
 set validate ;
 count+1 ;
 if count < (1/20)*&nv1  then dec=5; else
 if count < (2/20)*&nv1  then dec=10 ; else
 if count < (3/20)*&nv1  then dec=15 ; else
 if count < (4/20)*&nv1  then dec=20 ; else
 if count < (5/20)*&nv1  then dec=25 ; else
 if count < (6/20)*&nv1  then dec=30 ; else
 if count < (7/20)*&nv1  then dec=35; else
 if count < (8/20)*&nv1  then dec=40 ; else
 if count < (9/20)*&nv1  then dec=45 ; else
 if count < (10/20)*&nv1 then dec=50 ; else
 if count < (11/20)*&nv1 then dec=55 ; else
 if count < (12/20)*&nv1 then dec=60 ; else
 if count < (13/20)*&nv1 then dec=65 ; else
 if count < (14/20)*&nv1 then dec=70 ; else
 if count < (15/20)*&nv1 then dec=75 ; else
 if count < (16/20)*&nv1 then dec=80 ; else
 if count < (17/20)*&nv1 then dec=85 ; else
 if count < (18/20)*&nv1 then dec=90 ; else
 if count < (19/20)*&nv1 then dec=95 ; else
 dec = 100 ;
 run ;
proc means data=validate noprint ;
  class dec ;
  var delbal ;
  output out= outsum sum(delbal) = sum_delbal mean(delbal)=mean_delbal;
run ;
data Total(keep=sum_delbal rename=(sum_delbal=Tot_delbal)) deciles ;
  set outsum ;
  if _TYPE_ = 0 then output Total;
  else output deciles ;
run ;
```

Display 6.37B (continued from Table 6.37A)

```
45 data total ;
46   set total ;
47   scale = &scale/tot_delbal ;
48 run ;
49 data tables ;
50  set deciles ;
51  if _N_ = 1 then set total ;
52 run;
53 data lib2.Lift_&model ;
54  set tables ;
55  retain cumsum 0 nobs 0;
56  sum_delbal= sum_delbal*scale ;
57  tot_delbal= tot_delbal*scale ;
58  mean_delbal = mean_delbal*scale;
59  cumsum + sum_delbal ;
60  capc = 100*cumsum/tot_delbal ;
61  gmean = tot_delbal/&nv1 ;
62  nobs+_freq_ ;
63  meanc = cumsum/nobs ;
64  liftc = meanc/gmean;
65 run ;
66 *%let model= Model_C1;
67 data _NULL_ ;
68   *file 'C:\TheBook\EM_5.2\EM_Datasets\Lift_Model_C1.dat' ;
69  file print ;
70  set lib2.Lift_&model ;
71  if _N_ = 1 then do ;
72  put @1 "&model" ;
73  put ' ' ;
74  put @27 'INCREASE IN DEPOSITS' ;
75  put ' ' ;
76  put @31 'Cumulative' @43 'Cumulative' @55 'Cumulative' @67 'Cumulative' ;
77  put @1 'Demi-' @11 'Sum' @22 'Mean' @34 'Sum' @46 'Mean' @58 'Lift' @68'Capture' ;
78  put @1 'Decile' @68 'Rate(%)'  ;
79  put ' ' ;
80   end ;
81  put @2 dec 4. @8 sum_delbal dollar10.0 @18 mean_delbal dollar10.0
82      @31 cumsum dollar11.0 @41 meanc dollar11.0 @56 liftc 6.2 @67 capc 6.1 ;
83 run ;
```

Model_C2

The process flow for Model_C2 is given in Display 6.30. The **Transform Variables** node creates three data sets: Trans_Train, Trans_Validate, and Trans_Test. These data sets include the transformed variables. The interval inputs are transformed by the Maximum Normal method, which is described in Section 2.4.7.1 of Chapter 2. In the **Regression** node, the effect selection is done by using the Stepwise method, and the **Selection Criterion** property is set to **Validation Error**.

Results

Display 6.38 shows the **Score Rankings** in the **Regression** node's **Results** window.

Display 6.38

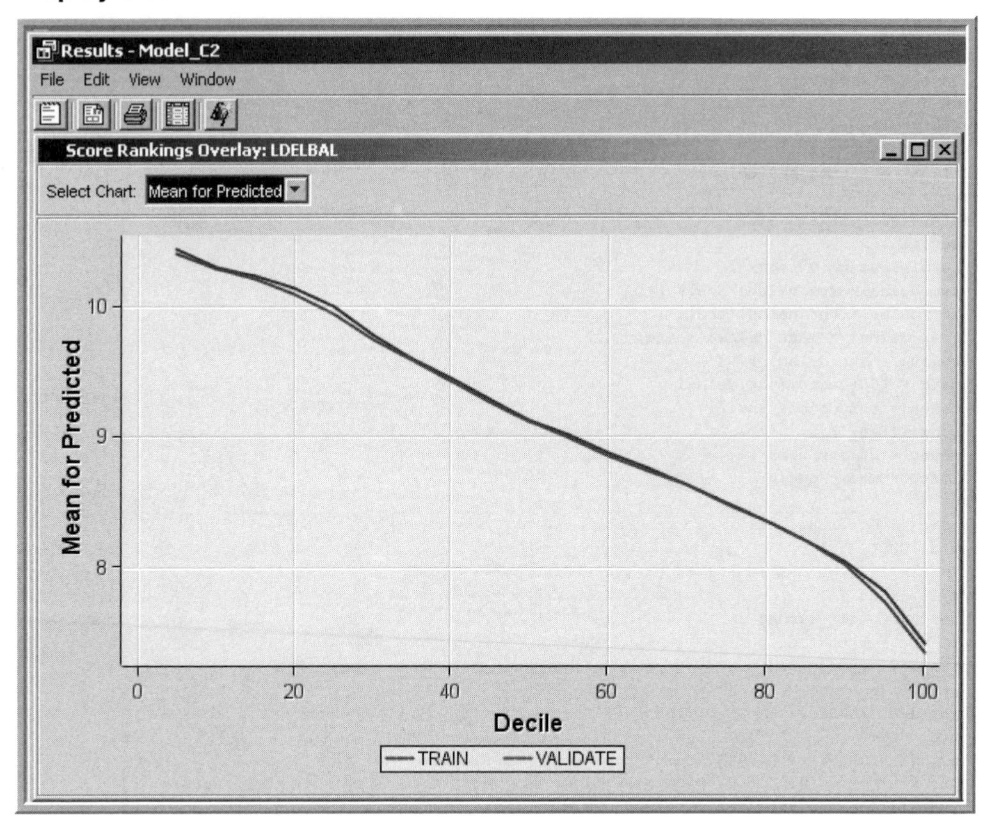

Table 6.6 shows the lift and capture rates of the Model_C2.

Table 6.6

Model_C2

INCREASE IN DEPOSITS

Demi-Decile	Sum	Mean	Cumulative Sum	Cumulative Mean	Cumulative Lift	Cumulative Capture Rate(%)
5	$1,107,718	$25,761	$1,107,71	$25,761	2.26	11.1
10	$699,998	$15,909	$1,807,71	$20,778	1.82	18.1
15	$753,936	$17,135	$2,561,65	$19,555	1.71	25.6
20	$815,392	$18,532	$3,377,04	$19,297	1.69	33.8
25	$565,195	$13,144	$3,942,24	$18,084	1.58	39.4
30	$1,209,389	$27,486	$5,151,62	$19,663	1.72	51.5
35	$1,033,859	$23,497	$6,185,48	$20,214	1.77	61.9
40	$382,551	$8,694	$6,568,03	$18,766	1.64	65.7
45	$867,015	$19,705	$7,435,05	$18,871	1.65	74.4
50	$365,129	$8,491	$7,800,18	$17,849	1.56	78.0
55	$324,400	$7,373	$8,124,58	$16,891	1.48	81.2
60	$230,007	$5,227	$8,354,59	$15,914	1.39	83.5
65	$207,036	$4,705	$8,561,62	$15,047	1.32	85.6
70	$170,232	$3,869	$8,731,86	$14,244	1.25	87.3
75	$251,222	$5,842	$8,983,08	$13,694	1.20	89.8
80	$150,610	$3,423	$9,133,69	$13,048	1.14	91.3
85	$281,679	$6,402	$9,415,37	$12,655	1.11	94.2
90	$296,148	$6,731	$9,711,51	$12,324	1.08	97.1
95	$180,100	$4,093	$9,891,61	$11,889	1.04	98.9
100	$108,382	$2,463	$10,000,00	$11,416	1.00	100.0

Model_C3

The process flow for Model_C3 is given in the lower region of Display 6.30. After the missing values were imputed, the data was passed to the **Decision Tree** node, which selected some important variables, and created a class variable called _NODE_. This variable indicates the node to which a record belongs. The selected variables and special class variables were then passed to the **Regression** node with their roles set to **Input**. The **Regression** node made further selections from these variables, and the final equation consisted of only the _NODE_ variable.

Results

Display 6.39 shows the **Score Rankings** in the **Regression** node's **Results** window.

Display 6.39

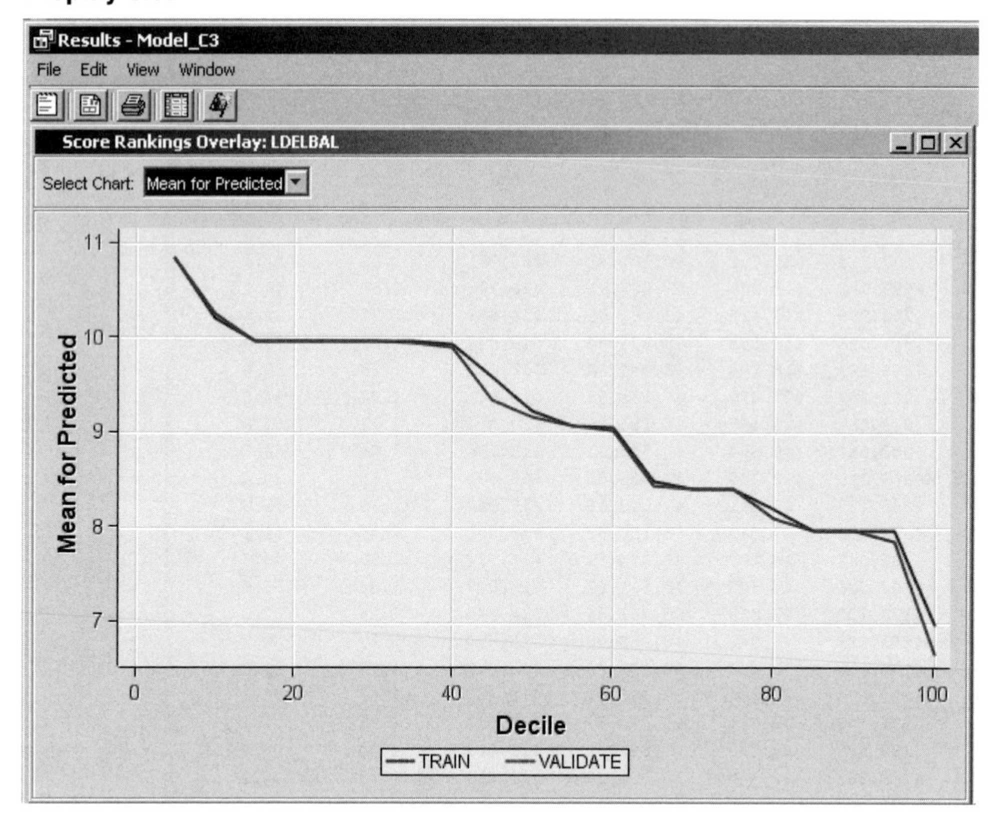

Table 6.7

Model_C3

INCREASE IN DEPOSITS

Demi-Decile	Sum	Mean	Cumulative Sum	Cumulative Mean	Cumulative Lift	Cumulative Capture Rate(%)
5	$869,727	$20,226	$869,72	$20,226	1.77	8.7
10	$664,529	$15,103	$1,534,25	$17,635	1.54	15.3
15	$1,344,183	$30,550	$2,878,43	$21,973	1.92	28.8
20	$776,870	$17,656	$3,655,30	$20,887	1.83	36.6
25	$697,180	$16,213	$4,352,48	$19,966	1.75	43.5
30	$946,223	$21,505	$5,298,71	$20,224	1.77	53.0
35	$817,983	$18,591	$6,116,69	$19,989	1.75	61.2
40	$330,245	$7,506	$6,446,94	$18,420	1.61	64.5
45	$408,095	$9,275	$6,855,03	$17,399	1.52	68.6
50	$279,353	$6,497	$7,134,38	$16,326	1.43	71.3
55	$248,261	$5,642	$7,382,64	$15,349	1.34	73.8
60	$286,250	$6,506	$7,668,89	$14,607	1.28	76.7
65	$816,808	$18,564	$8,485,70	$14,913	1.31	84.9
70	$306,491	$6,966	$8,792,19	$14,343	1.26	87.9
75	$289,684	$6,737	$9,081,88	$13,844	1.21	90.8
80	$255,411	$5,805	$9,337,29	$13,339	1.17	93.4
85	$167,508	$3,807	$9,504,80	$12,775	1.12	95.0
90	$164,483	$3,738	$9,669,28	$12,271	1.07	96.7
95	$154,688	$3,516	$9,823,97	$11,808	1.03	98.2
100	$176,028	$4,001	$10,000,00	$11,416	1.00	100.0

A Comparison of Model_C1, Model_C2, and Model_C3 is shown in Table 6.8

Table 6.8

Demi-Decile	Cumulative Mean			Cumulative Lift			Cumulative Capture Rate		
	Model_C1	Model_C2	Model_C3	Model_C1	Model_C2	Model_C3	Model_C1	Model_C2	Model_C3
5	$21,706	$25,761	$20,226	1.90	2.26	1.77	9.3	11.1	8.7
10	$21,260	$20,778	$17,635	1.86	1.82	1.54	18.5	18.1	15.3
15	$22,079	$19,555	$21,973	1.93	1.71	1.92	28.9	25.6	28.8
20	$19,289	$19,297	$20,887	1.69	1.69	1.83	33.8	33.8	36.6
25	$20,461	$18,084	$19,966	1.79	1.58	1.75	44.6	39.4	43.5
30	$19,158	$19,663	$20,224	1.68	1.72	1.77	50.2	51.5	53.0
35	$20,497	$20,214	$19,989	1.80	1.77	1.75	62.7	61.9	61.2
40	$20,461	$18,766	$18,420	1.79	1.64	1.61	71.6	65.7	64.5
45	$18,693	$18,871	$17,399	1.64	1.65	1.52	73.7	74.4	68.6
50	$17,453	$17,849	$16,326	1.53	1.56	1.43	76.3	78.0	71.3
55	$16,508	$16,891	$15,349	1.45	1.48	1.34	79.4	81.2	73.8
60	$15,820	$15,914	$14,607	1.39	1.39	1.28	83.1	83.5	76.7
65	$15,216	$15,047	$14,913	1.33	1.32	1.31	86.6	85.6	84.9
70	$14,773	$14,244	$14,343	1.29	1.25	1.26	90.6	87.3	87.9
75	$14,096	$13,694	$13,844	1.23	1.20	1.21	92.5	89.8	9[illegible].8
80	$13,421	$13,048	$13,339	1.18	1.14	1.17	93.9	91.3	93.4
85	$12,910	$12,655	$12,775	1.13	1.11	1.12	96.1	94.2	95.0
90	$12,380	$12,324	$12,271	1.08	1.08	1.07	97.6	97.1	96.7
95	$11,891	$11,889	$11,808	1.04	1.04	1.03	98.9	98.9	98.2
100	$11,416	$11,416	$11,416	1.00	1.00	1.00	100.0	100.0	100.0

6.5 Appendix to Chapter 6

Displays 6.40 through 6.44 show the SAS code used to generate Table 6.1.

Display 6.40

Program Editor

```
41  data train(rename=(liftc=liftc_train capc=capc_train))
42       validate(rename=(liftc=liftc_valid capc=capc_valid)) ;
43     *set emws5.reg2_emrank (keep=decile bin datarole liftc capc);
44     *set emws5.reg4_emrank (keep=decile bin datarole liftc capc);
45      set emws5.reg5_emrank (keep=decile bin datarole liftc capc);
46    if upcase(dataRole)='TRAIN'  then output train; else
47    if upcase(dataRole) = 'VALIDATE' then output validate;
48    if decile = 0 then delete ;
49    drop datarole ;
50  run;
51
52  data both ;
53    *retain decile bin liftc_train  liftc_valid capc_train capc_valid ;
54    merge train validate ;
55    by decile ;
56     if decile = 0 then delete ;
57    label liftc_train = "Cumulative Lift Training"
58          capc_train = "Cumulative $ Captured Response Training"
59          liftc_valid = "Cumulative Lift Validation"
60          capc_valid = "Cumulative $ Captured Response Validation"
61          decile = "Demi-Decile" ;
62
63  run ;
64
65  proc print data=both noobs label  ;
66    var bin decile liftc_train  liftc_valid capc_train capc_valid ;
67     Title "Model_B3" ;
68  run ;
```

Display 6.41

```
1  %let AGE = 50 ;
2  %let NPRVIO = 2 ;
3  %let CRED=670;
4
5  DATA ORDINAL ;
6  AGE = &AGE ;
7  NPRVIO = &NPRVIO ;
8  CRED=&CRED;
9  drop _TEMP;
10 drop _LP0;
11 _LP0 = 0;
12 *** Compute Linear Predictor;
13 ***  Effect: AGE ;
14 _TEMP = AGE ;
15 _LP0 = _LP0 + (    -0.02307201027961 * _TEMP);
16
17 ***  Effect: CRED ;
18 _TEMP = CRED ;
19 _LP0 = _LP0 + (    -0.00498225050304 * _TEMP);
20
21 ***  Effect: NPRVIO ;
22 _TEMP = NPRVIO ;
23 _LP0 = _LP0 + (     0.22524384228134 * _TEMP);
24
25 drop _MAXP _IY _P0 _P1 _P2;
26 _TEMP =     -2.88399896897927 + _LP0;
27 ETA2 = _TEMP ;
28 if (_TEMP < 0) then do;
29    _TEMP = exp(_TEMP);
30    _P0 = _TEMP / (1 + _TEMP);
31 end;
32 else _P0 = 1 / (1 + exp(-_TEMP));
33 _TEMP =      0.21333987443108 + _LP0;
34 ETA1=_TEMP ;
35 if (_TEMP < 0) then do;
36    _TEMP = exp(_TEMP);
37    _P1 = _TEMP / (1 + _TEMP);
38 end;
39 else _P1 = 1 / (1 + exp(-_TEMP));
40 _P2 = 1.0 - _P1;
41 _P1 = _P1 - _P0;
```

Display 6.42

```
42
43 *** Posterior Probabilities and Predicted Level;
44 label P_LOSSFRQ2 = 'Predicted: LOSSFRQ=2' ;
45 label P_LOSSFRQ1 = 'Predicted: LOSSFRQ=1' ;
46 label P_LOSSFRQ0 = 'Predicted: LOSSFRQ=0' ;
47 P_LOSSFRQ2 = _P0;
48 _MAXP = _P0;
49 P_LOSSFRQ1 = _P1;
50 if (_P1 - _MAXP > 1e-8) then do;
51    _MAXP = _P1;
52 end;
53 P_LOSSFRQ0 = _P2;
54 if (_P2 - _MAXP > 1e-8) then do;
55    _MAXP = _P2;
56 end;
57 run ;
58 ods html file='C:\TheBook\EM_5.2\Nov2006\Ordinal.html' ;
59 Proc print data=ORDINAL label noobs;
60  title  "ORDINAL TARGET" ;
61  title2 "AGE=50,CRED=670 and NPRVIO=2" ;
62  var P_LOSSFRQ2 P_LOSSFRQ1 P_LOSSFRQ0 ;
63 run ;
64 ods html close ;
```

Display 6.43

```
65 data NOMINAL ;
66 AGE = &AGE ;
67 NPRVIO = &NPRVIO ;
68 CRED=&CRED;
69 *** Compute Linear Predictor;
70 drop _TEMP;
71 drop _LP0  _LP1;
72 _LP0 = 0;
73 _LP1 = 0;
74 ***  Effect: AGE ;
75 _TEMP = AGE ;
76 _LP0 = _LP0 + (    -0.0323912354634 * _TEMP);
77 _LP1 = _LP1 + (    -0.0228181055783 * _TEMP);
78
79 ***  Effect: CRED ;
80 _TEMP = CRED ;
81 _LP0 = _LP0 + (   -0.01335715577514 * _TEMP);
82 _LP1 = _LP1 + (   -0.00448783417963 * _TEMP);
83 ***  Effect: NPRVIO ;
84 _TEMP = NPRVIO ;
85 _LP0 = _LP0 + (    0.21864487608312 * _TEMP);
86 _LP1 = _LP1 + (    0.22396412290576 * _TEMP);
87 *** Naive Posterior Probabilities;
88 drop _MAXP _IY _P0 _P1 _P2;
89 drop _LPMAX;
90 _LPMAX= 0;
91 _LP0 =      1.59463711057944 + _LP0;
92 if _LPMAX < _LP0 then _LPMAX = _LP0;
93 _LP1 =    -0.11855123381699 + _LP1;
94 ETA2 = _LP0 ;
95 ETA1 = _LP1 ;
96 if _LPMAX < _LP1 then _LPMAX = _LP1;
97 _LP0 = exp(_LP0 - _LPMAX);
98 _LP1 = exp(_LP1 - _LPMAX);
99 _LPMAX = exp(-_LPMAX);
100 _P2 = 1 / (_LPMAX + _LP0 + _LP1);
101 _P0 = _LP0 * _P2;
102 _P1 = _LP1 * _P2;
103 _P2 = _LPMAX * _P2;
```

Display 6.44

```
104 *** Posterior Probabilities and Predicted Level;
105 label P_LOSSFRQ2 = 'Predicted: LOSSFRQ=2' ;
106 label P_LOSSFRQ1 = 'Predicted: LOSSFRQ=1' ;
107 label P_LOSSFRQ0 = 'Predicted: LOSSFRQ=0' ;
108 P_LOSSFRQ2 = _P0;
109 _MAXP = _P0;
110 P_LOSSFRQ1 = _P1;
111 if (_P1 - _MAXP > 1e-8) then do;
112    _MAXP = _P1;
113  end;
114 P_LOSSFRQ0 = _P2;
115 if (_P2 - _MAXP > 1e-8) then do;
116    _MAXP = _P2;
117  end;
118 run ;
119 ods html file='C:\TheBook\EM_5.2\Nov2006\Nominal.html' ;
120
121 proc print data=NOMINAL label noobs;
122   title "NOMINAL TARGET" ;
123   title2 "AGE=50,CRED=670 and NPRVIO=2" ;
124    var P_LOSSFRQ2 P_LOSSFRQ1 P_LOSSFRQ0 ;
125 run ;
126 ods html close ;
```

Comparison of Different Models

7.1 Introduction

This chapter compares the output and performance of three modeling tools examined in the previous three chapters—**Decision Tree**, **Regression**, and **Neural Network**—by using all three tools to develop two types of models—one with a binary target and one with an ordinal target. I hope this chapter will help you decide what approach to take.

The model with a binary target is developed for a hypothetical bank that wants to predict the likelihood of a customer's attrition so that it can take suitable action to prevent the attrition if necessary. The details of this model will be presented later in this chapter.

The model with an ordinal target is developed for a hypothetical insurance company that wants to predict the probability of a given frequency of insurance claims for each of its existing customers and then use the resulting model to further profile customers who are most likely to have accidents.

7.2 Models for Binary Targets: An Example of Predicting Attrition

Attrition can be modeled as either a binary or a continuous target. When you model attrition as a binary target, you predict the probability of a customer "attriting" during the next few days or months, given the customer's general characteristics and change in the pattern of the customer's transactions. With a continuous target, you predict the expected time of attrition, or the residual lifetime, of a customer. For example, if a bank or credit card company wants to identify customers who are likely to terminate their accounts at *any* point within a pre-defined interval of time in the future, the company can model attrition as a binary target. If, on the other hand, they are interested in predicting the *specific* time at which the customer is likely to attrit, then the company should model attrition as a continuous target, and use techniques such as survival analysis.

In this chapter, I present an example of a hypothetical bank that wants to develop an early warning system to identify customers who are most likely to close their investment accounts in the next three months. To meet this business objective, an attrition model with a binary target is developed.

The customer record in the modeling data set for developing an attrition model consists of three types of variables (fields):

- Variables indicating the customer's past transactions (such as deposits, withdrawals, purchase of equities, etc.) by month for several months.
- Customer characteristics (demographic and socio-economic) such as age, income, lifestyle, etc.
- Target variable indicating whether a customer attrited during a pre-specified interval or not.

Assume that the model was developed during December 2006, and it was used to predict attrition for the period January1, 2007, through March 31, 2007.

Display 7.1 shows a chronological view of a data record in the data set used for developing an attrition model.

Display 7.1

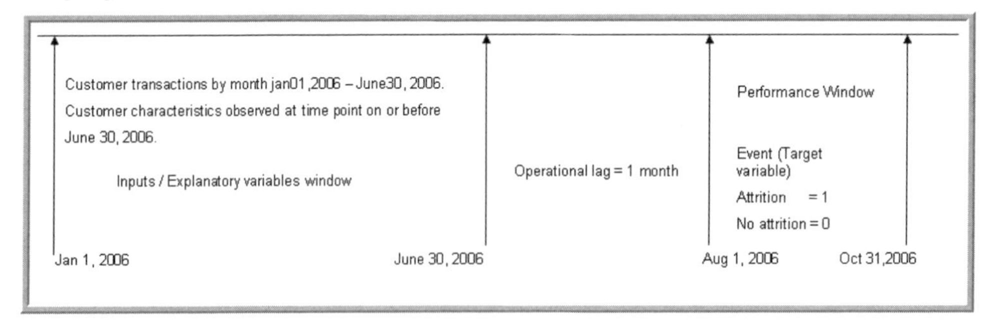

Here are some key input design definitions that are labeled in Display 7.1:

- Inputs/explanatory variables window

  This window refers to the period from which the transaction data is collected.

- Operational lag

  The model excludes any data for the month of July 2006 to allow for the operational lag that comes into play when the model is used for forecasting in real time. This lag may represent the period between the time at which a customer's transaction takes place and the time at which it is recorded in the database and becomes available to be used in the model.

- Performance window

  This is the time interval in which the customers in the sample data set are observed for attrition. If a customer attrited during this interval then the target variable ATTR takes the value 1; if not, it takes the value 0.

The model should exclude all data from any of the inputs that have been captured during the performance window.

If the model is developed using the data shown in Display 7.1, and used for predicting, or scoring, attrition propensity for the period January 1, 2007, through March 2007, then the inputs window, the operation lag, and the prediction window will look like what is shown in Display 7.2 when the scoring or prediction is done.

Display 7.2

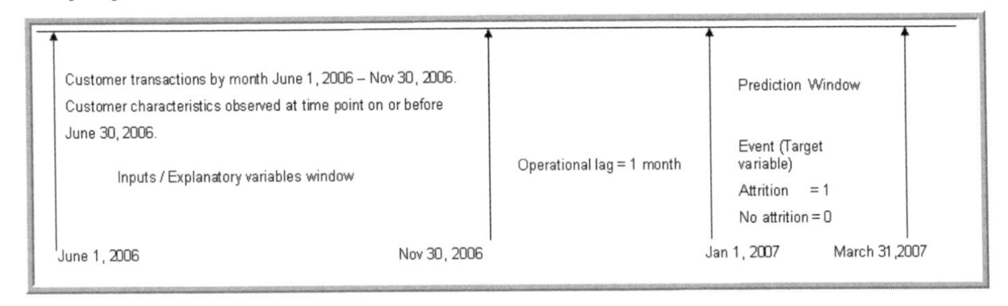

At the time of prediction (say on December 31, 2006), all data in the inputs/explanatory variables window would be available. In this hypothetical example, however, the inputs were not available for the month of December 2006 because of the time lag in collecting the input data. In addition, no data will be available for the prediction window, because it resides in the future (see Display 7.2).

As mentioned earlier, we assume that the model was developed during December 2006. The modeling data set was created by observing a sample of current customers with investment accounts during a time interval of three months starting August 1, 2006, through October 31, 2006. This is the performance window, as shown in Display 7.1. The fictitious bank is interested in predicting the probability of attrition, specifically among their customers' investment accounts. Accordingly, if an investment account is closed during the performance window, an indicator value of 1 is entered on the customer's record; otherwise, a value of 0 is given. Thus, for the target variable ATTR, 1 represents attrition and 0 represents no attrition.

Each record in the modeling data set includes the following details:

- monthly transactions
- balances
- monthly rates of change of customer balances
- specially constructed trend variables.

These trend variables indicate patterns in customer transaction behavior, etc., for all accounts held by the customer during the inputs/ explanatory variables window, which spans the period of January 1, 2006, through June 30, 2006, as shown in Display 7.1. The customer's records also show how long the customer has held an investment account with the bank (investment account tenure). In addition, each record is appended with the customer's age, marital status, household income, home equity, life-stage indicators, and number of months the customer has been on the books (customer tenure), etc. All of these details are candidates for inclusion in the model.

In the process flow shown in Display 7.3, the data source is created first. The **Input Data** node reads this data. The data is then partitioned such that 45% of the records are allocated to training, 35% to validation, and 25% for testing. I allocated more records for training and validation than for testing so that the models are estimated as accurately as possible. The value of the **Partitioning Method** property is set to **Default**, which performs stratification with respect to the target variable if the target variable is categorical. I also imputed missing observations using the **Impute** node. Display 7.3 shows the process flow for modeling the binary target.

Display 7.3

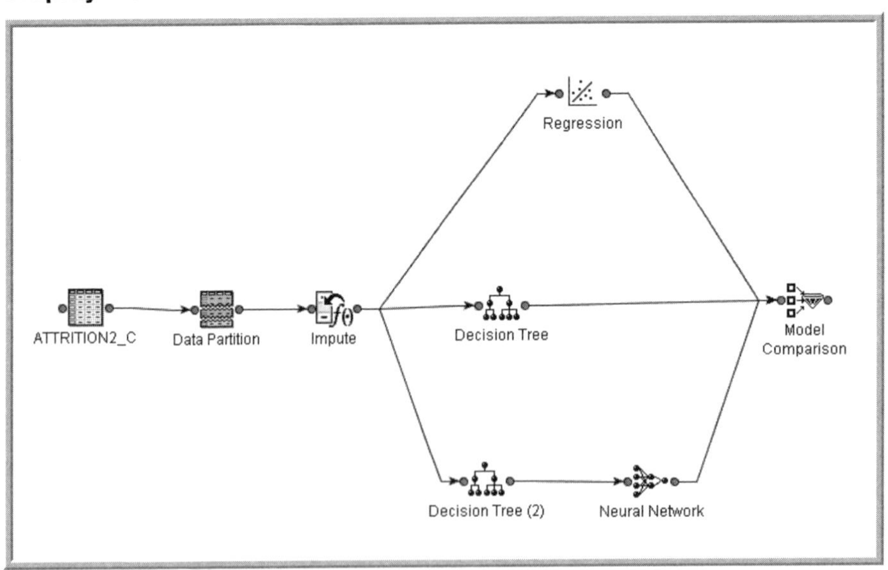

7.2.1 Logistic Regression for Predicting Attrition

The top segment of Display 7.3 shows the process flow for a logistic regression model of attrition. In the **Regression** node, I set the **Selection Model** property to **Stepwise**, as this method was found to produce the most parsimonious model. Before making a choice of value for the **Selection Criterion** property, I tested the model using different values. In this example, the validation error criterion produced a model that made the most business sense. Display 7.4 shows the model that is estimated by the **Regression** node with these property settings.

Display 7.4

Analysis of Maximum Likelihood Estimates

Parameter	DF	Estimate	Standard Error	Wald Chi-Square	Pr > ChiSq
Intercept	1	-1.3239	0.1381	91.90	<.0001
btrend	1	-0.1505	0.00898	280.89	<.0001
duration	1	-0.0294	0.00743	15.63	<.0001

The variables that are selected by the **Regression** node are BTREND and DURATION. The variable BTREND measures the trend in the customer's balances during the six-month period prior to the operational lag period and the operational lag period. If there is a *downward trend* in the balances over the period, e.g., a *decline* of one percent, then the odds of attrition will increase by $100 * \{e^{0.1505} - 1\} = 16.24\%$. This may seem a bit high, but the direction of the result does make sense. (Also, keep in mind that these estimates are based on simulated data. It is best not to consider them as general results.)

The variable DURATION measures the customer's investment account tenure. Longer tenure corresponds to lower probability of attrition. This can be interpreted as the positive effect of customer loyalty.

Display 7.5 shows the lift charts from the **Results** window of the **Regression** node.

Display 7.5

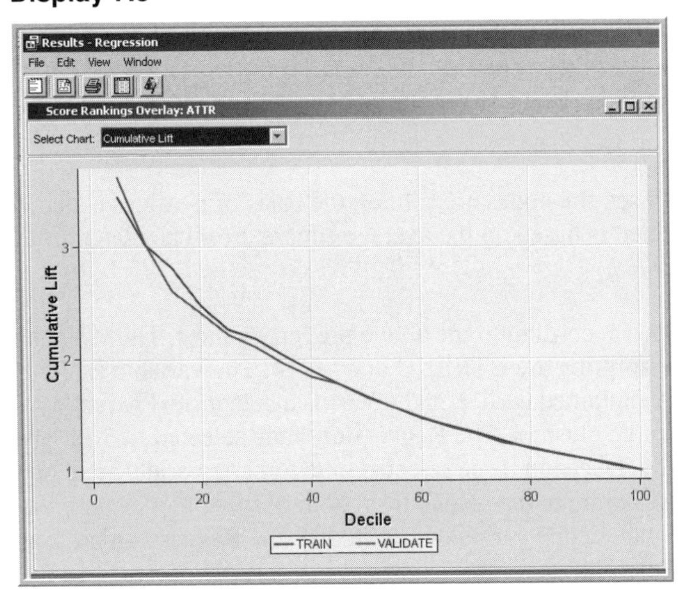

Table 7.1 shows the cumulative lift and capture rates for the Train, Validate, and Test data sets.

Table 7.1 [1]

REG

	Train		Validate		Test	
Demi-decile	Cumulative Lift	Cumulative Capture Rate(%)	Cumulative Lift	Cumulative Capture Rate(%)	Cumulative Lift	Cumulative Capture Rate(%)
5	3.44	17.2	3.63	18.1	3.49	17.4
10	3.03	30.3	3.03	30.3	2.93	29.3
15	2.82	42.3	2.60	39.1	2.85	42.7
20	2.48	49.5	2.41	48.1	2.69	53.9
25	2.26	56.6	2.21	55.3	2.47	61.7
30	2.18	65.3	2.10	63.1	2.31	69.2
35	2.03	71.1	1.96	68.4	2.15	75.1
40	1.90	75.8	1.84	73.4	2.01	80.4
45	1.78	80.0	1.78	80.0	1.86	83.5
50	1.67	83.3	1.66	82.8	1.70	85.0
55	1.57	86.2	1.56	85.9	1.57	86.6
60	1.46	87.8	1.47	88.4	1.47	88.2
65	1.40	90.8	1.39	90.6	1.39	90.7
70	1.33	93.2	1.33	92.8	1.33	93.1
75	1.26	94.4	1.25	93.4	1.27	95.0
80	1.19	95.3	1.20	95.6	1.21	96.9
85	1.14	96.9	1.14	97.2	1.15	97.8
90	1.10	98.8	1.09	98.4	1.10	99.4
95	1.05	99.3	1.05	99.7	1.05	99.7
100	1.00	100.0	1.00	100.0	1.00	100.0

7.2.2 Decision Tree Model for Predicting Attrition

The middle section of Display 7.3 shows the process flow for the decision tree model of attrition.

I set the **Splitting Rule Criterion** property to **ProbChisq** and the **Subtree Method** property to **Average Square Error**. My choice of these property values is somewhat arbitrary, but it resulted in a tree that can serve as an illustration. You can set alternative values for these properties and examine the trees produced.

According to the property values I set, the nodes are split on the basis of *p*-values of the Pearson Chi-Square, and the sub-tree selected is based on the average square error calculated from the validation data set.

Display 7.6 shows the tree produced according to the above property values. The variables selected by the **Decision Tree** for splitting are BTREND and CV14. The variable BTREND is the trend in the customer balances, as explained earlier, and CV14 is a categorical variable, which reflects certain hypothetical economic clusters. The **Regression** node selected two variables, which were both numeric, while the **Decision Tree** selected one numeric variable and one nominal variable. One could either combine the results from both of these models or create a larger set of variables that would include the variables selected by the **Regression** node and the **Decision Tree** node.

[1] In Tables 7.1 through 7.8, I used the term *demi-decile* when I referred to the equal sized bins of the lift tables, since there are 20 of them. A more accurate term, perhaps, would be *centiles*, because the description found in the Demi-decile column is actually a 5-percentile range indicator, e.g., 5, 10, 15… 90, 95,100. In the **Score Rankings Overlay** window (shown in Displays 7.5, 7.7, 7.8, 7.9, 7.10, etc.) and in the **Store Rankings Overlay** tables, Enterprise Miner refers to them as Deciles.

Forecast combination is an important topic, and it deserves an in-depth study on its own. However, it is beyond the scope of the present discussion.[2]

Display 7.6

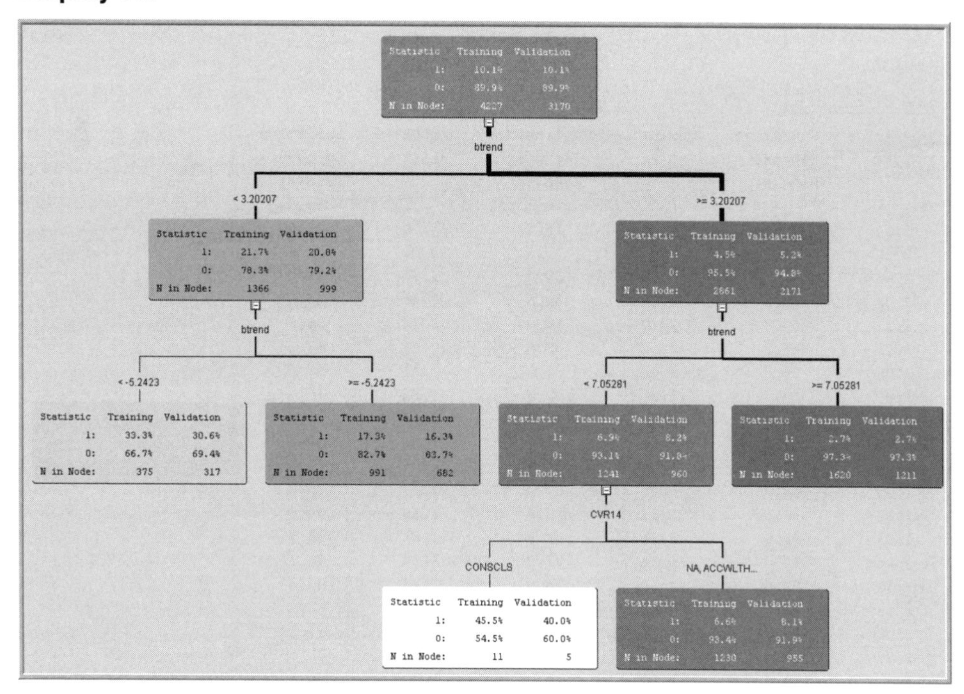

Display 7.7 shows the lift and cumulative lift of the decision tree model.

Display 7.7

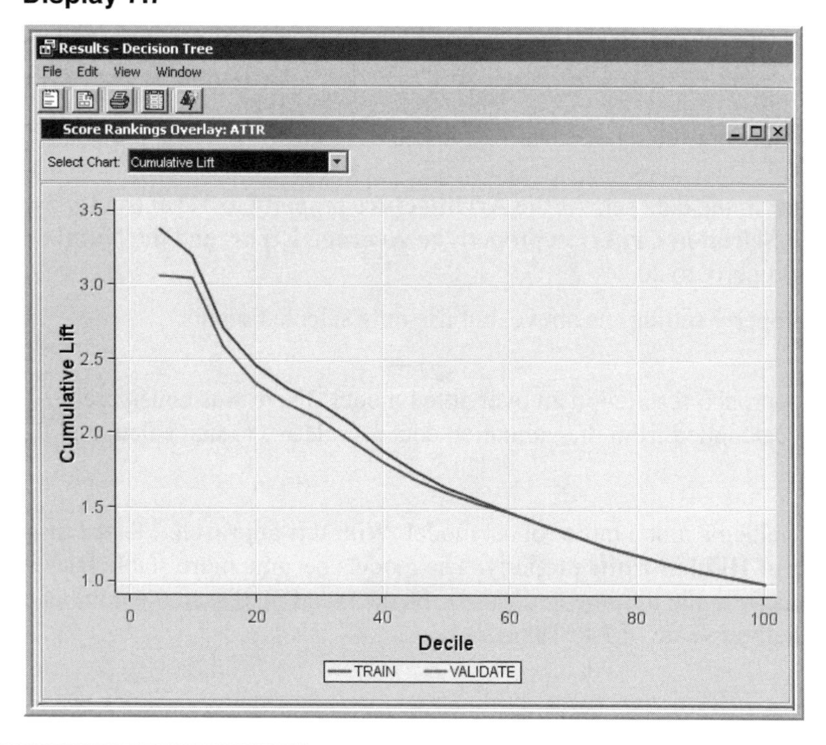

---

[2] For more information on forecast combination, see Breiman (1996a and 1996b), Sarma (2005), Eklund and Karlsson (2005), and Guidolin and Na (2007).

The cumulative lift and cumulative capture rates for Train, Validate, and Test data are shown in Table 7.2.

Table 7.2

TREE

	Train		Validate		Test	
Demi-decile	Cumulative Lift	Cumulative Capture Rate(%)	Cumulative Lift	Cumulative Capture Rate(%)	Cumulative Lift	Cumulative Capture Rate(%)
5	3.37	16.9	3.06	15.3	3.00	15.0
10	3.20	32.0	3.05	30.5	3.00	30.0
15	2.70	40.6	2.58	38.7	2.63	39.5
20	2.46	49.1	2.34	46.8	2.45	49.0
25	2.31	57.7	2.19	54.9	2.34	58.5
30	2.21	66.2	2.10	62.9	2.27	68.1
35	2.06	72.2	1.95	68.3	2.11	73.9
40	1.89	75.5	1.81	72.3	1.92	76.7
45	1.75	78.8	1.69	76.3	1.77	79.6
50	1.64	82.0	1.61	80.3	1.65	82.4
55	1.55	85.3	1.53	84.3	1.55	85.3
60	1.48	88.6	1.47	88.3	1.47	88.2
65	1.39	90.6	1.39	90.6	1.39	90.3
70	1.31	91.9	1.31	91.9	1.31	91.7
75	1.24	93.3	1.24	93.3	1.24	93.1
80	1.18	94.6	1.18	94.6	1.18	94.5
85	1.13	96.0	1.13	96.0	1.13	95.9
90	1.08	97.3	1.08	97.3	1.08	97.2
95	1.04	98.7	1.04	98.7	1.04	98.6
100	1.00	100.0	1.00	100.0	1.00	100.0

7.2.3 A Neural Network Model for Predicting Attrition

The bottom segment of Display 7.3 shows the process flow for the neural network model of attrition.

In developing a neural network model of attrition, I tested the following two approaches:

- Use all the inputs in the data set, set the **Architecture** property to **MLP** (multi/layer perceptron), the **Selection Criterion** property to **Average Error**, and the **Number of Hidden Units** property to 20.
- Use the same property settings as above, but use only selected inputs.

As expected, the first approach resulted in an over-fitted model. There was considerable deterioration in the lift calculated from the validation data set relative to that calculated from the training data set.

The second approach yielded a much more robust model. With this approach, I tested different values for the **Number of Hidden Units** property. The models became more stable (as seen by comparing the lift charts from the training and validation data) as I increased the number of hidden units from the default value of 3 to 10 or 20.

Using only a selected number of inputs enables you to test a variety of architectural specifications fast. For purposes of illustration, I present below the results from a model with the following property settings:

- **Architecture** property: MLP
- **Selection Criterion** property: 20
- **Number of Hidden Units** property: 20

The bottom segment of Display 7.3 shows the process flow for the neural network model of attrition. In this process flow I used the **Decision Tree** node with the **Splitting Rule Criterion** property set to **ProbChisq** and **Assessment Measure** property set to **Average Square Error** to select the inputs for use in the **Neural Network** node.

The lift charts for this model are shown in Display 7.8.

Display 7.8

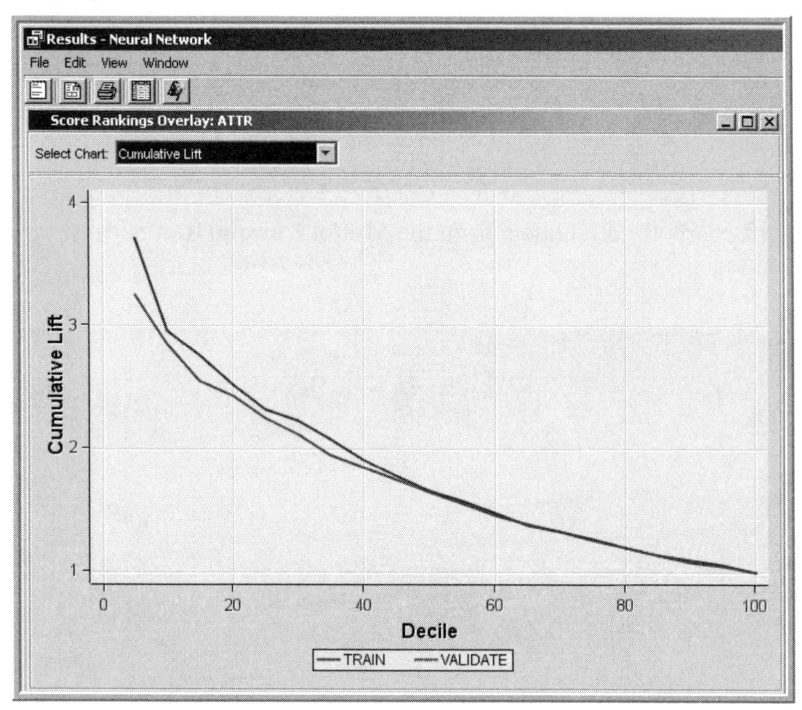

Table 7.3 shows the cumulative lift and cumulative capture rates for the Train, Validate, and Test data sets.

Table 7.3

NEURAL

	Train		Validate		Test	
Demi-decile	Cumulative Lift	Cumulative Capture Rate(%)	Cumulative Lift	Cumulative Capture Rate(%)	Cumulative Lift	Cumulative Capture Rate(%)
5	3.71	18.5	3.25	16.3	3.12	15.6
10	2.95	29.5	2.84	28.4	3.03	30.3
15	2.75	41.3	2.54	38.1	2.80	42.1
20	2.51	50.2	2.42	48.4	2.76	55.3
25	2.32	58.0	2.25	56.3	2.45	61.4
30	2.22	66.7	2.11	63.4	2.32	69.5
35	2.07	72.5	1.94	68.0	2.16	75.7
40	1.90	76.1	1.84	73.8	1.99	79.4
45	1.78	80.0	1.74	78.4	1.84	82.9
50	1.66	82.9	1.64	82.2	1.69	84.7
55	1.57	86.6	1.55	85.3	1.59	87.2
60	1.47	88.3	1.45	87.2	1.49	89.4
65	1.38	89.7	1.39	90.3	1.40	91.0
70	1.32	92.5	1.33	92.8	1.32	92.5
75	1.26	94.6	1.25	94.1	1.27	95.0
80	1.20	95.8	1.20	95.6	1.21	96.6
85	1.14	96.7	1.14	96.9	1.16	98.4
90	1.09	98.4	1.08	97.2	1.10	99.1
95	1.05	99.8	1.04	99.1	1.05	99.4
100	1.00	100.0	1.00	100.0	1.00	100.0

Display 7.9 shows the lift charts for all models from the **Model Comparison** node.

Display 7.9

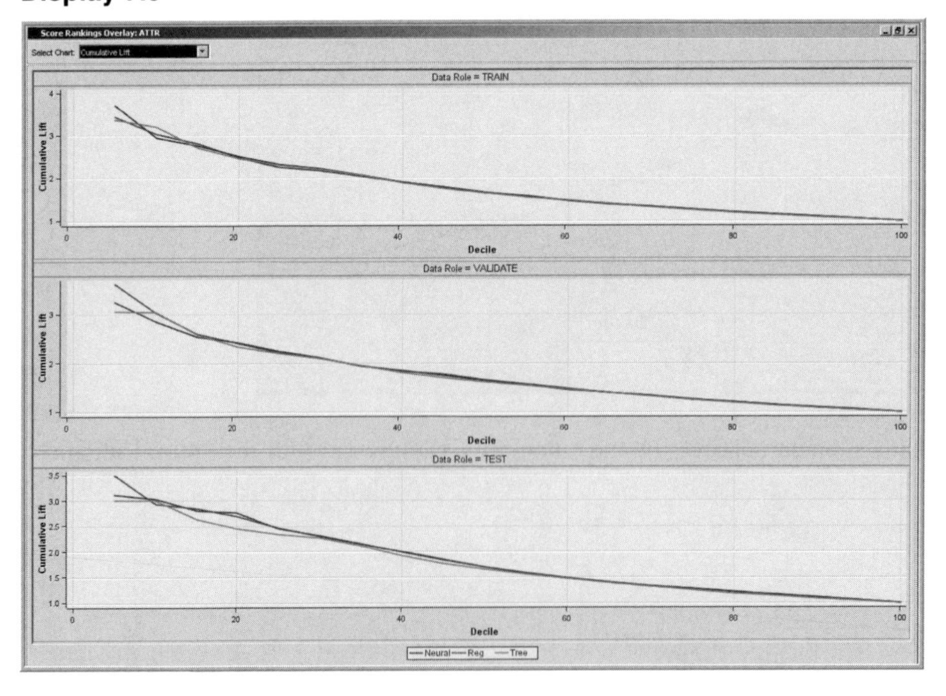

Display 7.10 shows the cumulative capture rates for the three models.

Display 7.10

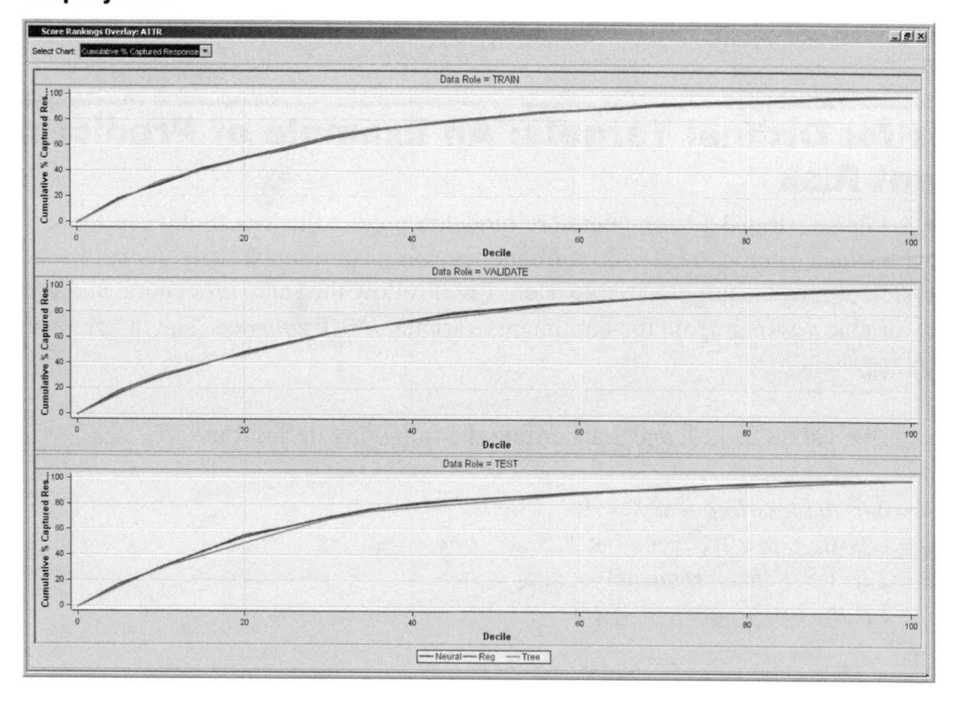

Table 7.4 shows a comparison of the three models using test data.

Table 7.4

	Regression		Neural Network		Decision Tree	
Demi-decile	Cumulative Lift	Cumulative Capture Rate(%)	Cumulative Lift	Cumulative Capture Rate(%)	Cumulative Lift	Cumulative Capture Rate(%)
5	3.49	17.4	3.12	15.6	3.00	15.0
10	2.93	29.3	3.03	30.3	3.00	30.0
15	2.85	42.7	2.80	42.1	2.63	39.5
20	2.69	53.9	2.76	55.3	2.45	49.0
25	2.47	61.7	2.45	61.4	2.34	58.5
30	2.31	69.2	2.32	69.5	2.27	68.1
35	2.15	75.1	2.16	75.7	2.11	73.9
40	2.01	80.4	1.99	79.4	1.92	76.7
45	1.86	83.5	1.84	82.9	1.77	79.6
50	1.70	85.0	1.69	84.7	1.65	82.4
55	1.57	86.6	1.59	87.2	1.55	85.3
60	1.47	88.2	1.49	89.4	1.47	88.2
65	1.39	90.7	1.40	91.0	1.39	90.3
70	1.33	93.1	1.32	92.5	1.31	91.7
75	1.27	95.0	1.27	95.0	1.24	93.1
80	1.21	96.9	1.21	96.6	1.18	94.5
85	1.15	97.8	1.16	98.4	1.13	95.9
90	1.10	99.4	1.10	99.1	1.08	97.2
95	1.05	99.7	1.05	99.4	1.04	98.6
100	1.00	100.0	1.00	100.0	1.00	100.0

From Table 7.4 it appears that both the logistic regression and neural network models have outperformed the decision tree model by a slight margin, and that the logistic regression model outperformed the neural network model by a slight margin. The top four deciles (the top eight demi-deciles), based on the logistic regression model, capture 80.4% of attritors, while the top four deciles based on the neural network model capture 79.4% of the attritors, while the top four deciles, based on the decision tree model, capture 76.7% of attritors. The cumulative lift for all the three models are monotonically declining with each successive demi-decile.

7.3 Models for Ordinal Targets: An Example of Predicting Accident Risk

In Chapter 5, I demonstrated a neural network model to predict the loss frequency for a hypothetical insurance company. The loss frequency was a continuous variable, but I used a discrete version of it as the target variable. Here I will follow the same procedure, and create a discretized variable *lossfrq* from the continuous variable *loss frequency*. The discretization is done as follows:

lossfrq takes the values 0, 1, 2, and 3 according the following definitions:

$$\begin{aligned}
&lossfrq = 0 \text{ if the loss frequency } = 0\\
&lossfrq = 1 \text{ if } 0 < loss\ frequency < 1.5\\
&lossfrq = 2 \text{ if } 1.5 \le loss\ frequency < 2.5\\
&lossfrq = 3 \text{ if the loss frequency } \ge 2.5.
\end{aligned}$$

The goal here is to develop three models using the **Regression**, **Decision Tree**, and **Neural Network** nodes to predict the following probabilities:

$$\begin{aligned}
&\Pr(lossfrq = 0 \mid X) = \Pr(loss\ frequency = 0 \mid X)\\
&\Pr(lossfrq = 1 \mid X) = \Pr(0 < loss\ frequency < 1.5 \mid X)\\
&\Pr(lossfrq = 2 \mid X) = \Pr(1.5 \le loss\ frequency < 2.5 \mid X)\\
&\Pr(lossfrq = 3 \mid X) = \Pr(the\, loss\ frequency \ge 2.5 \mid X)
\end{aligned}$$

where X is a vector of inputs or explanatory variables.

When you are using a target variable such as *lossfrq*, which is a discrete version of a continuous variable, the variable becomes ordinal if it has more than two levels.

The first model I develop for predicting the above probabilities is a proportional odds model (or logistic regression with cumulative logits link) using the **Regression** node. As explained in Section 6.2.2 of Chapter 6, the **Regression** node produces such a model if the measurement level of the target is set to **Ordinal**.

The second model I develop is a decision tree model using the **Decision Tree** node. The **Decision Tree** node does not produce any equations, but it does provide certain rules for portioning the data set into disjoint groups (leaf nodes). The rules are stated in terms of input ranges (or definitions). For each group, the **Decision Tree** node gives the predicted probabilities:

$Pr(lossfrq = 0 \mid X), Pr(lossfrq = 1 \mid X), Pr(lossfrq = 2 \mid X)$ and $Pr(lossfrq = 3 \mid X)$. The **Decision Tree** node also assigns a target level such as 0, 1, 2, and 3 to each group, and hence to all the records belonging to that group.

The third model will be developed by using the **Neural Network** node, which produces a Proportional Odds type model, provided I set the **Target Activation Function** property to **Logistic**, as explained in Section 5.5.1.1 of Chapter 5.

Display 7.11 shows the process flow for comparing the ordinal target.

Display 7.11

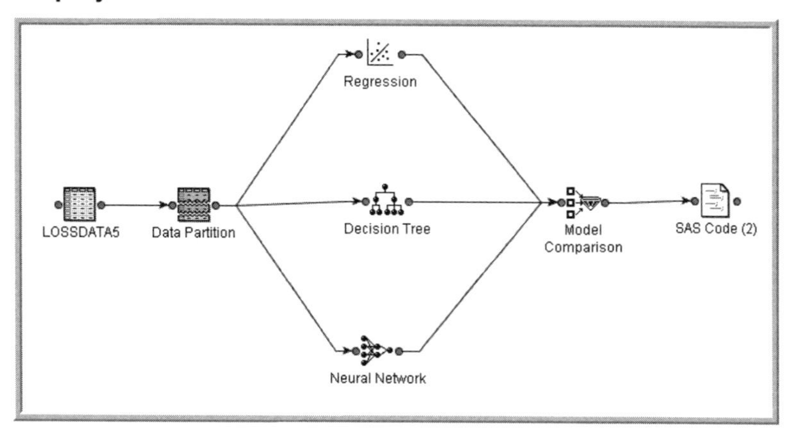

7.3.1 Lift Charts and Capture Rates for Models with Ordinal Targets

Method Used by SAS Enterprise Miner

When the target is ordinal, Enterprise Miner creates lift charts based on the probability of the highest level of the target variable. The highest level in this example is 3. For each record in the Test data set, Enterprise Miner computes the predicted, or posterior, probability $Pr(lossfrq = 3 \mid X_i)$ from the model. Then it sorts the data set in descending order of the predicted probability and divides the data set into 20 deciles, which I refer to alternatively as demi-deciles. Within each decile Enterprise Miner calculates the proportion of cases, as well as the total number of cases, with *lossfrq*= 3. The lift for a decile is the ratio of the proportion of cases where *lossfrq* =3 in the decile to the proportion of cases with *lossfrq*=3 in the entire data set. The capture rate is the ratio of the total number of cases with *lossfrq*=3 within the decile to the total number of cases with *lossfrq* =3 in the entire data set.

An Alternative Approach Using Expected Lossfrq

As pointed out earlier in Section 5.5.1.4, it is sometimes required to calculate lift and capture rate based on expected value of the target variable. For each model, I shall present lift tables and capture rates based on the expected value of the target variable.

For each record in the Test data set, I calculate the expected *lossfrq* as

$$E(lossfrq \mid X_i) = Pr(lossfrq = 0 \mid X_i)*0 + Pr(lossfrq = 1 \mid X_i)*1 + Pr(lossfrq = 2 \mid X_i)*2 + Pr(lossfrq = 3 \mid X_i)*3.$$

Next, I sort the records of the data set in descending order of $E(lossfrq \mid X_i)$, divide the data set into 20 demi-deciles (bins), and within each demi-decile, I calculate the mean of the actual or observed value of the target variable *lossfrq*. The ratio of the mean of actual *lossfrq* in a demi-decile to the overall mean of actual *lossfrq* for the data set is the lift for the demi-decile. Likewise, the capture rate for a demi-decile is the ratio of the sum of the actual *lossfrq* for all the records within the demi-decile to the sum of the actual *lossfrq* of all the records in for the entire data set.

7.3.2 Logistic Regression with Proportional Odds for Predicting Risk in Auto Insurance

Given that the measurement level of the target variable is set to **Ordinal**, a proportional odds model is estimated by the **Regression** node using the property settings shown in Display 7.12.

Display 7.12

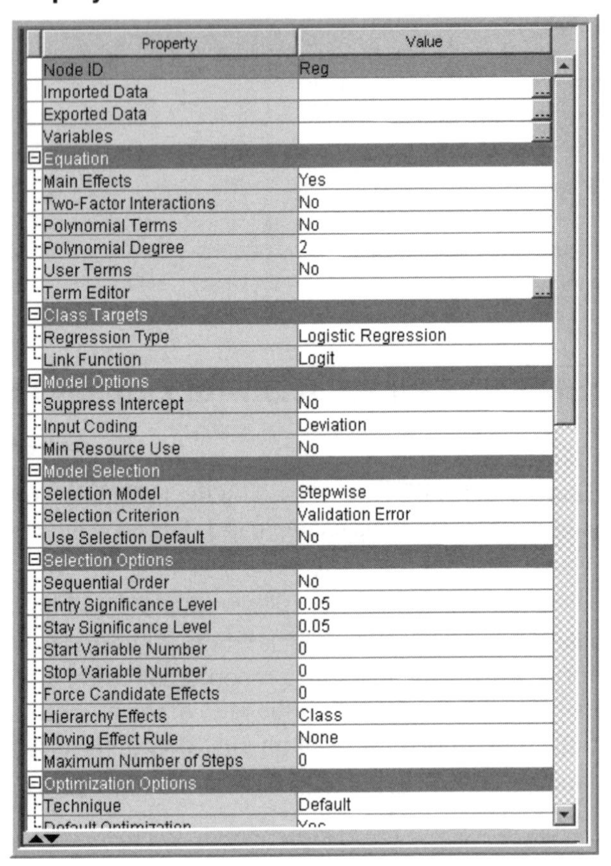

Property	Value
Node ID	Reg
Imported Data	
Exported Data	
Variables	
Equation	
Main Effects	Yes
Two-Factor Interactions	No
Polynomial Terms	No
Polynomial Degree	2
User Terms	No
Term Editor	
Class Targets	
Regression Type	Logistic Regression
Link Function	Logit
Model Options	
Suppress Intercept	No
Input Coding	Deviation
Min Resource Use	No
Model Selection	
Selection Model	Stepwise
Selection Criterion	Validation Error
Use Selection Default	No
Selection Options	
Sequential Order	No
Entry Significance Level	0.05
Stay Significance Level	0.05
Start Variable Number	0
Stop Variable Number	0
Force Candidate Effects	0
Hierarchy Effects	Class
Moving Effect Rule	None
Maximum Number of Steps	0
Optimization Options	
Technique	Default

For the reasons I cited in the previous sections, I set the **Selection Model** property to **Stepwise** and the **Selection Criterion** property to **Validation Error**. The choice of **Validation Error** is also prompted by the lack of an appropriate profit matrix.

Display 7.13 shows the estimated equations for the cumulative logits.

Display 7.13

Analysis of Maximum Likelihood Estimates

Parameter		DF	Estimate	Standard Error	Wald Chi-Square	Pr > ChiSq	Standardized Estimate	Exp(Est)
Intercept	3	1	-0.1100	0.3948	0.08	0.7806		0.896
Intercept	2	1	1.8692	0.3278	32.52	<.0001		6.483
Intercept	1	1	3.6020	0.3219	125.18	<.0001		36.671
AGE		1	-0.0344	0.00334	105.69	<.0001	-0.2977	0.966
CRED		1	-0.00594	0.000407	213.28	<.0001	-0.3083	0.994
NPRVIO	0	1	-1.9172	0.1910	100.78	<.0001		0.147
NPRVIO	1	1	-1.4764	0.2050	51.87	<.0001		0.228
NPRVIO	2	1	-0.7174	0.2360	9.24	0.0024		0.488
NPRVIO	3	1	-1.2199	0.4144	8.66	0.0032		0.295
NPRVIO	4	1	0.0259	0.4004	0.00	0.9484		1.026
NPRVIO	5	1	0.4649	0.9918	0.22	0.6392		1.592
NPRVIO	6	1	2.4145	0.4774	25.57	<.0001		11.184

It can be seen from Display 7.13 that the estimated model is logistic with a cumulative logits link. Accordingly, the slopes (coefficients of the explanatory variables) for the three logit equations are the same, but the intercepts are different. There are only three equations in Display 7.13, while there are four events represented by the four levels (0, 1, 2, and 3) of the target variable $lossfrq$. The three equations given in Display 7.13 plus the fourth equation, given by the identity

$$\Pr(lossfrq = 0) + \Pr(lossfrq = 1) + \Pr(lossfrq = 2) + \Pr(lossfrq = 3) = 1$$

are sufficient to solve for the probabilities of the four events. (See Section 6.2.2 of Chapter 6 to review how these probabilities are calculated by the **Regression** node.) In the model presented in Section 6.2.2, the target variable has only three levels, while the target variable in the model presented in Display 7.13 has four levels: 0, 1, 2, and 3.

The variables selected by the **Regression** node are AGE (age of the insured), CRED (credit score of the insured), and NPRVIO (number of prior violations). Because NPRVIO has only six levels, Enterprise Miner has treated it as a class input.

The cumulative lift charts from the **Results** window of the **Regression** node are shown in Display 7.14.

Display 7.14

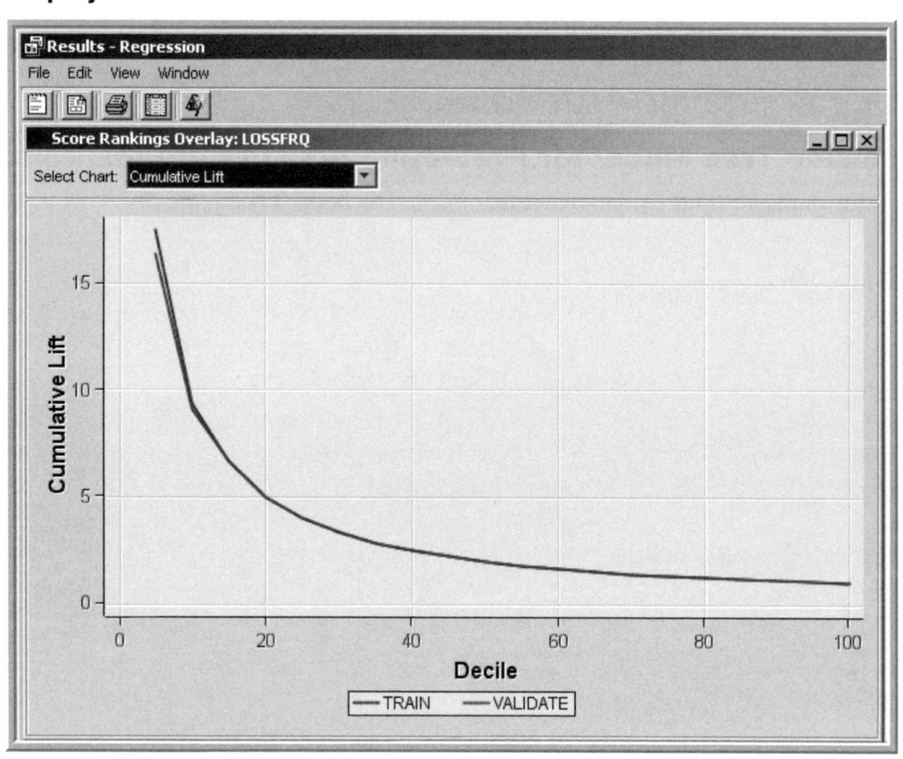

The cumulative lift charts shown in Display 7.14 are based on the highest level of the target variable, as described in Section 7.3.1. Using test data, alternative cumulative lift and capture rates based on $E(lossfrq)$, as described in Section 7.3.1, were computed and are shown in Table 7.5.

Table 7.5

reg

Demi-Decile	Number of Policies	Loss Frequency Total	Mean	Cumulative Total	Cumulative Mean	Cumulative Lift	Cumulative Capture Rate(%)
5	450	126	0.280	126	0.280	5.09	25.4%
10	451	57	0.126	183	0.203	3.69	36.9%
15	451	60	0.133	243	0.180	3.27	49.0%
20	450	39	0.087	282	0.156	2.84	56.9%
25	451	43	0.095	325	0.144	2.62	65.5%
30	451	24	0.053	349	0.129	2.35	70.4%
35	451	19	0.042	368	0.117	2.12	74.2%
40	450	22	0.049	390	0.108	1.97	78.6%
45	451	12	0.027	402	0.099	1.80	81.0%
50	451	14	0.031	416	0.092	1.68	83.9%
55	451	13	0.029	429	0.087	1.57	86.5%
60	450	13	0.029	442	0.082	1.49	89.1%
65	451	7	0.016	449	0.077	1.39	90.5%
70	451	7	0.016	456	0.072	1.31	91.9%
75	451	6	0.013	462	0.068	1.24	93.1%
80	450	10	0.022	472	0.065	1.19	95.2%
85	451	10	0.022	482	0.063	1.14	97.2%
90	451	9	0.020	491	0.061	1.10	99.0%
95	451	5	0.011	496	0.058	1.05	100.0%
100	451	0	0.000	496	0.055	1.00	100.0%

7.3.3 Decision Tree Model for Predicting Risk in Auto Insurance

Display 7.15 shows the settings of the properties of the **Decision Tree** node.

Display 7.15

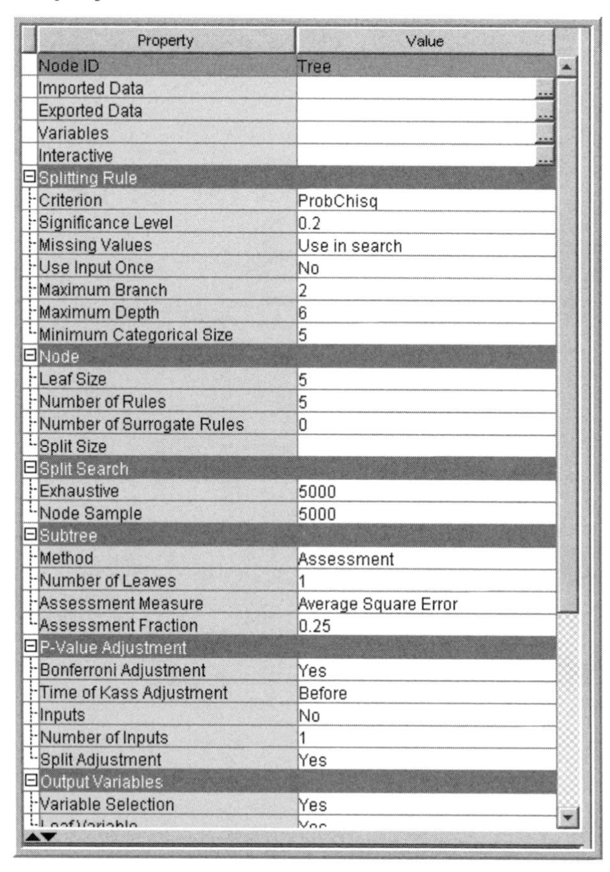

Property	Value
Node ID	Tree
Imported Data	
Exported Data	
Variables	
Interactive	
Splitting Rule	
Criterion	ProbChisq
Significance Level	0.2
Missing Values	Use in search
Use Input Once	No
Maximum Branch	2
Maximum Depth	6
Minimum Categorical Size	5
Node	
Leaf Size	5
Number of Rules	5
Number of Surrogate Rules	0
Split Size	
Split Search	
Exhaustive	5000
Node Sample	5000
Subtree	
Method	Assessment
Number of Leaves	1
Assessment Measure	Average Square Error
Assessment Fraction	0.25
P-Value Adjustment	
Bonferroni Adjustment	Yes
Time of Kass Adjustment	Before
Inputs	No
Number of Inputs	1
Split Adjustment	Yes
Output Variables	
Variable Selection	Yes

Note that I set the **Splitting Rule Criterion** property to **ProbChisq**, the **Subtree Method** property to **Assessment**, and the **Assessment Measure** property to **Average Square Error**. These choices were arrived at after trying out different values. The settings chosen end up yielding a reasonably sized tree—one that is not too small or too large. In order to find general rules for making these choices, you need to experiment with different values for these properties with many replications.

Displays 7.16A, 7.16B, and 7.16C show the tree produced by the property settings given in Display 7.15.

Display 7.16A

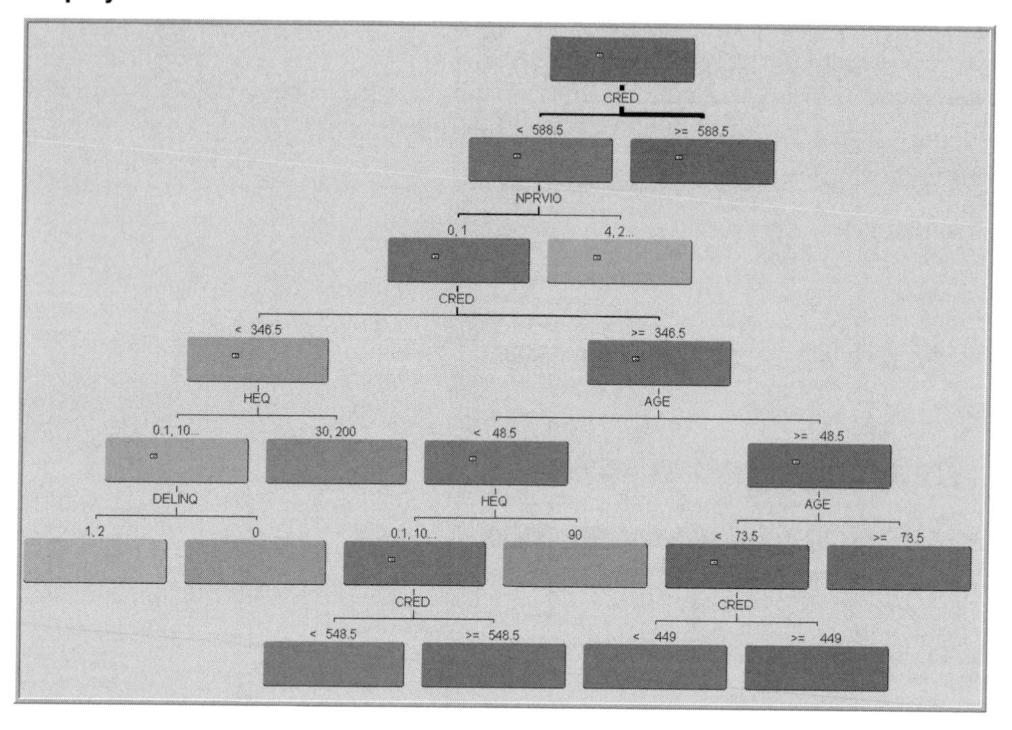

Display 7.16B

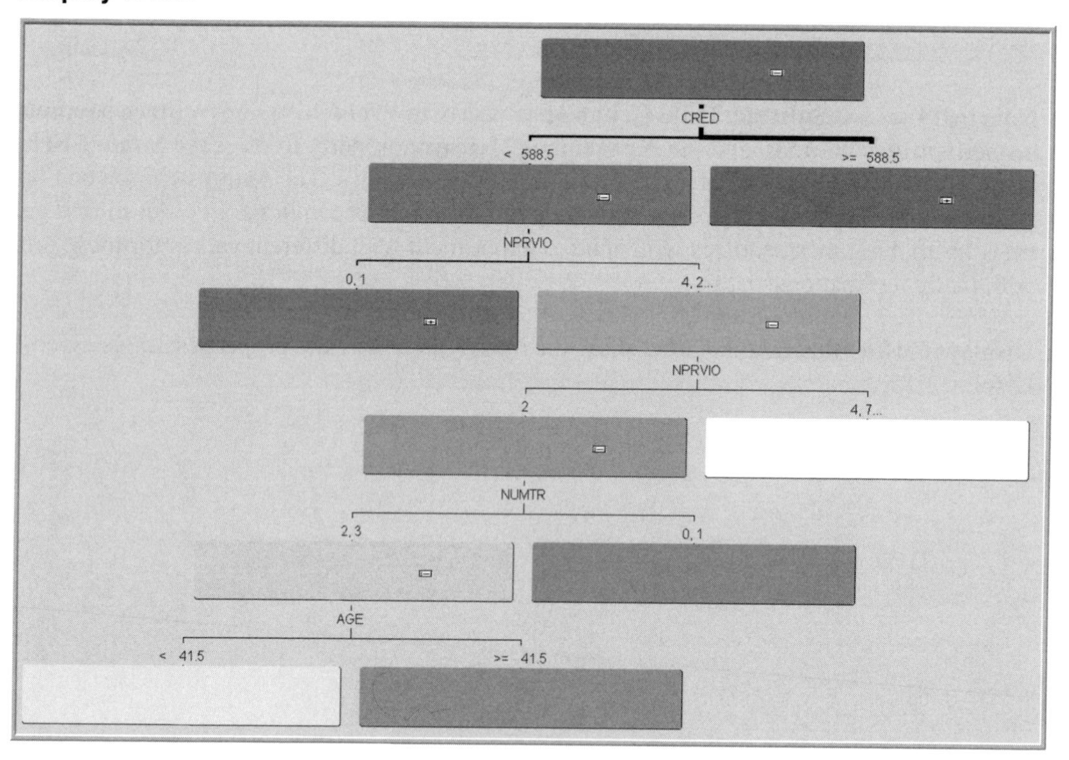

Display 7.16C

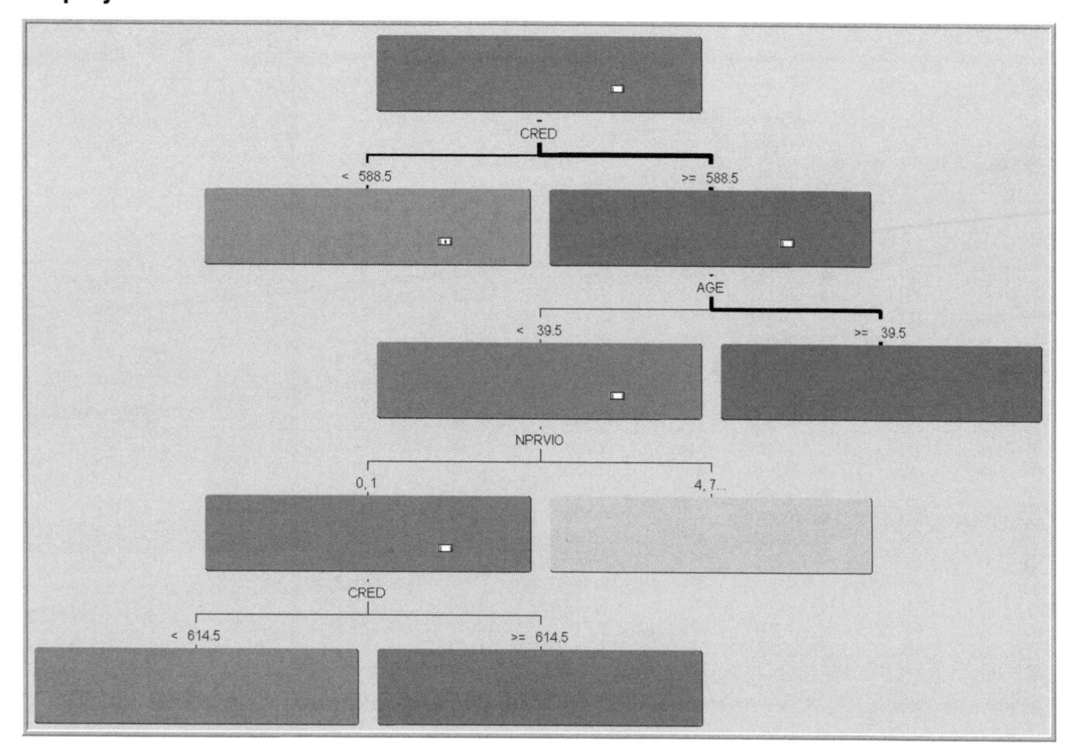

The **Decision Tree** node has selected the three variables AGE, CRED, and NPRVIO, which were also selected by the **Regression** node. In addition, the **Decision Tree** node selected two more variables NUMTR (number of credit cards owned by the policyholder) and HEQ (value of home equity).

Display 7.17 shows the lift charts for the decision tree model.

Display 7.17

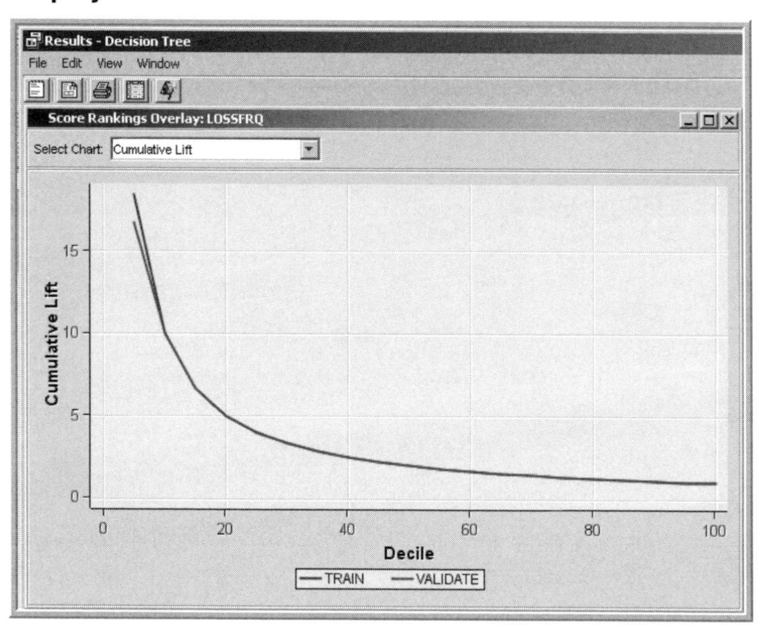

Table 7.6 shows the lift charts calculated using the expected loss frequency.

Table 7.6

tree

Demi-Decile	Number of Policies	Loss Frequency Total	Mean	Cumulative Total	Cumulative Mean	Cumulative Lift	Cumulative Capture Rate(%)
5	450	108	0.240	108	0.240	4.36	21.8%
10	451	48	0.106	156	0.173	3.15	31.5%
15	451	78	0.173	234	0.173	3.15	47.2%
20	450	25	0.056	259	0.144	2.61	52.2%
25	451	29	0.064	288	0.128	2.32	58.1%
30	451	36	0.080	324	0.120	2.18	65.3%
35	451	13	0.029	337	0.107	1.94	67.9%
40	450	25	0.056	362	0.100	1.83	73.0%
45	451	15	0.033	377	0.093	1.69	76.0%
50	451	15	0.033	392	0.087	1.58	79.0%
55	451	17	0.038	409	0.082	1.50	82.5%
60	450	9	0.020	418	0.077	1.40	84.3%
65	451	9	0.020	427	0.073	1.32	86.1%
70	451	8	0.018	435	0.069	1.25	87.7%
75	451	8	0.018	443	0.066	1.19	89.3%
80	450	12	0.027	455	0.063	1.15	91.7%
85	451	18	0.040	473	0.062	1.12	95.4%
90	451	5	0.011	478	0.059	1.07	96.4%
95	451	3	0.007	481	0.056	1.02	97.0%
100	451	15	0.033	496	0.055	1.00	100.0%

7.3.4 Neural Network Model for Predicting Risk in Auto Insurance

The property settings for the **Neural Network** node are shown in Display 7.18.

Display 7.18

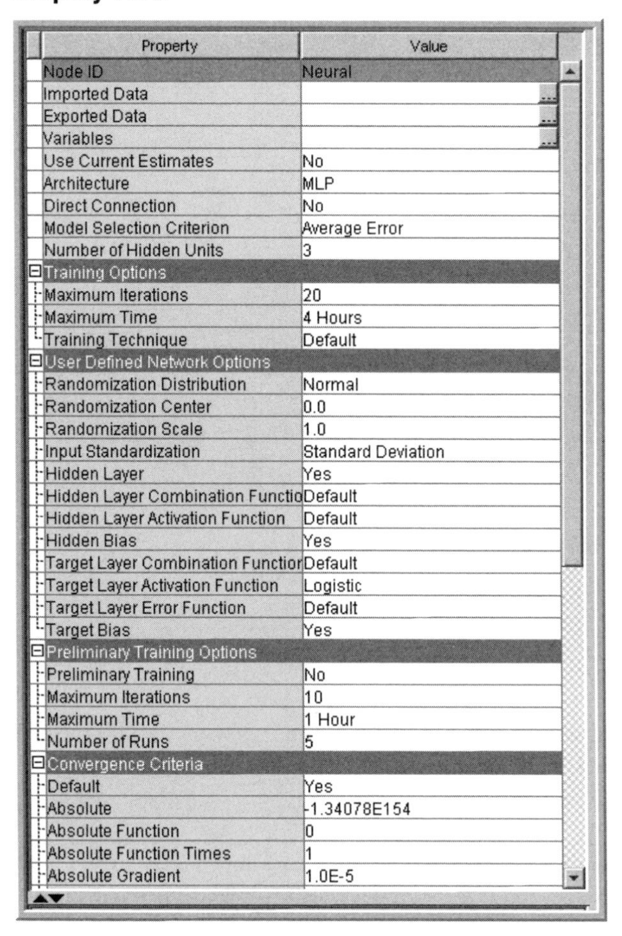

Property	Value
Node ID	Neural
Imported Data	...
Exported Data	...
Variables	...
Use Current Estimates	No
Architecture	MLP
Direct Connection	No
Model Selection Criterion	Average Error
Number of Hidden Units	3
⊟ Training Options	
Maximum Iterations	20
Maximum Time	4 Hours
Training Technique	Default
⊟ User Defined Network Options	
Randomization Distribution	Normal
Randomization Center	0.0
Randomization Scale	1.0
Input Standardization	Standard Deviation
Hidden Layer	Yes
Hidden Layer Combination Functio	Default
Hidden Layer Activation Function	Default
Hidden Bias	Yes
Target Layer Combination Functior	Default
Target Layer Activation Function	Logistic
Target Layer Error Function	Default
Target Bias	Yes
⊟ Preliminary Training Options	
Preliminary Training	No
Maximum Iterations	10
Maximum Time	1 Hour
Number of Runs	5
⊟ Convergence Criteria	
Default	Yes
Absolute	-1.34078E154
Absolute Function	0
Absolute Function Times	1
Absolute Gradient	1.0E-5

Because the number of inputs available in the data set is small, I passed all of them into the **Neural Network** node. I set the **Number of Hidden Units** property to its default value of 3, since any increase in this value did not improve the results significantly.

Display 7.19 shows the lift charts for the neural networks model.

Display 7.19

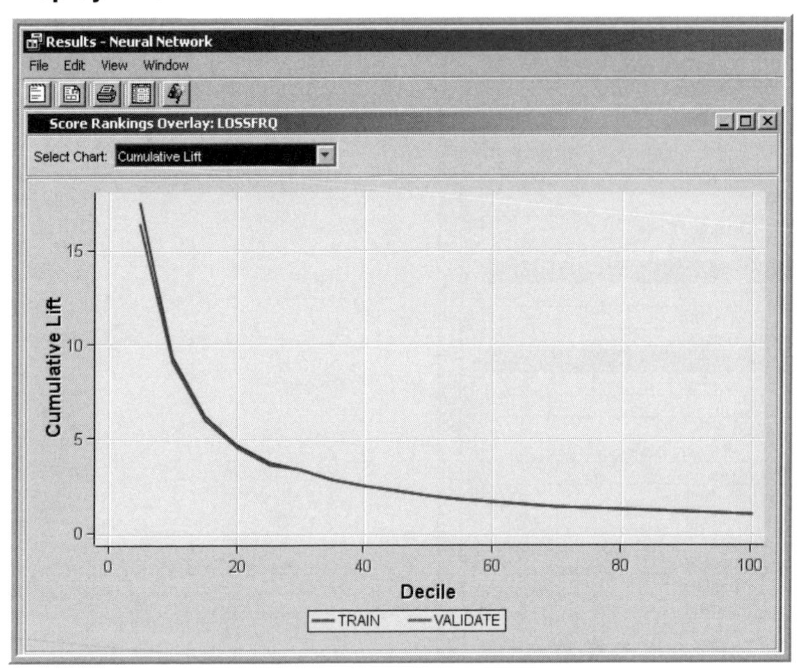

Table 7.7 shows the cumulative lift and capture rates for the neural network model based on the expected loss frequency using the test data.

Table 7.7

neural

Demi-Decile	Number of Policies	Loss Frequency Total	Mean	Cumulative Total	Cumulative Mean	Cumulative Lift	Cumulative Capture Rate(%)
5	450	118	0.262	118	0.262	4.77	23.8%
10	451	63	0.140	181	0.201	3.65	36.5%
15	451	45	0.100	226	0.167	3.04	45.6%
20	450	38	0.084	264	0.147	2.66	53.2%
25	451	35	0.078	299	0.133	2.41	60.3%
30	451	33	0.073	332	0.123	2.23	66.9%
35	451	19	0.042	351	0.111	2.02	70.8%
40	450	19	0.042	370	0.103	1.87	74.6%
45	451	19	0.042	389	0.096	1.74	78.4%
50	451	20	0.044	409	0.091	1.65	82.5%
55	451	19	0.042	428	0.086	1.57	86.3%
60	450	10	0.022	438	0.081	1.47	88.3%
65	451	7	0.016	445	0.076	1.38	89.7%
70	451	15	0.033	460	0.073	1.32	92.7%
75	451	10	0.022	470	0.070	1.26	94.8%
80	450	7	0.016	477	0.066	1.20	96.2%
85	451	7	0.016	484	0.063	1.15	97.6%
90	451	6	0.013	490	0.060	1.10	98.8%
95	451	4	0.009	494	0.058	1.05	99.6%
100	451	2	0.004	496	0.055	1.00	100.0%

7.4 Comparison of All Three Accident Risk Models

Table 7.8 shows the lift and capture rates calculated for the test data set ranked by $E(lossfrq)$, as outlined in Section 7.3.

Table 7.8

	Regression		Neural Network		Decision Tree	
Demi-Decile	Cumulative Lift	Cumulative Capture Rate(%)	Cumulative Lift	Cumulative Capture Rate(%)	Cumulative Lift	Cumulative Capture Rate(%)
5	5.09	25.4%	4.77	23.8%	4.36	21.8%
10	3.69	36.9%	3.65	36.5%	3.15	31.5%
15	3.27	49.0%	3.04	45.6%	3.15	47.2%
20	2.84	56.9%	2.66	53.2%	2.61	52.2%
25	2.62	65.5%	2.41	60.3%	2.32	58.1%
30	2.35	70.4%	2.23	66.9%	2.18	65.3%
35	2.12	74.2%	2.02	70.8%	1.94	67.9%
40	1.97	78.6%	1.87	74.6%	1.83	73.0%
45	1.80	81.0%	1.74	78.4%	1.69	76.0%
50	1.68	83.9%	1.65	82.5%	1.58	79.0%
55	1.57	86.5%	1.57	86.3%	1.50	82.5%
60	1.49	89.1%	1.47	88.3%	1.40	84.3%
65	1.39	90.5%	1.38	89.7%	1.32	86.1%
70	1.31	91.9%	1.32	92.7%	1.25	87.7%
75	1.24	93.1%	1.26	94.8%	1.19	89.3%
80	1.19	95.2%	1.20	96.2%	1.15	91.7%
85	1.14	97.2%	1.15	97.6%	1.12	95.4%
90	1.10	99.0%	1.10	98.8%	1.07	96.4%
95	1.05	100.0%	1.05	99.6%	1.02	97.0%
100	1.00	100.0%	1.00	100.0%	1.00	100.0%

From Table 7.8, it is clear that the logistic regression with a cumulative logits link is the winner in terms of lift and capture rates. The table shows that, for the logistic regression, the cumulative lift at the eighth demi-decile (eighth row of the table) is 1.97. This means that the actual average loss frequency of the customers who are in the top eight demi-deciles (centiles 1 to 40) based on the logistic regression is 1.97 times the overall average loss frequency. The corresponding numbers for the neural networks and decision tree models are 1.87 and 1.83, respectively.

Customer Profitability

8.1 Introduction

This chapter presents a general framework for calculating the profitability of different groups of customers using a simplified example. The methodology can be extended to calculate profits at the individual customer level as well.

My goal here is to illustrate how costs of acquisition and the costs associated with risk factors can affect the decision to acquire new customers using the example of a credit card company.

For example, suppose you have a population of 10,000 prospects. A response model is used to score each of these 10,000 prospects. The prospects are arranged in descending order of the predicted probability of response and divided into deciles. A risk model is then used to calculate the risk rate for each prospect. In this simplified example, the *risk rate* is the probability of a customer defaulting on payment. Assume that a default results in a net loss of $10 for the credit card company on the average. If the customer did not default, the company would have made $100. (These numbers are fictitious, and I use them for demonstration purposes only.) Table 8.1 shows the estimated response and risk rates for the deciles of the prospect population.

Table 8.1

Decile	Response Rate	Risk Rate
1	0.050	0.100
2	0.045	0.092
3	0.040	0.080
4	0.030	0.067
5	0.020	0.059
6	0.010	0.042
7	0.005	0.038
8	0.004	0.029
9	0.003	0.018
10	0.002	0.010

From Table 8.1 you can see that response rate and risk rate move in the same direction. The response rate is highest in the first decile and declines with succeeding higher numbered deciles. A similar pattern is observed for the risk rate also. One reason why this might occur is that when a credit card company solicits applications for credit cards, the groups that respond most are likely to be those who cannot get credit elsewhere because of their relatively high risk rates.

In Table 8.1, the term *risk rate* is used in a general sense. A risk rate of 10% in decile 1 in Table 8.1 means that 10% of the persons who responded and are in decile 1 tend to default on their payments. (Here I am assuming that each person who responded is issued a credit card, but this assumption can be easily relaxed without violating the logic.) The response rate in a particular decile is the proportion of individuals in the decile who are responders. Display 8.1 graphs the response and risk rates for each decile.

Display 8.1

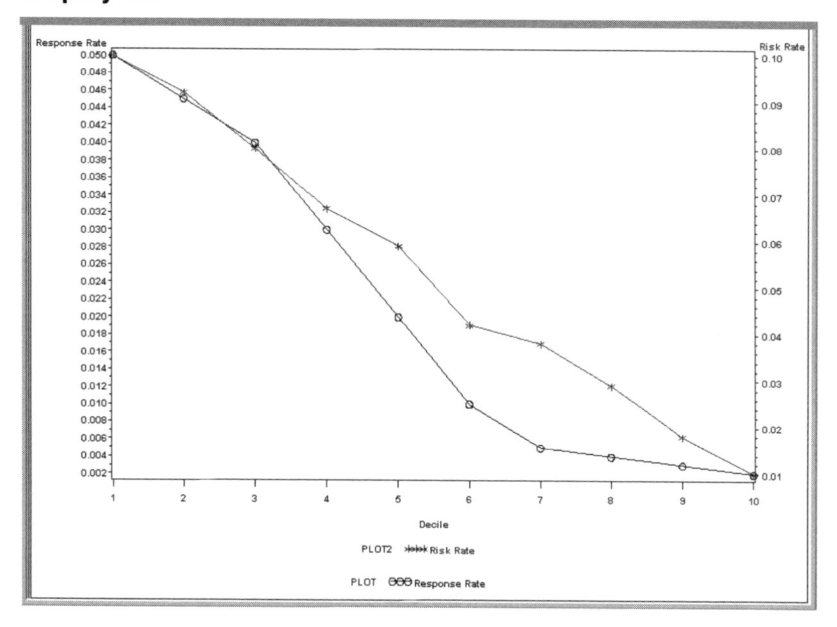

In Display 8.1, the horizontal axis shows the decile number. Decile 1 is the top decile, and decile 10 is the lowest decile.

Note: The example presented in this chapter is hypothetical and greatly simplified for the purpose of exposition; the costs and revenue figures are also quite arbitrary. The example refers to a credit card company but it can be extended to insurance and other companies as well. Details of the methodology should be modified according to the specific situation being analyzed. In many situations, for example, customers with a high response rate to a direct mailing campaign might also pose higher risks to the company that is soliciting business.

8.2 Acquisition Cost

New customers are acquired through channels such as newspaper or Internet advertisements, radio and TV broadcasting, direct mail, etc. If a company spends X dollars on a particular channel, and as a result it acquires n customers, then the average cost of acquisition per customer is X / n. For direct mail this can be calculated easily. Suppose the response rate is r in a segment of the target population and suppose the cost of sending one mail piece is m dollars. If mail is sent to N customers from that segment, then the average cost of acquiring a customer is $\frac{N.m}{rN} = \frac{m}{r}$.

This means that, as the response rate decreases, the average acquisition cost increases. Display 8.1A illustrates this relationship.

Display 8.1A

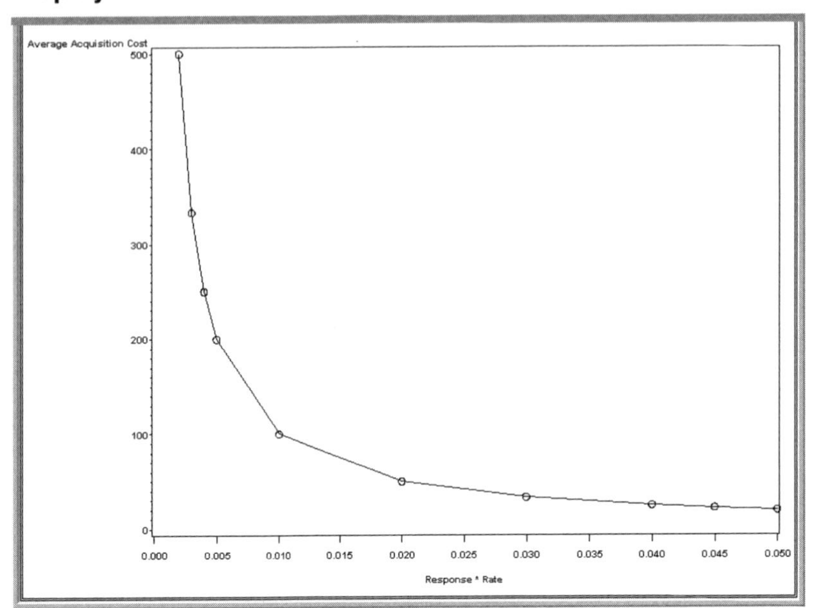

Table 8.2 shows the acquisition cost by decile. Since the response rate declines with succeeding higher-numbered deciles, the average acquisition cost increases. When the company sends mail to the 1,000 prospects in the top decile (decile 1) inviting them to purchase a product, at a cost of $1 per piece of mail, the total cost is $1,000. In return, the company acquires 50 customers, as the response rate in the top decile is 0.05. The average cost of acquisition is therefore $1,000/50 = $20. Similarly, the average cost of acquisition for the second decile is $1,000/45 = $22.22. The last column shows the cumulative acquisition cost. If all the prospects in the top two deciles are sent a mail piece, the total cost would be $2,000. This is the cumulative acquisition cost, and it is an important item in determining the optimum cut-off point for mailing.

Table 8.2

Decile	Number of Customers	Cost per Mail-piece	Average Acquisition Cost	Total Acquisition Cost	Cumulative Acquisition Cost
1	1000	$1	$20	$1,000	$1,000
2	1000	$1	$22	$1,000	$2,000
3	1000	$1	$25	$1,000	$3,000
4	1000	$1	$33	$1,000	$4,000
5	1000	$1	$50	$1,000	$5,000
6	1000	$1	$100	$1,000	$6,000
7	1000	$1	$200	$1,000	$7,000
8	1000	$1	$250	$1,000	$8,000
9	1000	$1	$333	$1,000	$9,000
10	1000	$1	$500	$1,000	$10,000

8.3 Cost of Default

If a customer defaults on payment, his credit card company incurs a loss. The credit card company might face other types of risk, but for the purpose of illustration only one type of risk, the risk of default, is considered here.

Table 8.3 illustrates the costs due to the risk of default. Assume, for the purpose of illustration, that each event (default) results in a cost of $10 to the credit card company. In general, you can include more events with associated probabilities and with more realistic costs in these calculations.

In this simplified example, the top decile has 50 responders. Of these responders, five people (the risk rate is 0.1) tend to default on their payments, resulting in an expected loss of $50 to the company. In the second decile there are 45 responders (since the response rate is 4.5%). Of these 45 customers, 9.2% are likely to default on their payments resulting in an expected loss of 45 × 0.092 × $10 = $41. If the company targets the top two deciles, it will experience an expected loss of $91. This is the cumulative loss[1] for the second decile. Similarly cumulative losses can be calculated for the remaining deciles.

Table 8.3

Decile	Number of Responders	Risk Rate	Loss per Event	Loss Amount	Cumulative Loss
1	50	0.100	$10	$50	$50
2	45	0.092	$10	$41	$91
3	40	0.080	$10	$32	$123
4	30	0.067	$10	$20	$144
5	20	0.059	$10	$12	$155
6	10	0.042	$10	$4	$160
7	5	0.038	$10	$2	$161
8	4	0.029	$10	$1	$163
9	3	0.018	$10	$1	$163
10	2	0.010	$10	$0	$163

[1] The losses are calculated using the estimated probabilities, and hence, strictly speaking, they should be called *expected losses*. But here I will use the terms *losses* and *expected losses* interchangeably.

8.4 Revenue

Revenue depends on the number of customers who do not default. These revenues are shown in Table 8.4. To simplify the example, I have assumed that a customer who defaults on payment generates no revenue. This assumption can easily be relaxed without violating the logic of my argument.

In the top decile, there are 50 responders, but only 45 of them are customers in good standing at the end of the year (assuming that our analysis is based on calculations for one year). Hence the revenue generated in the top decile is 45 × \$100 = \$4,500. A similar calculation yields revenue of \$4,086 for the second decile. Hence, the cumulative revenue for the first two deciles is \$8,586. Similarly, cumulative revenue calculated for all the remaining deciles is shown in Table 8.4.

Table 8.4

Decile	Number of Responders	Risk Rate	Price	Revenue	Cumulative Revenue
1	50	0.100	$100	$4,500	$4,500
2	45	0.092	$100	$4,086	$8,586
3	40	0.080	$100	$3,680	$12,266
4	30	0.067	$100	$2,799	$15,065
5	20	0.059	$100	$1,882	$16,947
6	10	0.042	$100	$958	$17,905
7	5	0.038	$100	$481	$18,386
8	4	0.029	$100	$388	$18,774
9	3	0.018	$100	$295	$19,069
10	2	0.010	$100	$198	$19,267

8.5 Profit

If, for example, the company targets the top two deciles, the expected revenue will be \$8,586, the expected acquisition cost will be \$2,000, and the expected losses will be \$91. Therefore, the expected cumulative profit will be \$8,586 – \$2,000 – \$91 = \$6,495. Table 8.5 shows the cumulative profit for all the deciles.

Table 8.5

Decile	Cumulative Revenue	Cumulative Cost	Cumulative Profit
1	$4,500	$1,050	$3,450
2	$8,586	$2,091	$6,495
3	$12,266	$3,123	$9,143
4	$15,065	$4,144	$10,922
5	$16,947	$5,155	$11,792
6	$17,905	$6,160	$11,746
7	$18,386	$7,161	$11,225
8	$18,774	$8,163	$10,612
9	$19,069	$9,163	$9,906
10	$19,267	$10,163	$9,104

The optimum cut-off point is where the cumulative profit peaks. In our example it occurs at decile 5. The company can maximize its profit by sending mail to only the 5,000 prospects who are in the top five deciles. This can be seen from Display 8.2.

Display 8.2

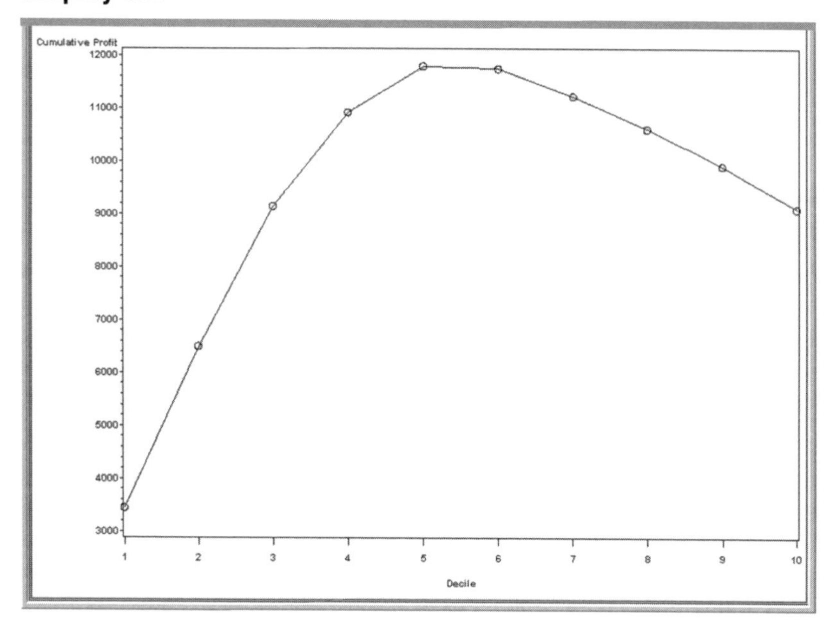

The cumulative profit peaks at the fifth decile because, beyond the fifth decile, the marginal profit earned by mailing to an additional decile is negative.

The *marginal cost* at the fifth decile is defined as the additional cost that the company would incur in acquiring the responders in the next (sixth) decile. The *marginal revenue* at the fifth decile is defined as the additional revenue the company would earn if it acquired all the responders in the next (sixth) decile. The *marginal profit* at the fifth decile is defined as the additional profit the company would make if it acquired the responders in the next (sixth) decile, which is the marginal revenue minus the marginal cost.

Beyond the fifth decile, the marginal cost outweighs the marginal revenue. Hence the marginal profit is negative. This is depicted in Display 8.3.

Display 8.3

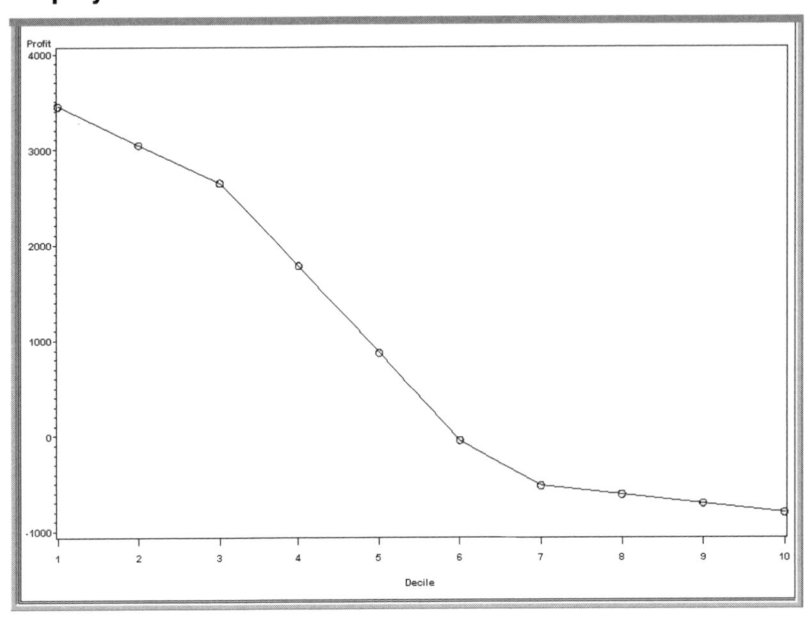

8.6 The Optimum Cut-off Point

The optimization problem can be analyzed in terms of marginal revenue and marginal cost. In this simplified example, marginal revenue (MR) at any decile is the additional revenue that is gained by mailing to the next decile, in addition to the previous deciles. Similarly, the marginal cost (MC) at any decile is the additional cost the company incurs by adding an additional decile for mailing. The marginal cost and marginal revenue, which are derived from the figures in Table 8.5, are shown in Table 8.6. From Table 8.5 it can be seen that, if the company mails to the top five deciles, the cumulative (total) revenue is $16,947 and cumulative (total) cost is $5,155. If the company wants to mail to the top six deciles, the revenue will be $17,905 and cost will be $6,160. Hence the marginal revenue at 5 is $17,905 – $16,947 = $958 and marginal cost is $6,160 – $5,155 = $1005. Since marginal cost exceeds marginal revenue, the cumulative profits will decline by adding the sixth decile to the mailing. The marginal revenues and marginal costs at different deciles shown in Table 8.6 are plotted in Display 8.4.

Table 8.6

Decile	Marginal Revenue	Marginal Cost	Profit by Decile
1	$4,500	$1,050	$3,450
2	$4,086	$1,041	$3,045
3	$3,680	$1,032	$2,648
4	$2,799	$1,020	$1,779
5	$1,882	$1,012	$870
6	$958	$1,004	$-46
7	$481	$1,002	$-521
8	$388	$1,001	$-613
9	$295	$1,001	$-706
10	$198	$1,000	$-802

Display 8.4

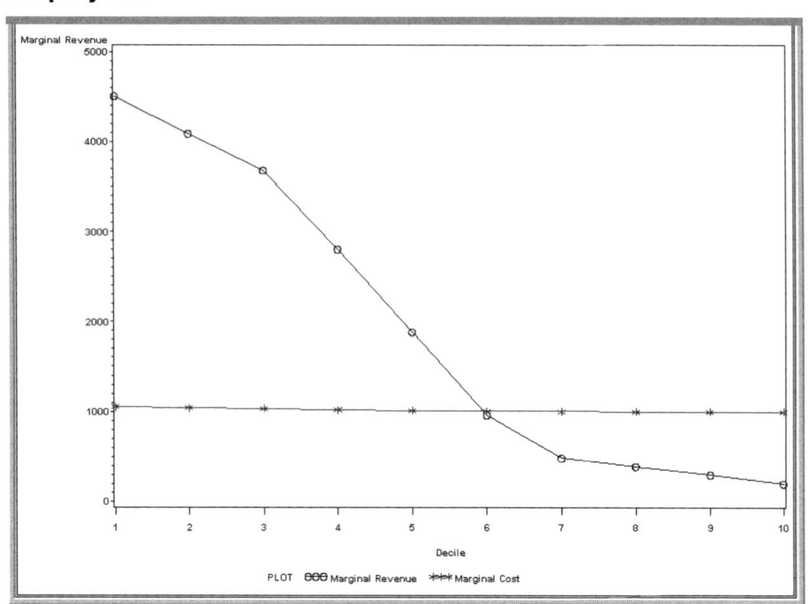

From Display 8.4 it can be seen that the point at which MR=MC is somewhere between the fifth and sixth deciles. As an approximation, one can stop mailing after the fifth decile.

In the above example the profits are calculated for one year only for the sake of simplicity. Alternatively, you can calculate profits over a longer period of time or add other complications to the analysis to make it more pertinent to the business problem that you are trying to solve. All these calculations can be made in the **SAS Code** node.

8.7 Alternative Scenarios of Response and Risk

In the example presented in Table 8.1, it is assumed that response rate and risk rate move in the same direction. The response rate is highest in the first decile, and declines with succeeding higher numbered deciles. A similar pattern is observed for the risk rate. However, there are many situations in which response rate and risk rate may not show this type of a pattern. In such situations, you can estimate two scores for each customer—one based on probability of response, and the other based on risk. Next, the prospects can be arranged in a 2x2 matrix, each cell of the matrix consisting of customers belonging to the same response-decile and the same risk-decile. Acquisition cost, cost of risk, revenue, and profit can be calculated for each cell, and acquisition decisions can be made for each group of customers based on the profitability of the cells.

8.8 Customer Lifetime Value

At the end of Section 8.6 I pointed out that the analysis of customer acquisition based on the marginal profit from mailing to additional prospects could be performed with a more distant horizon than the one-year analysis shown in my example here. Taking this to its logical conclusion, calculations of profitability at the level of an individual customer can be extended further by calculating the customer's lifetime value. For this you need answers to three questions:

- What is the expected residual lifetime of each customer?
- What is the flow of revenue that the customer is expected to generate in his "lifetime"?
- What are the future costs to the company of acquiring and retaining each customer?

Proper analysis of these elements can yield valuable insights into the type of actions that a company can take for extending the residual lifetime or customer lifetime value or both.

8.9 Suggestions for Extending Results

In the analysis presented in this chapter, revenue, cost, and profit are calculated for each decile. Alternatively, you can do the same calculations for each percentile in your database of prospects, rather than for each decile. In the above example, only one type of risk is identified. If, however, there is more than one type of risk, you can model competing risks by means of logistic hazard functions using either the **Regression** node or the **Neural Network** node in Enterprise Miner.

Glossary

assessment
the process of determining how well a model computes good outputs from input data that is not used during training. Assessment statistics are automatically computed when you train a model with a modeling node. By default, assessment statistics are calculated from the validation data set.

association analysis rule
in association analyses, an association between two or more items. An association analysis rule should not be interpreted as a direct causation. An association analysis rule is expressed as follows: If item A is part of an event, then item B is also part of the event X percent of the time.

association discovery
the process of identifying items that occur together in a particular event or record. This technique is also known as market basket analysis. Association discovery rules are based on frequency counts of the number of times items occur alone and in combination in the database.

binary variable
a variable that contains two discrete values (for example, PURCHASE: Yes and No).

branch
a subtree that is rooted in one of the initial divisions of a segment of a tree. For example, if a rule splits a segment into seven subsets, then seven branches grow from the segment.

CART (classification and regression trees)
a decision tree technique that is used for classifying or segmenting a data set. The technique provides a set of rules that can be applied to new data sets in order to predict which records will have a particular outcome. It also segments a data set by creating 2-way splits. The CART technique requires less data preparation than CHAID.

case
a collection of information about one of many entities that are represented in a data set. A case is an observation in the data set.

CHAID (chi-squared automatic interaction detection)
a technique for building decision trees. The CHAID technique specifies a significance level of a chi-square test to stop tree growth.

champion model
the best predictive model that is chosen from a pool of candidate models in a data mining environment. Candidate models are developed using various data mining heuristics and algorithm configurations. Competing models are compared and assessed using criteria such as training, validation, and test data fit and model score comparisons.

clustering
the process of dividing a data set into mutually exclusive groups such that the observations for each group are as close as possible to one another, and different groups are as far as possible from one another.

confidence
in association analyses, a measure of the strength of the association. In the rule A --> B, confidence is the percentage of times that event B occurs after event A occurs. See also **association analysis rule**.

cost variable
a variable that is used to track cost in a data mining analysis.

data mining database (DMDB)
a SAS data set that is designed to optimize the performance of the modeling nodes. DMDBs enhance performance by reducing the number of passes that the analytical engine needs to make through the data. Each DMDB contains a meta catalog, which includes summary statistics for numeric variables and factor-level information for categorical variables.

data source
a data object that represents a SAS data set in the Java-based Enterprise Miner GUI. A data source contains all the metadata for a SAS data set that Enterprise Miner needs in order to use the data set in a data mining process flow diagram. The SAS data set metadata that is required to create an Enterprise Miner data source includes the name and location of the data set, the SAS code that is used to define its library path, and the variable roles, measurement levels, and associated attributes that are used in the data mining process.

data subdirectory
a subdirectory within the Enterprise Miner project location. The data subdirectory contains files that are created when you run process flow diagrams in an Enterprise Miner project.

decile
any of the nine points that divide the values of a variable into ten groups of equal frequency, or any of those groups.

dependent variable
a variable whose value is determined by the value of another variable or by the values of a set of variables.

depth
the number of successive hierarchical partitions of the data in a tree. The initial, undivided segment has a depth of 0.

diagram
See **process flow diagram**.

expected confidence
in association analyses, the number of consequent transactions divided by the total number of transactions. For example, suppose that 100 transactions of item B were made, and that the data set includes a total of 10,000 transactions. Given the rule A-->B, the expected confidence for item B is 100 divided by 10,000, or one percent.

format
a pattern or set of instructions that SAS uses to determine how the values of a variable (or column) should be written or displayed. SAS provides a set of standard formats and also enables you to define your own formats.

generalization
the computation of accurate outputs, using input data that was not used during training.

hidden layer
in a neural network, a layer between input and output to which one or more activation functions are applied. Hidden layers are typically used to introduce nonlinearity.

hidden neuron
in a feed-forward, multilayer neural network, a neuron that is in one or more of the hidden layers that exist between the input and output neuron layers. The size of a neural network depends largely on the number of layers and on the number of hidden units per layer. See also **hidden layer**.

hold-out data
a portion of the historical data that is set aside during model development. Hold-out data can be used as test data to benchmark the fit and accuracy of the emerging predictive model. See also **model**.

imputation
the computation of replacement values for missing input values.

input variable
a variable that is used in a data mining process to predict the value of one or more target variables.

internal node
in a tree, a segment that has been further segmented. See also **node**.

interval variable
a continuous variable that contains values across a range. For example, a continuous variable called Temperature could have values such as 0, 32, 34, 36, 43.5, 44, 56, 80, 99, 99.9, and 100.

Kohonen network
any of several types of competitive networks that were invented by Teuvo Kohonen. Kohonen vector quantization networks and self-organizing maps (SOMs) are two types of Kohonen network that are commonly used in data mining. See also **Kohonen vector quantization network, SOM (self-organizing map)**.

Kohonen vector quantization network
a type of competitive network that can be viewed either as an unsupervised density estimator or as an autoassociator. The Kohonen vector quantization algorithm is closely related to the k-means cluster analysis algorithm. See also **Kohonen network**.

leaf
in a tree diagram, any segment that is not further segmented. The final leaves in a tree are called terminal nodes.

level
a successive hierarchical partition of data in a tree. The first level represents the entire unpartitioned data set. The second level represents the first partition of the data into segments, and so on.

libref (library reference)
a name that is temporarily associated with a SAS library. The complete name of a SAS file consists of two words, separated by a period. The libref, which is the first word, indicates the library. The second word is the name of the specific SAS file. For example, in VLIB.NEWBDAY, the libref VLIB tells SAS which library contains the file NEWBDAY. You assign a libref with a LIBNAME statement or with an operating system command.

lift
in association analyses and sequence analyses, a calculation that is equal to the confidence factor divided by the expected confidence. See also **confidence**, **expected confidence**.

logistic regression
a form of regression analysis in which the target variable (response variable) represents a binary-level or ordinal-level response.

macro variable
a variable that is part of the SAS macro programming language. The value of a macro variable is a string that remains constant until you change it. Macro variables are sometimes referred to as symbolic variables.

measurement
the process of assigning numbers to an object in order to quantify, rank, or scale an attribute of the object.

measurement level
a classification that describes the type of data that a variable contains. The most common measurement levels for variables are nominal, ordinal, interval, log-interval, ratio, and absolute. See also **interval variable**, **nominal variable**, **ordinal variable**.

metadata
a description or definition of data or information.

metadata sample
a sample of the input data source that is downloaded to the client and that is used throughout SAS Enterprise Miner to determine meta information about the data, such as number of variables, variable roles, variable status, variable level, variable type, and variable label.

model
a formula or algorithm that computes outputs from inputs. A data mining model includes information about the conditional distribution of the target variables, given the input variables.

multilayer perceptron (MLP)
a neural network that has one or more hidden layers, each of which has a linear combination function and executes a nonlinear activation function on the input to that layer. See also **hidden layer**.

neural networks
a class of flexible nonlinear regression models, discriminant models, data reduction models, and nonlinear dynamic systems that often consist of a large number of neurons. These neurons are usually interconnected in complex ways and are often organized into layers. See also **neuron**.

neuron
a linear or nonlinear computing element in a neural network. Neurons accept one or more inputs. They apply functions to the inputs, and they can send the results to one or more other neurons. Neurons are also called nodes or units.

node
(1) in the SAS Enterprise Miner user interface, a graphical object that represents a data mining task in a process flow diagram. The statistical tools that perform the data mining tasks are called nodes when they are placed on a data mining process flow diagram. Each node performs a mathematical or graphical operation as a component of an analytical and predictive data model.
(2) in a neural network, a linear or nonlinear computing element that accepts one or more inputs, computes a function of the inputs, and optionally directs the result to one or more other neurons. Nodes are also known as neurons or units. (3) a leaf in a tree diagram. The terms leaf, node, and segment are closely related and sometimes refer to the same part of a tree. See also **process flow diagram**, **internal node**.

nominal variable
a variable that contains discrete values that do not have a logical order. For example, a nominal variable called Vehicle could have values such as car, truck, bus, and train.

numeric variable
a variable that contains only numeric values and related symbols, such as decimal points, plus signs, and minus signs.

observation
a row in a SAS data set. All of the data values in an observation are associated with a single entity such as a customer or a state. Each observation contains either one data value or a missing-value indicator for each variable.

ordinal variable
a variable that contains discrete values that have a logical order. For example, a variable called Rank could have values such as 1, 2, 3, 4, and 5.

partition
to divide available data into training, validation, and test data sets.

perceptron
a linear or nonlinear neural network with or without one or more hidden layers.

predicted value
in a regression model, the value of a dependent variable that is calculated by evaluating the estimated regression equation for a specified set of values of the explanatory variables.

process flow diagram
a graphical representation of the various data mining tasks that are performed by individual Enterprise Miner nodes during a data mining analysis. A process flow diagram consists of two or more individual nodes that are connected in the order in which the data miner wants the corresponding statistical operations to be performed.

profit matrix
a table of expected revenues and expected costs for each decision alternative for each level of a target variable.

project
a collection of Enterprise Miner process flow diagrams. See also **process flow diagram**.

root node
the initial segment of a tree. The root node represents the entire data set that is submitted to the tree, before any splits are made.

rule
See **association analysis rule, sequence analysis rule, tree splitting rule**.

sampling
the process of subsetting a population into n cases. The reason for sampling is to decrease the time required for fitting a model.

SAS data set
a file whose contents are in one of the native SAS file formats. There are two types of SAS data sets: SAS data files and SAS data views. SAS data files contain data values in addition to descriptor information that is associated with the data. SAS data views contain only the descriptor information plus other information that is required for retrieving data values from other SAS data sets or from files whose contents are in other software vendors' file formats.

scoring
the process of applying a model to new data in order to compute outputs. Scoring is the last process that is performed in data mining.

seed
an initial value from which a random number function or CALL routine calculates a random value.

segmentation
the process of dividing a population into sub-populations of similar individuals. Segmentation can be done in a supervisory mode (using a target variable and various techniques, including decision trees) or without supervision (using clustering or a Kohonen network). See also **Kohonen network**.

self-organizing map
See **SOM (self-organizing map)**.

SEMMA
the data mining process that is used by Enterprise Miner. SEMMA stands for Sample, Explore, Modify, Model, and Assess.

sequence analysis rule
in sequence discovery, an association between two or more items, taking a time element into account. For example, the sequence analysis rule A --> B implies that event B occurs after event A occurs.

sequence variable
a variable whose value is a time stamp that is used to determine the sequence in which two or more events occurred.

SOM (self-organizing map)
a competitive learning neural network that is used for clustering, visualization, and abstraction. A SOM classifies the parameter space into multiple clusters, while at the same time organizing the clusters into a map that is based on the relative distances between clusters. See also **Kohonen network**.

subdiagram
in a process flow diagram, a collection of nodes that are compressed into a single node. The use of subdiagrams can improve your control of the information flow in the diagram.

target variable
a variable whose values are known in one or more data sets that are available (in training data, for example) but whose values are unknown in one or more future data sets (in a score data set, for example). Data mining models use data from known variables to predict the values of target variables.

test data
currently available data that contains input values and target values that are not used during training, but which instead are used for generalization and to compare models.

training
the process of computing good values for the weights in a model.

training data
currently available data that contains input values and target values that are used for model training.

transformation
the process of applying a function to a variable in order to adjust the variable's range, variability, or both.

tree
the complete set of rules that are used to split data into a hierarchy of successive segments. A tree consists of branches and leaves, in which each set of leaves represents an optimal segmentation of the branches above them according to a statistical measure.

tree splitting rule
in decision trees, a conditional mathematical statement that specifies how to split segments of a tree's data into subsegments.

validation data
data that is used to validate the suitability of a data model that was developed using training data. Both training data sets and validation data sets contain target variable values. Target variable values in the training data are used to train the model. Target variable values in the validation data set are used to compare the training model's predictions to the known target values, assessing the model's fit before using the model to score new data.

variable
a column in a SAS data set or in a SAS data view. The data values for each variable describe a single characteristic for all observations. Each SAS variable can have the following attributes: name, data type (character or numeric), length, format, informat, and label.

variable attribute

any of the following characteristics that are associated with a particular variable: name, label, format, informat, data type, and length.

variable level

the set of data dimensions for binary, interval, or class variables. Binary variables have two levels. A binary variable CREDIT could have levels of 1 and 0, Yes and No, or Accept and Reject. Interval variables have levels that correspond to the number of interval variable partitions. For example, an interval variable PURCHASE_AGE might have levels of 0-18, 19-39, 40-65, and >65. Class variables have levels that correspond to the class members. For example, a class variable HOMEHEAT might have four variable levels: Coal/Wood, FuelOil, Gas, and Electric. Data mining decision and profit matrixes are composed of variable levels.

References

Afifi, A., V. Clark, and S. May. 2004. *Computer Aided Multivariate Analysis*. 4th ed. London: Chapman & Hall/CRC Press.

Agresti, A. 2002. *Categorical Data Analysis*. 2d ed. New York: John Wiley & Sons.

Allison, P. D. 2005. *Fixed Effects Regression Methods for Longitudinal Data Using SAS*. Cary, NC: SAS Institute Inc.

Allison, P. D. 1999. *Logistic Regression Using SAS: Theory and Application*. Cary, NC: SAS Institute Inc.

Bishop, C. M. 1995. *Neural Networks for Pattern Recognition*. New York: Oxford University Press.

Breiman, L. 1996a. "Stacking Regressions." *Machine Learning* 24: 49–64.

Breiman, L. 1996b. "Bagging Predictors." *Machine Learning* 24: 123–140.

Cerrito, P. B. 2006. *Introduction to Data Mining Using SAS Enterprise Miner*. Cary, NC: SAS Institute Inc.

Cody, R. P. 1999. *Cody's Data Cleaning Techniques Using SAS Software*. Cary, NC: SAS Institute Inc.

deVille Barry. 2006. *Decision Trees for Business Intelligence and Data Mining: Using SAS Enterprise Miner*. Cary, NC: SAS Institute Inc.

Eklund, J., and S. Karlsson. 2005. "Forecast Combination and Model Averaging Using Predictive Measures." Working Paper Series 191, Sveriges Riksbank (Central Bank of Sweden).

Friendly, M. 2000. *Visualizing Categorical Data.* Cary, NC: SAS Institute Inc.

Freund, R. J., and R. C. Littell. 2000. *SAS System for Regression.* 3d ed. Cary, NC: SAS Institute Inc.

Guidolin, M., and C. F. Na. 2007. "The Economic and Statistical Value of Forecast Combinations under Regime Switching: An Application to Predictable U.S. Returns," Working Paper Series 2006-059B, Federal Reserve Bank of St. Louis.

Littell, R. C., W. W. Stroup, and R. J. Freund. 2002. *SAS for Linear Models.* 4th ed. Cary, NC: SAS Institute Inc.

Maddala, G. S. 1986. "Limited-Dependent and Qualitative Variables in Econometrics." *Econometric Society Monographs*. New York: Cambridge University Press.

Muller, K. E. and B. A. Fetterman. 2002. *Regression and ANOVA: An Integrated Approach Using SAS Software*. Cary, NC: SAS Institute Inc.

Rawlings, J. O., S. G. Pantula, and D. A. Dickey. 2001. *Applied Regression Analysis: A Research Tool*. 2d ed. New York: Springer-Verlag.

Ripley, B.D. 1996. *Pattern Recognition and Neural Networks*. New York: Cambridge University Press.

Sarma, K. S. 2005. "Combining Decision Trees with Regression in Predictive Modeling with SAS Enterprise Miner." *Proceedings of the Thirtieth Annual SAS Users Group International Conference*. Cary, NC: SAS Institute.

Sarma, K. S. 2001. "Using SAS Enterprise Miner for Forecasting." *Proceedings of the Twenty-sixth Annual SAS Users Group International Conference*. Cary, NC: SAS Institute.

Sarma, K. S. 2001. "Using SAS Enterprise Miner for Forecasting Response and Risk." *Proceedings of the Ninth Annual Conference of Western Users of SAS Software.* Cary, NC: SAS Institute Inc.

SAS Institute Inc. 2004. *Advanced Predictive Modeling Using SAS Enterprise Miner 5.1 Course Notes*. Cary, NC: SAS Institute Inc.

SAS Institute Inc. 2007. *Applied Analytics Using SAS Enterprise Miner 5 Course Notes*. Cary, NC: SAS Institute Inc.

SAS Institute Inc. 2007. *Applying Data Mining Techniques Using SAS Enterprise Miner Course Notes.* Cary, NC: SAS Institute Inc.

SAS Institute Inc. 2005. *Categorical Data Analysis Using Logistic Regression Course Notes*. Cary, NC: SAS Institute Inc.

SAS Institute Inc. 2004. *Data Preparation for Data Mining Using SAS Software Course Notes*. Cary, NC: SAS Institute Inc.

SAS Institute Inc. 2001. *Decision Tree Modeling Course Notes*. Cary, NC: SAS Institute Inc.

SAS Institute Inc. 2004. *Extending SAS Enterprise Miner 5.1 Course Notes*. Cary, NC: SAS Institute Inc.

SAS Institute Inc. 2005. *Neural Network Modeling Course Notes*. Cary, NC: SAS Institute Inc.

SAS Institute Inc. 2004. *Predictive Modeling Using SAS Enterprise Miner 5.1 Course Notes*. Cary, NC: SAS Institute Inc.

SAS Institute Inc. 2004. *SAS/STAT 9.1 User's Guide.* Cary, NC: SAS Institute Inc.

Stokes, M. E., C. S. Davis, and G. G. Koch. 2000. *Categorical Data Analysis Using the SAS System.* 2d ed. Cary, NC: SAS Institute Inc.

Svolba, G. 2006. *Data Preparation for Analytics Using SAS*. Cary, NC: SAS Institute Inc.

Westfall, P. H., R. D. Tobias, D. Rom, R. D. Wolfinger, and Y. Hochberg. 1999. *Multiple Comparisons and Multiple Tests Using SAS*. Cary, NC: SAS Institute Inc.

Index

A

B

C

D

E

S

T

W

Books Available from SAS Press

Advanced Log-Linear Models Using SAS®
by **Daniel Zelterman**

Analysis of Clinical Trials Using SAS®: A Practical Guide
by **Alex Dmitrienko, Geert Molenberghs, Walter Offen,** *and* **Christy Chuang-Stein**

Analyzing Receiver Operating Characteristic Curves with SAS®
by **Mithat Gönen**

Annotate: Simply the Basics
by **Art Carpenter**

Applied Multivariate Statistics with SAS® Software, Second Edition
by **Ravindra Khattree**
and **Dayanand N. Naik**

Applied Statistics and the SAS® Programming Language, Fifth Edition
by **Ronald P. Cody**
and **Jeffrey K. Smith**

An Array of Challenges — Test Your SAS® Skills
by **Robert Virgile**

Building Web Applications with SAS/IntrNet®: A Guide to the Application Dispatcher
by **Don Henderson**

Carpenter's Complete Guide to the SAS® Macro Language, Second Edition
by **Art Carpenter**

Carpenter's Complete Guide to the SAS® REPORT Procedure
by **Art Carpenter**

The Cartoon Guide to Statistics
by **Larry Gonick**
and **Woollcott Smith**

Categorical Data Analysis Using the SAS® System, Second Edition
by **Maura E. Stokes, Charles S. Davis,**
and **Gary G. Koch**

Cody's Data Cleaning Techniques Using SAS® Software
by **Ron Cody**

Common Statistical Methods for Clinical Research with SAS® Examples, Second Edition
by **Glenn A. Walker**

The Complete Guide to SAS® Indexes
by **Michael A. Raithel**

CRM Segmentation and Clustering Using SAS® Enterprise Miner™
by **Randall S. Collica**

Data Management and Reporting Made Easy with SAS® Learning Edition 2.0
by **Sunil K. Gupta**

Data Preparation for Analytics Using SAS®
by **Gerhard Svolba**

Debugging SAS® Programs: A Handbook of Tools and Techniques
by **Michele M. Burlew**

Decision Trees for Business Intelligence and Data Mining: Using SAS® Enterprise Miner™
by **Barry de Ville**

Efficiency: Improving the Performance of Your SAS® Applications
by **Robert Virgile**

The Essential Guide to SAS® Dates and Times
by **Derek P. Morgan**

The Essential PROC SQL Handbook for SAS® Users
by **Katherine Prairie**

Fixed Effects Regression Methods for Longitudinal Data Using SAS®
by **Paul D. Allison**

Genetic Analysis of Complex Traits Using SAS®
Edited by **Arnold M. Saxton**

A Handbook of Statistical Analyses Using SAS®, Second Edition
by **B.S. Everitt**
and **G. Der**

Health Care Data and SAS®
by **Marge Scerbo, Craig Dickstein,**
and **Alan Wilson**

The How-To Book for SAS/GRAPH® Software
by **Thomas Miron**

In the Know ... SAS® Tips and Techniques From Around the Globe, Second Edition
by **Phil Mason**

Instant ODS: Style Templates for the Output Delivery System
by **Bernadette Johnson**

Integrating Results through Meta-Analytic Review Using SAS® Software
by **Morgan C. Wang**
and **Brad J. Bushman**

Introduction to Data Mining Using SAS® Enterprise Miner™
by **Patricia B. Cerrito**

Learning SAS® by Example: A Programmer's Guide
by **Ron Cody**

Learning SAS® in the Computer Lab, Second Edition
by **Rebecca J. Elliott**

The Little SAS® Book: A Primer
by **Lora D. Delwiche**
and **Susan J. Slaughter**

The Little SAS® Book: A Primer, Second Edition
by **Lora D. Delwiche**
and **Susan J. Slaughter**
(updated to include SAS 7 features)

The Little SAS® Book: A Primer, Third Edition
by **Lora D. Delwiche**
and **Susan J. Slaughter**
(updated to include SAS 9.1 features)

The Little SAS® Book for Enterprise Guide® 3.0
by **Susan J. Slaughter**
and **Lora D. Delwiche**

The Little SAS® Book for Enterprise Guide® 4.1
by **Susan J. Slaughter**
and **Lora D. Delwiche**

Logistic Regression Using the SAS® System: Theory and Application
by **Paul D. Allison**

Longitudinal Data and SAS®: A Programmer's Guide
by **Ron Cody**

Maps Made Easy Using SAS®
by **Mike Zdeb**

Measurement, Analysis, and Control Using JMP®: Quality Techniques for Manufacturing
by **Jack E. Reece**

Models for Discrete Data
by **Daniel Zelterman**

Multiple Comparisons and Multiple Tests Using SAS® Text and Workbook Set
(books in this set also sold separately)
by **Peter H. Westfall, Randall D. Tobias, Dror Rom, Russell D. Wolfinger,**
and **Yosef Hochberg**

Multiple-Plot Displays: Simplified with Macros
by **Perry Watts**

Multivariate Data Reduction and Discrimination with SAS® Software
by **Ravindra Khattree**
and **Dayanand N. Naik**

Output Delivery System: The Basics
by **Lauren E. Haworth**

Painless Windows: A Handbook for SAS® Users, Third Edition
by **Jodie Gilmore**
(updated to include SAS 8 and SAS 9.1 features)

Pharmaceutical Statistics Using SAS®: A Practical Guide
Edited by **Alex Dmitrienko, Christy Chuang-Stein,**
and **Ralph D'Agostino**

The Power of PROC FORMAT
by **Jonas V. Bilenas**

Predictive Modeling with SAS® Enterprise Miner™: Practical Solutions for Business Applications
by **Kattamuri S. Sarma, Ph.D.**

PROC SQL: Beyond the Basics Using SAS®
by **Kirk Paul Lafler**

PROC TABULATE by Example
by **Lauren E. Haworth**

Professional SAS® Programmer's Pocket Reference, Fifth Edition
by **Rick Aster**

Professional SAS® Programming Shortcuts, Second Edition
by **Rick Aster**

Quick Results with SAS/GRAPH® Software
by **Arthur L. Carpenter**
and **Charles E. Shipp**

Quick Results with the Output Delivery System
by **Sunil K. Gupta**

Reading External Data Files Using SAS®: Examples Handbook
by **Michele M. Burlew**

Regression and ANOVA: An Integrated Approach Using SAS® Software
by **Keith E. Muller**
and **Bethel A. Fetterman**

SAS® For Dummies®
by **Stephen McDaniel**
and **Chris Hemedinger**

SAS® for Forecasting Time Series, Second Edition
by **John C. Brocklebank**
and **David A. Dickey**

SAS® for Linear Models, Fourth Edition
by **Ramon C. Littell, Walter W. Stroup,**
and **Rudolf J. Freund**

SAS® for Mixed Models, Second Edition
by **Ramon C. Littell, George A. Milliken, Walter W. Stroup, Russell D. Wolfinger,** *and* **Oliver Schabenberger**

SAS® for Monte Carlo Studies: A Guide for Quantitative Researchers
by **Xitao Fan, Ákos Felsővályi, Stephen A. Sivo,**
and **Sean C. Keenan**

SAS® Functions by Example
by **Ron Cody**

SAS® Graphics for Java: Examples Using SAS® AppDev Studio™ and the Output Delivery System
by **Wendy Bohnenkamp**
and **Jackie Iverson**

SAS® Guide to Report Writing, Second Edition
by **Michele M. Burlew**

SAS® Macro Programming Made Easy, Second Edition
by **Michele M. Burlew**

SAS® Programming by Example
by **Ron Cody**
and **Ray Pass**

SAS® Programming for Researchers and Social Scientists, Second Edition
by **Paul E. Spector**

SAS® Programming in the Pharmaceutical Industry
by **Jack Shostak**

SAS® Survival Analysis Techniques for Medical Research, Second Edition
by **Alan B. Cantor**

SAS® System for Elementary Statistical Analysis, Second Edition
by **Sandra D. Schlotzhauer**
and **Ramon C. Littell**

SAS® System for Regression, Third Edition
by **Rudolf J. Freund**
and **Ramon C. Littell**

SAS® System for Statistical Graphics, First Edition
by **Michael Friendly**

The SAS® Workbook and *Solutions* Set
(books in this set also sold separately)
by **Ron Cody**

Saving Time and Money Using SAS®
by **Philip R. Holland**

Selecting Statistical Techniques for Social Science Data: A Guide for SAS® Users
by **Frank M. Andrews, Laura Klem, Patrick M. O'Malley, Willard L. Rodgers, Kathleen B. Welch,**
and **Terrence N. Davidson**

Statistical Quality Control Using the SAS® System
by **Dennis W. King**

Statistics Using SAS® Enterprise Guide®
by **James B. Davis**

A Step-by-Step Approach to Using the SAS® System for Factor Analysis and Structural Equation Modeling
by **Larry Hatcher**

A Step-by-Step Approach to Using SAS® for Univariate and Multivariate Statistics, Second Edition
by **Norm O'Rourke, Larry Hatcher,**
and **Edward J. Stepanski**

Step-by-Step Basic Statistics Using SAS®: Student Guide and *Exercises*
(books in this set also sold separately)
by **Larry Hatcher**

Survival Analysis Using SAS®: A Practical Guide
by **Paul D. Allison**

Tuning SAS® Applications in the OS/390 and z/OS Environments, Second Edition
by **Michael A. Raithel**

Univariate and Multivariate General Linear Models: Theory and Applications Using SAS® Software
by **Neil H. Timm**
and **Tammy A. Mieczkowski**

Using SAS® in Financial Research
by **Ekkehart Boehmer, John Paul Broussard,**
and **Juha-Pekka Kallunki**

Using the SAS® Windowing Environment: A Quick Tutorial
by **Larry Hatcher**

Visualizing Categorical Data
by **Michael Friendly**

Web Development with SAS® by Example, Second Edition
by **Frederick E. Pratter**

Your Guide to Survey Research Using the SAS® System
by **Archer Gravely**

JMP® Books

Elementary Statistics Using JMP®
by **Sandra D. Schlotzhauer**

JMP® for Basic Univariate and Multivariate Statistics: A Step-by-Step Guide
by **Ann Lehman, Norm O'Rourke, Larry Hatcher,**
and **Edward J. Stepanski**

JMP® Start Statistics, Third Edition
by **John Sall, Ann Lehman,**
and **Lee Creighton**

Regression Using JMP®
by **Rudolf J. Freund, Ramon C. LIttell,**
and **Lee Creighton**

support.sas.com/pubs